Geospatial Technologies and Homeland Security

The GeoJournal Library

Volume 94

The titles published in this series are listed at the end of this volume.

This book is dedicated to the loving memory of

Dr. Chor-Pang (C.P.) Lo, 1939–2007, *whose devotion to research and education provides an endless source of inspiration for generations to come;*

Dr. Richard E. Ewing, 1946–2007, *whose unwavering support for geospatial research at Texas A&M University has made this book possible.*

Foreword

Homeland security and context

In the *Geographical Dimensions of Terrorism* (GDOT) (Cutter et al. 2003), the first book after 9/11 to address homeland security and geography, we developed several thematic research agendas and explored intersections between geographic research and the importance of *context*, both geographical and political, in relationship to the concepts of terrorism and security. It is good to see that a great deal of new thought and research continues to flow from that initial research agenda, as illustrated by many of the papers of this new book, entitled *Geospatial Technologies and Homeland Security: Research Frontiers and Future Challenges.*

Context is relevant not only to understanding homeland security issues broadly, but also to the conduct of research on geospatial technologies. It is impossible to understand the implications of a homeland security strategy, let alone hope to make predictions, conduct meaningful modeling and research, or assess the value and dangers of geospatial technologies, without consideration of overarching political, social, economic, and geographic contexts within which these questions are posed.

Witness, as an example, the reversals of context for the analysis of homeland security issues that have occurred since September 11, 2001. Immediately after 9/11 there was a genuine outpouring of international sympathy and support for the United States and a historic window of opportunity and international good will to work together to create a safer, more peaceful, and more just world. That context is gone today, either ineptly squandered or purposely undermined, depending on the analysis. In the few short years since 9/11, we have seen a rapid and near universal deterioration of global confidence in the credibility, motives, and competence of the United States government as a force for security, peace, or international understanding. Domestically, we have seen legitimate public concerns about security following 9/11 cynically hijacked to enrich private interests closely allied with government officials, and extraordinary justifications advanced for broad new programs of surveillance and even for the use of torture. How does one conduct geographic research on homeland security issues or evaluate the role of geospatial technologies for security purposes in these two different contexts?

Researchers, both academic and in other sectors, have long been concerned about the ultimate ends to which governments or others put their research, insight, or inventions. Among the most cogent explorations of these issues are Noam Chomsky's classic *American Power and the New Mandarins*, and Heinar Kipphardt's (1968) play, *In the Matter of J. Robert Oppenheimer*.

Personally, as one who has helped to create some of our core geographic technologies, specifically real-time interactive Global Positioning System/Geographic Information System (GPS/GIS) technologies that are now so widespread in societal applications, I am deeply concerned about some of the uses to which these powerful technologies are being put. While there are nearly endless beneficial uses of interactive GPS/GIS systems, ranging from planning and economic development to health research, environmental protection, and even legitimate defense and security applications, the usurpation of this science and technology for surveillance and control and warfare in contexts less than legitimate is a disturbing prospect.

While geographers and GIScientists obviously have a great deal to contribute to legitimate security needs ("homeland" or otherwise), we also have an obligation to work to ensure that the uses of our work, as well as the context in which they are employed is legitimate, responsible, and furthers the common good. As I wrote in the *Geographical Dimensions of Terrorism*, a crucial challenge is

> How can we safely develop and implement the powerful capabilities of the advanced new geographic technologies we are creating—which hold so much promise for individual and scientific benefit—when, as with so many other advanced technologies, they have inherent within them a risk for potential abuse? As part of this policy equation, we must examine the social responsibilities of those employing spatial technologies. How can we ensure that individual rights and locational privacy are protected from inadvertent or willful misuse of such technology? What is appropriate in terms of legal or regulatory safeguards regarding their use? (Cutter et al. 2003: 205)

Geographers are particularly well-suited and positioned to develop and understand both the technological as well as the contextual aspects of legitimate national and international security needs. Our revolutionary new geographic technologies are an outgrowth of a discipline which has long and deep traditions of integrating the social sciences with the humanities and the natural sciences, and also of critical analyses of research and its applications in society.

I am pleased to see that so many of these integrative and self-reflective ideas, ranging from scientific and technical challenges to those of political and social context that were identified in the research agendas during the GDOT process and research agendas in 2002–2003, are represented and being further explored in this current book, *Geospatial Technologies and Homeland Security*.

The academic and public debate on topics of security in the context of our current political and economic climate is understandably highly charged and often acrimonious. As a result, many researchers, geographers among them, are often tempted to avoid research on the topic altogether. However, security issues and concerns are not going to disappear in our world. Even with the most humane, competent, honest, good-willed and democratically responsive government in place, the issues surrounding national and international security will still be with us, and will require the reasoned and thoughtful attention of geographers, and all citizens.

I commend the editor and authors of this book for grappling with these issues, and at the same time urge them and the readers of this book to consider carefully the context in which our geospatial technologies are employed. Hopefully, the knowledge, caring, and engagement of geographers in the full range of these issues can contribute to both the better world and the better context we all desire.

Douglas Richardson
November, 2007

References

Cutter, S.L., Richardson, D.B. & Wilbanks, T.J. (Eds.) (2003). *The geographical dimensions of terrorism* (New York: Routledge)

Chomsky, N. (1969). *American power and the new mandarins* (New York: Pantheon Books)

Kipphardt, H. (1968). *In the matter of J. Robert Oppenheimer* (translated by Ruth Speirs. New York: Hill and Wang)

Acknowledgements

Financial support from the Office of Vice President for Research at Texas A&M University for the TAMU GIST Symposium series is gratefully acknowledged. Heartfelt thanks are also due to Gina Lane and Gayle Willis who provided extensive editorial assistance for this book project. The editor would also like to thank Susan Cutter and all the reviewers for their timely reviewing of the chapters in this book.

Contents

Biographies of the Editor and Contributors

Editor

Daniel Z. Sui

Daniel Z. Sui is a professor of geography and holder of the Reta A. Haynes endowed chair in the College of Geosciences at Texas A&M University. He also holds an adjunct professorship in epidemiology and biostatistics at the School of Rural Public Health, Texas A&M System Health Science Center. Since 2004, Dr. Sui has also been serving as assistant vice president for research and director for Geospatial Information Science and Technology (GIST) at Texas A&M. Dr. Sui is the current editor-in-chief of *GeoJournal* and the North American editor for the international journal *Computers, Environment, and Urban Systems* (CEUS). Daniel Sui holds a BS (1986) and MS (1989) from Peking University and Ph.D. from the University of Georgia (1993). His research areas cover geospatial information science & technology and their applications to homeland security, public health, urban studies, and sustainable development. He is author/co-author of six books and over 140 scientific publications. He became an AT&T Faculty Fellow in 2000, an ORISE Fellow in 2004 and he won the Michael Breheny Prize for the best paper in *Environment and Planning B* in 2007.Daniel Sui is also recipient of two outstanding teaching awards at Texas A&M University. More details can be found at: http://geog.tamu.edu/~sui

Contributors

David I. Ashby

David Ashby is Head of Local Government at Dr Foster Research Ltd. and is an expert in neighborhood profiling, geodemographics and spatial analysis. David has a Ph.D. from University College London, an MSc from the University of Toronto, and a First Class B.Sc. (Hons) from the University of Nottingham. He is an Honorary Research Fellow at the Department of Geography, UCL. David's research interests lie in GIS, spatial statistics, and advanced spatial analysis. His research experience in the area is diverse; focusing on the geographical profiling of crime for the Police, Home Office, and Audit Commission to the profiling of inner city health in Canada. At Dr Foster, he leads on the development of new management information products and services, originating in the fields of health and social care and expanding into all other areas of local government and including Policing. Before joining Dr Foster, David was an academic, consultant, and Business Manager at University College London, where his remit covered all Geospatial Technologies and the wider Faculty of Social Sciences.

Robert S. Bednarz

Robert S. Bednarz is a professor of geography at Texas A&M University in College Station, TX. He earned his AB from Dartmouth College in Geography (1968), his MA from Northwestern University in Geography (1969), and his Ph.D. from the University of Chicago in Geography (1974). Professor Bednarz is interested in spatial thinking, learning, and cognition; how geo-technologies can support geographic education; and urban geography and economic geography. He serves as North American editor of the *Journal of Geography in Higher Education*, is a member of the editorial boards of *Focus and Research in Geographic Education*, and is past-president of the National Council for Geographic Education.

Sarah Witham Bednarz

Sarah Witham Bednarz (AB Geography, Mount Holyoke College, 1973 magna cum laude; MAT Geography, University of Chicago, 1974; Ph.D. Educational Curriculum and Instruction, Texas A&M University, 1992) is Professor of Geography, Texas A&M University. Bednarz specializes in geosciences education with a focus on learning and teaching geography, particularly in the ways people learn to think spatially using geospatial technologies. She has worked for more than 20 years to improve the quality of geoscience and environmental education in elementary, middle, and high schools, working with educators and other stakeholders to develop content standards, curriculum support materials, and research-based strategies to improve student learning. As one of the primary authors of the National Geography Standards, she developed the sections on geographic skills. Bednarz has experience in coordinating large educational projects such as Mission Geography, a NASA-funded project to develop curriculum materials linking the National Geography Standards with NASA's missions and results, and the Texas Alliance for Geographic Education, a coalition of teachers and geographers sponsored by the National Geographic Society. She is currently PI for an NSF-funded program Advancing Geospatial Skills in Science and Social Science, which links geospatially skilled graduate and advanced undergraduate fellows with science and social science teachers, grades 6–12, in a collaborative, three year cycle to enhance teacher and student knowledge and skills in spatial thinking.

Aaron Binns

Aaron Binns is a Ph.D. candidate studying GIS and Remote Sensing integration at Florida State University. Aaron's dissertation research is focused on the use of GIS and photogrammetric techniques, specifically aerial photography, satellite imagery, and LiDAR data, to locate and plot the now-extinct river channels of Mesoamerica's greatest ancient cultures: including the Maya and the Olmec civilizations. Once complete, this data can be merged with the current literature and has the potential for future archaeological site prediction using existing site-specific electromagnetic signatures and LiDAR based Digital Elevation Models.

Samuel D. Brody
Samuel D. Brody is an Associate Professor of Environmental Planning in the Department of Landscape Architecture and Urban Planning at Texas A&M University. He is the Director of the Environmental Planning and Sustainability Research Unit, Co-Director of the Center for Texas Beaches and Shores, and a faculty fellow in the Hazard Reduction & Recovery Center. Dr. Brody's research focuses on environmental planning, spatial analysis, environmental dispute resolution, climate change policy, and natural hazards mitigation. He has several projects within these research areas funded by organizations such as the National Science Foundation, NOAA, Texas Sea Grant, and the National Park Service. Dr. Brody teaches graduate courses in environmental planning, sustainable development, and dispute resolution. He has also worked in both the public and private sectors to help local coastal communities to draft land use and environmental plans.

Christian J. E. Castle
Christian J. E. Castle is currently the GIS Manager at the Transport for London, UK. Christian Castle received his Bachelor of Science degree in Geography from the University of Leicester, where he also undertook a Master of Science degree in GIS. Christian has subsequently completed his Doctorate of Philosophy in GIS at the Centre for Advanced Spatial Analysis, University College London. The focus of his research during this period was pedestrian evacuation modelling, culminating in the development of an agent-based simulation model to evaluate the egress of people from London's King's Cross Underground station. Christian is motivated by real-world problems, whether it is analyzing the evacuation process of buildings to facilitate emergency management strategies, calculating new exposures of emerging insurance risks (e.g. terrorism), or determining an efficient means to effectively manage London's Congestion Charging Scheme. Since completing his Ph.D., Christian is now GIS manager for a local UK government body responsible for transport within London.

Spencer Chainey
Spencer Chainey is Director of Geographical Information Science at the Jill Dando Institute of Crime Science (JDI), University College London. He is one of the pioneers in the field of crime mapping (working with police forces, community safety partnerships, CENTREX and the Home Office in the UK, and with law enforcement and justice departments in the USA, Brazil, Australia and New Zealand). Prior to joining JDI, Spencer spent several years working in the private sector and in local government on GIS, community safety, information sharing, housing development and regeneration projects.

Xuwei Chen

Xuwei Chen is an Assistant Professor in the Department of Geography at Northern Illinois University. She obtained her BS and MS in Geography from Nanjing University in China. She holds a Ph.D. in Geographic Information Science from Texas State University-San Marcos. Her research interests include transportation simulation and modeling, emergency evacuation, spatial analysis, geovisualization, and GIScience. She is currently involved in projects that examine how different variables, e.g. evacuation zoning structures, evacuation sequences, spatial-temporal variations of evacuation population, human behaviors, affect the effectiveness of evacuation strategies at a micro-scale level. Her publications have appeared in peer-reviewed journals such as *Social Science Research, Natural Hazards, Biomass & Bioenergy* etc.

Richard S. Courtney

Richard S. Courtney is Associate Professor of Geography at Kutztown University, Pennsylvania, USA. He has a Ph.D. in Geography from Ohio State University and his research interests include residential segregation, spatial population redistribution, and urban systems population dynamics. Dr. Courtney is also committed to development of cartography in the GIS environment.

Jeremy W. Crampton

Jeremy W. Crampton is an Associate Professor at Georgia State University and obtained his Ph.D. from The Pennsylvania State University. Crampton's research interests focus on the politics of identity, especially how identity is produced through geographical representations such as maps and GIS. Using critical approaches to cartography and GIS as informed by the work of Michel Foucault, he has examined the cartographic biopolitics of governmentality both in contemporary settings and historical. Crampton's recent publications include articles in *Progress in Human Geography*, *Environment and Planning D*, *Social and Cultural Geography*, the *Geographical Review*, and chapters in *Multimedia Cartography* (Springer-Verlag, edited by William Cartwright, Mike Peterson & Georg Gartner) and *The Hyperlinked Society* (University of Michigan Press, edited by Joseph Turow & Lokman Tsui). His book *The Political Mapping of Cyberspace* was published by the University of Chicago Press in 2004. Crampton's latest book is *Space, Knowledge and Power: Foucault and Geography* (Ashgate Press, 2007) co-edited with Stuart Elden. Crampton is the Section Editor for Cartography of the forthcoming *International Encyclopedia of Human Geography* which will be published by Elsevier. He is also Editor of *Cartographica: The International Journal of Geographic Information and Visualization*.

Joni Downs

Joni Downs is a Ph.D. candidate in the Department of Geography at Florida State University. She received both BS and MS degrees in Natural Resources from the Ohio State University. Her research interests include GIScience, spatial analysis, and location-allocation modelling, with an emphasis on applications in wildlife ecology and management. Joni has worked on a variety of research projects, ranging from wildlife home range estimation and habitat suitability modelling to network models and hurricane disaster relief planning.

Anthony M. Filippi
Anthony M. Filippi received the BA (summa cum laude) degree from Kansas State University, Manhattan, KS, USA, in 1995, and the MS and Ph.D. degrees from the University of South Carolina, Columbia, SC, USA, in 1998 and 2003, respectively, all in geography. He is currently an Assistant Professor in the Department of Geography, Texas A&M University, College Station, TX, USA. He was an Office of Naval Research (ONR)/NASA summer fellow at Friday Harbor Laboratories, University of Washington, in 1998, and he has been a faculty fellow at Oak Ridge National Laboratory (ORNL) during the summers of 2005, 2006, and 2007. His research interests include hyperspectral remote sensing, environmental and ocean optics, machine learning, and GIS-based modeling. Dr. Filippi is a Fellow of the American Geographical Society (AGS), and he is a member of the American Geophysical Union (AGU), American Society of Limnology and Oceanography (ASLO), American Society for Photogrammetry and Remote Sensing (ASPRS), Association of American Geographers (AAG), and SPIE. He is also a member of Phi Beta Kappa, Phi Kappa Phi, Golden Key National Honor Society, and Gamma Theta Upsilon.

Michael F. Goodchild
Michael F. Goodchild is Professor of Geography at the University of California, Santa Barbara, and Director of spatial@ucsb. He received his BA degree from Cambridge University in Physics in 1965 and his Ph.D. in Geography from McMaster University in 1969. He was elected member of the National Academy of Sciences and Foreign Fellow of the Royal Society of Canada in 2002, and member of the American Academy of Arts and Sciences in 2006, and in 2007 he received the Prix Vautrin Lud. He was editor of *Geographical Analysis* between 1987 and 1990 and editor of the Methods, Models, and Geographic Information Sciences section of the *Annals of the Association of American Geographers* from 2000 to 2006. He serves on the editorial boards of ten other journals and book series, and has published over 15 books and 400 articles. He was Chair of the National Research Council's Mapping Science Committee from 1997 to 1999, and currently chairs the Advisory Committee on Social, Behavioral, and Economic Sciences of the National Science Foundation. His current research interests center on geographic information science, spatial analysis, and uncertainty in geographic data.

Peter Gordon
Peter Gordon is a Professor in the University of Southern California's School of Policy, Planning and Development. He is also attached to USC's Center for Risk and Economic Analysis of Terrorist Events (CREATE). Gordon's research interests are in applied urban economics and policy analysis. This includes the spatial and institutional evolution of cities. Gordon and his colleagues have developed various economic impact models which they apply to the study of the effects of infrastructure

investments or disruptions from natural events or terrorist attacks. Peter Gordon has published in most of the major urban planning, urban transportation, and urban economics journals. His recent papers are at www-rcf.usc.edu/~pgordon. He has acted as a consultant for local, state, and federal agencies, the World Bank, the United Nations, and many private groups. Gordon received the Ph.D. from the University of Pennsylvania in 1971. Gordon maintains a blog at http://www-rcf.usc.edu/~pgordon/blog/

Matthew Hannah

Matthew Hannah is Chair in Human Geography at the University of Wales, Aberystwyth. His research interests center on relations between power, knowledge and territory in the modern West. In addition to his book *Governmentality and the Mastery of Territory in Nineteenth Century America* (Cambridge 2000), he has published numerous articles on the history of censuses and administration of space in the US. He is now researching census boycott movements in the 1980s West Germany.

Wesley 'Clint' Hoffmann

Wesley 'Clint' Hoffmann is an Agricultural Engineer with the USDA-Agricultural Research Service (ARS) in College Station, TX. After receiving a MS from the University of Florida and a Ph.D. from Texas A&M University in Agricultural Engineering, Dr. Hoffmann joined ARS with a focus on aerial application or 'crop dusting' research, where he serves as the Lead Scientist of the Aerial Application Technology project. His research efforts are focused on effects of physical properties and nozzle operational parameters on spray atomization, spray evaluation and development, and sampling methodologies for measuring spray droplet transport in the environment. He has published over 30 articles related to application technology and serves on several technical committees within professional societies.

Yanbo Huang

Yanbo Huang is a General Engineer of the Areawide Pest Management Research Unit for the US Department of Agriculture, Agricultural Research Service at College Station, TX. He earned a BS degree in Industrial Automation at Beijing University of Science and Technology in 1983, an MS degree in Industrial Automation at Chinese Academy of Mechanics and Electronics Sciences in 1986, and a Ph.D. degree in Agricultural Engineering at Texas A&M University in 1995. In the last decade Dr. Huang has been working in research, extension and teaching on information and electronics technologies for biological and agricultural engineering. He has extensive theoretical and applied knowledge in information systems, sensors, instrumentation, process control, mathematical modeling, pattern recognition, statistics and decision support for food engineering, irrigation engineering, pest management and manufacturing industry.

Andrew G. Klein

Andrew G. Klein received a BA in Geology and Environmental Studies from Macalester College in 1990 and a Ph.D. in Geological Sciences from Cornell University in 1997. He is currently an Associate Professor in the Department of

Geography at Texas A&M University. Dr. Klein's primary research interests lie in the application of Geographic Information Science (GISci) and Remote Sensing to the cryosphere. He is actively monitoring the retreat of glaciers throughout the tropics using remote sensing, and is interested in developing and validating algorithms to retrieve properties of snow and ice from satellite images. Dr. Klein is actively involved in employing geospatial technologies to help monitor the impact of science and operations in Antarctica as part of the efforts of the United States Antarctic Program to comply with the International Antarctic Treaty. Dr. Klein is also interested in spatial thinking and how to best educate students in Geospatial Technologies. He has been actively involved in several educational programs that are designed to improve spatial thinking and the use of geospatial technologies in the K-12 classroom.

Yubin Lan

Yubin Lan works as an Agricultural Engineer with Aerial Application Technology Group, Areawide Pest Management Research Unit, USDA-ARS at College Station. He is also an adjunct professor and graduate faculty with the Department of Biological and Agricultural Engineering, Texas A&M University, College Station, TX. Dr. Lan received his BS (1982) and MS (1987) from Jilin University of Technology and a Ph.D. (1994) from Texas A&M University (TAMU). Dr. Lan's current research interests include precision agriculture in aerial application, decision support system for precision areawide pest management, sensor and controls development, remote sensing/GIS/GPS, remote sensing with crop modeling, multisensor data fusion, and electronic nose/biosensor.

Paul A. Longley

Paul A Longley is a Professor of Geographic Information Science at University College London, UK. Paul Longley's research interests are grouped around the development and application of geographic information systems and geographic information science for socioeconomic analysis. They include: information integration within GIS (predominantly socioeconomic applications but also remote sensing–GIS integration); geodemographics; public service delivery (specifically health, education and policing); Internet GIS applications and e-social science; housing and retail market analysis; fractal geometry; and social survey research practice. These interests have been funded to date through a wide portfolio of research grants, including extensive knowledge transfer/exchange activities. In addition to contributing to the research literature (Paul has published 11 books and over 100 refereed articles), his work has been linked to important pedagogic and outreach activities, particularly through the website www.spatial-literacy.org.

Praveen Maghelal

Praveen Maghelal, Assistant Professor in the Department of Urban and Regional Planning in the College of Architecture, Urban, and Public Affairs at Florida Atlantic University, has educational and professional experience in Planning, Architecture, and Civil Engineering. He received his Ph.D. in Urban and Regional Sciences from the Texas A&M University, Masters in Infrastructure Planning from

New Jersey Institute of Technology, and Bachelors in Architecture and Diploma in Civil Engineering from India. His research interest includes improving sustainability community planning through sustainable transportation. Dr. Maghelal's specialization includes transportation planning, spatial planning, urban form assessment, and land-use planning. Currently, Dr. Maghelal is working on developing indices to measure the suitability and walkability of built-up environment that affects walking and biking in communities. He is also working with members at Texas A&M on developing an index that measures the critical risk in communities due to the presence of sex offenders. He has published in journals including Journal of American Planning Association and Institute of Transportation Engineers.

David H. McIntyre
David H. McIntyre is the Director of the Integrative Center for Homeland Security at Texas A&M University. He also teaches homeland security and terrorism at the Bush School of Government and Public Service at Texas A&M, where he directs the graduate Certificate for Homeland Security program. He hosts the weekly radio short 'Just a Minute for Homeland Security' and co-hosts the weekly program 'Homeland Security: Inside and Out.' Dr. McIntyre is a Fellow at the Homeland Security Policy Institute, The George Washington University, Washington, DC, and on the Steering Committee, Homeland Security/Defense Educational Consortium, USNORTHCOM. A nationally recognized analyst, writer, and teacher specializing in national and homeland security, Dr. McIntyre is a 30 year Army veteran, who has been designing and teaching national security and homeland security strategy at senior levels of government for 19 years. His specialty is security analysis and teaching through experiential learning. He holds a BS in Engineering from West Point, an MA in English and American Literature from Auburn University, and a Ph.D. in Political Science from the University of Maryland. He is a graduate of the US Army War College and the National War College.

Victor Mesev
Victor Mesev is Professor and Chair of Geography at Florida State University, Tallahassee, USA. He completed his doctoral studies at the University of Bristol, England and his primary research is summarized in two books, the '*Remotely Sensed Cities*' and the '*Integration of GIS and Remote Sensing*.' Dr. Mesev was formerly at the University of Ulster where research on conflict resolution and residential segregation was funded by the European Union Interreg program and the Community Relations section of the Office of the First Minister and Deputy First Minister of Northern Ireland.

Mark Monmonier
Mark Monmonier is Distinguished Professor of Geography in Syracuse University's Maxwell School of Citizenship and Public Affairs. He received a BA from Johns Hopkins University in 1964 and a Ph.D. from Pennsylvania State University in 1969. He has served as editor of *The American Cartographer* and president of the American Cartographic Association, and has been a research geographer for the US Geological Survey. Monmonier's awards include a Guggenheim Fellowship (1984),

the Association of American Geographers' Media Achievement Award (2000), and the American Geographical Society's O. M. Miller Cartographic Medal (2001). He has published numerous papers on map design, automated map analysis, cartographic generalization, the history of cartography, statistical graphics, and mass communications, and is author of fifteen books, including *How to Lie with Maps* (1991, 1996), *Cartographies of Danger* (1997), and *Coast Lines: How Mapmakers Frame the World and Chart Environmental Change* (2008). Monmonier has served on the National Research Council's Mapping Science Committee and the NRC Panel on Planning for Catastrophe, and is editor of Volume Six (the Twentieth Century) of the general history of cartography published by the University of Chicago Press. He is a member of the EPA's Coastal Elevations and Sea Level Rise Advisory Committee (CESLAC).

James Elliott Moore II

James Elliott Moore II, Ph.D., has been with University of Southern California (USC) since 1988. He is director of the Transportation Engineering program in the Department of Civil and Environmental Engineering, and co-director of the Master of Construction Management (MCM) program, which is jointly sponsored by Civil Engineering and the School of Policy, Planning, and Development. Professor Moore conducts fundamental and applied research on the engineering economic aspects of large-scale transportation and land use systems. His research interests include risk management of infrastructure networks subject to natural hazards and terrorist threats; economic impact modeling; transportation network performance and control; large scale computational models of metropolitan land use/transport systems, especially in California; evaluation of new technologies; and infrastructure investment and pricing policies. He has been published extensively in transportation planning and engineering literature. Prior to joining USC, he was on the faculty of Northwestern University's McCormick School of Engineering and Applied Science.

Miriam Olivares

Miriam Olivares is a doctoral candidate of Urban and Regional Science at Texas A&M University. She holds a Masters in Land Development from Texas A&M and a Bachelor of Architecture degree from Monterrey Tech, Mexico. She has more than 12 years experience of interdisciplinary work in architecture and urban planning and has her own firm based in Mexico. Her research interests are sustainable communities, sex crime risk management, and urban strategic planning. Current research efforts focus on sex crime and sustainable communities.

Qisheng Pan

Qisheng Pan is an Associate Professor in the Department of Urban Planning and Environmental Policy at Texas Southern University. He teaches courses at the Master and Ph.D. levels in quantitative analysis, advanced planning analysis, GIS, and transportation planning. He earned a Masters Degree in Computer Sciences and a Ph.D. in Urban and Regional Planning from the University of Southern California (USC), Los Angeles. Professor Pan had worked for the National Center for Metropolitan

Transportation Research (METRANS) at USC from 2000 to 2003 on METRANS-funded research projects. He was a key member on a team that evaluated CALTRANS (California Department of Transportation) projects about the socio-economic impacts of Caltrans I-5 and SR91 expansion plan on the Los Angeles metropolitan region. He was also involved in two NSF (National Science Foundation) digital government projects and worked on several DHS (Department of Homeland Security) research projects as consultant. He served as Principal Investigator on multiple research projects from the Texas Department of Transportation and other state and federal funding agencies. Recently, he worked as consultant in a RAND Study on the Western Riverside County Multiple Species Habitat Conservation Plan.

Harry W. Richardson
Harry Richardson is the James Irvine Chair of Urban and Regional Planning in the School of Policy, Planning and Development and a Professor of Economics at the University of Southern California. He is focusing much of his current research efforts on the economics of terrorism at CREATE (Center for Risk and Economic Analysis of Terrorism Events), USC's terrorist research center. He is the author of 26 books and about 200 research papers. He was an Overseas Visiting Fellow at Churchill College, Cambridge University in the United Kingdom in Fall 2004 and Spring 2006, pursuing research on urban regeneration and London's congestion pricing scheme. He was given the Walter Isard Award for Scholarly Achievement in Regional Science by the Regional Science Association International in 2004. He is also doing research on Korean reunification, and was a Posco Visiting Fellow in the East–West Center at the University of Hawaii in Fall 2006.

Peter Shirlow
Peter Shirlow is Senior Lecturer in Law at Queen's University, Belfast, Northern Ireland. He is co-author of the books *'Beyond the Wire,' Former Prisoners and Conflict Transformation in Northern Ireland* (with Kieran McEvoy), and *Belfast: Segregation, Violence and the City* (both University of Michigan Press). He has written widely on conflict in Northern Ireland in journals such as *Political Geography*, *Urban Studies*, *Antipode*, *Area*, *Capital*, and *Class and Environment and Planning A*.

John Ford Shroder, Jr.
Dr. John 'Jack' F. Shroder, Jr., is currently Professor of Geography and Geology. He has plentiful research experience in South Asia (Afghanistan and Pakistan) and is editor of 11 books and monographs and over 100 professional papers. Dr. Shroder and another faculty member initiated the Afghanistan Studies Center in 1972 at the University of Nebraska at Omaha, as well as a project with Kabul University to produce a National Atlas of Afghanistan. His Fulbright award there was terminated by the communist takeover in late 1978, so Dr. Shroder worked thereafter during the Afghan/Soviet war on issues related to the ongoing confrontation and in helping the Afghan mujahideen. Awarded a Fulbright to Peshawar, Pakistan, in 1983–84, Dr. Shroder renewed his work on geology, geography, and resources in South Asia,

as well as lecturing to Afghan refugees on clean water sources and sewage sanitation. In the 1980s he worked to assess agriculture in Afghanistan, as well surveying mineral resources and hydrocarbons for use in post-war redevelopment. Dr. Shroder is currently Co-Director of the GLIMS (Global Land Ice Measurements from Space) Regional Center on Afghanistan and Pakistan, which is actively assessing ice, snow, and meltwater resources in drought-torn South Asia.

Michael Ward

Michael Ward (B.V.Sc., MSc., MPVM, Ph.D.) is a Professor of Epidemiology, Department of Veterinary Integrative Biosciences, College of Veterinary Medicine and Biomedical Sciences, Texas A&M University [http://www.cvm.tamu.edu/vaph/vetepi/people.htm]. Dr. Ward is a veterinary epidemiologist. His undertook his veterinary medical training in Australia and graduate training at the University of California, Davis. Prior to joining Texas A&M University, Dr. Ward was employed at Purdue University. He has 20 years experience in conducting research on diseases of livestock in Australia, the US, Argentina, and Romania. He teaches infectious disease epidemiology and spatial epidemiology, and has taught GIS short courses in Texas, Minnesota, Chile, Denmark, Romania and Argentina. Dr. Ward's areas of expertise include spatial epidemiology and simulation modeling. Current areas of research include forecasting systems for mosquito-borne diseases, such as West Nile and St. Louis Encephalitis, and spatial diffusion models of foot-and-mouth disease and avian influenza. In 2008, Dr. Ward will be taking up the inaugural Sesquicentennial Chair of Veterinary Public health at the University of Sydney, Australia.

John K. Westbrook

John K. Westbrook serves as Research Leader of the Areawide Pest Management Research Unit for the US Department of Agriculture, Agricultural Research Service at College Station, TX. He earned a BS degree in Meteorology at San Jose State University in 1977, and MS and Ph.D. degrees in Biometeorology at Utah State University in 1980 and 1982, respectively. Dr. Westbrook investigates atmospheric impacts on insect population dynamics and migratory flights, and incorporates GIS for spatial analysis and mapping of pest populations. He has developed weather balloon tracking instrumentation and pioneered the use of superpressure balloons for insect migration research. Throughout his career, he has contributed to interdisciplinary research on areawide insect pest management strategies.

Douglas F. Wunneburger

Douglas Wunneburger, Research Scientist and Lecturer, directs the GeoInformatics Studio in the College of Architecture at Texas A&M University. He received his Ph.D. in Forestry from Texas A&M University having studied remote sensing and spatial information systems. Earlier work includes a Masters in Forest Science from Stephen F. Austin State University, and Bachelor of Arts in Economics from the University of Texas. Dr. Wunneburger teaches courses in GIS for the Landscape Architecture and Urban Planning Department at Texas A&M. General research interests include the integration of spatial and information technology for research, planning and management of urban landscapes. Current efforts focus on spatial impacts of governmental policies and legal statutes, particularly with regard to the relationship between school attendance zones and academic performance, and to the viability of laws directed at the reduction of adverse impacts of sex offenders in communities.

Sammy Zahran
Sammy Zahran holds a BA from University of Windsor (Canada) and Ph.D. from University of Tennessee. He is an assistant professor of sociology at Colorado State University. Dr. Zahran's areas of research include population and demography, spatial science, and environmental risk, planning and policy. Prior to joining Colorado State University as a faculty member, he was a post-doctoral research fellow at the Institute for Science, Technology, and Public Policy at the George Bush School of Government, Texas A & M University. He is currently involved in many funded projects, including advancing the resilience of coastal localities in Texas and Louisiana, developing, implementing and sustaining the use of coastal resilience indicators, and policy making for climate change & drought in the US South, among others.

F. Benjamin Zhan
F. Benjamin Zhan is professor and director of the Texas Center for Geographic Information Science in the Department of Geography at Texas State University-San Marcos. He obtained his Bachelor of Engineering from Wuhan Technical University of Surveying and Mapping in China, MSc from the International Institute for Geo-Information Science and Earth Observation (ITC) in the Netherlands, and Ph.D. from the Department of Geography and the US National Center for Geographic Information and Analysis (NCGIA) at the University at Buffalo, the State University of New York (SUNY). Professor Zhan's major research interests are in Geographic Information Science and its applications, focusing on three topical areas: (1) spatial analysis and modeling, (2) transportation and network science, and (3) health and the environment. He has written or co-authored more than 90 journal articles and other professional publications. Professor Zhan's papers have been cited by researchers in at least 32 countries.

Chapter 1
Geospatial Technologies and Homeland Security: An Overview

Daniel Z. Sui

1.1 Introduction

In the aftermath of the terrorist attacks of September 11, homeland security has not only become a top priority of US government policies at all levels (National Governors' Association 2006; Homeland Security Council 2007), it also has become an emerging field of study (Bullock et al. 2006). Homeland security presents an enormous opportunity but also poses daunting challenges for higher education (National Research Council 2005). Despite the diverse interpretation of the term 'homeland security,' existing literature can be grouped in two categories: either narrowly defined as dealing with intentional, human-induced terrorist attacks, or more broadly defined as dealing with any disasters of technological, natural, and human origin. This book takes the broader view.

Regardless of how homeland security is defined, the tasks it deals with typically follow more or less the same cycles of response in a typical emergency situation: detection, preparedness, prevention, protection, rescue, relief, recovery, and reconstruction (Cutter 2003). Although data needed in each phase of an emergency response may differ in spatial and temporal scales, government agencies and citizens need access to real-time, multiple types of general information, as well as accurate geospatial information in order to accomplish many of the tasks during an emergency response situation (Briggs et al. 2002; National Commission on Terrorist Attacks upon the United States 2004; Patterson and Apostolakis 2007).

Geospatial technologies encompass a suite of technologies that deal with where, what, when, and how information can be obtained explicitly covering a particular locale, increasingly on multiple scales. Immediately following the 9/11 attacks, government, industry and academy have recognized the need for a tighter coupling between geospatial technologies and homeland security missions, as the former can provide crucial information for accomplishing most of the tasks of the latter (ESRI 2001; FGDC 2001; Beck 2003; Katoaka 2007). Along with bio- and nanotechnology, geospatial technologies are widely regarded as one of the three leading technological areas in the 21st century. Generally speaking, geospatial technologies include,

Texas A&M University

D.Z. Sui (ed.) *Geospatial Technologies and Homeland Security*,

but are not limited to, geographic information systems (GIS), remote sensing, global positioning systems (GPS), and location-based systems (LBS). The emerging geographic information science (GIScience) addresses the fundamental issues behind spatial data handling efforts, such as spatial data representation, analysis, modeling, and visualization (National Research Council 2006a). Although the various chapters in this volume are primarily related to applications of geospatial technologies in the homeland security context, we believe that this dynamic field also poses new challenges and questions that are better dealt with within GIScience.

1.2 Synopsis of this Book

This book is based upon papers presented during the one-day symposium on 'Geospatial Technologies and Homeland Security,' which took place at Texas A&M University on Nov. 16, 2006. The symposium was sponsored by the Office of Vice President for Research at Texas A&M University, with support from the Integrative Center for Homeland Security, National Center for Foreign Animal and Zoonotic Disease Defense (FAZD), and Department of Geography in the College of Geosciences at Texas A&M. The TAMU geospatial information science and technology (GIST) symposium series are devoted to discussing the recent developments in geospatial technologies and their applications in areas of great public interest by leading researchers and government program officers. Homeland security was chosen as the inaugural topic for the TAMU GIST symposium series.

All the chapters in the book have gone through a rigorous reviewing process—similar to that of refereed journal articles. Chapters 2–18 cover a variety of topics related to the scientific, technical, and social aspects of geospatial technologies for homeland security applications. In terms of applications, we have tried to maintain a balance between chapters devoted to issues related to terrorist attacks vs. applications on natural disasters.

Chapter 2 by Jack Shroder takes us right to the frontline of the on-going war against terrorism—Afghanistan. According to Shroder, Afghanistan was created in the 19th century as a classic buffer state to avoid confrontation between major powers, but in the 21st century it has become a focal point for armed conflict. Due to its unique topographic characteristics, many of Afghanistan's problems can be studied through geospatial technologies. Shroder argues that instead of the present weakness and ineptitude, a Marshall-Plan style of post-war rebuilding could have led to a robust Islamic democracy instead of re-energizing the Taliban and worsening other problems. He also warns that winning every battle could still result in losing the overall war for Afghanistan unless stronger redevelopment measures are promptly undertaken.

Chapter 3 by Qisheng Pan, Peter Gordon, James Moore, II, and Harry Richardson examines the economic impacts of terrorist attacks and natural disasters using two case studies from Los Angeles and Houston. Pan and his colleagues developed the

Southern California Planning Model (SCPM)—a GIS-based regional planning model capable of endogenizing freight and passenger flows, and allocating impacts spatially via unexpected impedances to trips and shipments through the regional highway network. Using SCPM as the primary simulation tool, Chapter 3 presents simulation results of a radiological bomb (or so-called 'dirty bomb') and conventional bomb attacks in Los Angeles, and a major hurricane striking the Houston-Galveston-Brazoria (HGB) region. Their results show that the model can allocate losses to various types of impact analysis zones or political jurisdictions.

Chapter 4 by David Ashby, Spencer Chainey, and Paul Longley reviews the use of geodemographics, and more specifically small area neighborhood profiles, in community policing and, by extension, in homeland security applications. They discuss the merits of a local focus in policing, and the data and analytical frameworks that are necessary to support this activity. Using case-study examples to illustrate how priorities for neighborhood policing may be developed, Ashby and his colleagues suggest that available public sector data may be used to drive improved bespoke classifications of neighborhoods. They further argue that better measures of local social capital and community cohesion can be used to develop interventions for specific local conditions in order to maintain and enhance community stability. Some ethical impediments to this type of homeland security application are also discussed in the chapter.

In Chapter 5, Victor Mesev, Richard Courtney, Joni Downs, Peter Shirlow, and Aaron Binns map the spatial distributions of conflict-related deaths in Belfast (Northern Ireland's capital) in an attempt to unravel the complex social, political, and ethno-religious underpinnings of the disputes between Irish republicans (mostly Catholics) and British unionists (mostly Protestants) in Northern Ireland, which have lasted for centuries and have claimed around 3,600 lives since the late 1960s. Religious segregation is claimed by many analysts to be a major contributory variable to explaining the pattern of conflict-related deaths. Mesev and his colleagues explore a modification of the spatial segregation index to examine the distribution of Catholic and Protestant neighborhoods in the city. This chapter concludes that while politically motivated attacks can be unpredictable, they seem to cluster within highly segregated and low social class neighborhoods located within close proximity to areas of interaction between Catholics and Protestants.

Chapter 6 by Douglas Wunneburger, Miriam Olivares, and Praveen Maghelal evaluates the effectiveness of geospatial sex offender legislation using a GIS-based spatial analysis approach. A number of laws have been passed in recent years to protect children from sexual predators in the US, most of which mandate creation and maintenance of registries for convicted offenders, establish safety zones from which known offenders are restricted, and require notification to residents of the presence of registered offenders in their neighborhoods. The results of this chapter cast serious doubts about the effectiveness of such sex offender laws. Although not directly linked to issues of homeland security, the results of this chapter nonetheless are informative on how geospatial technologies can be deployed to track potential terrorist activities.

Anthony Filippi's Chapter 7 offers an overview of remote sensing-based damage-assessment. Filippi paid particular attention to remote sensing-based detection of vegetation damage and soil contamination, including a discussion of the remotely-sensed effects of artificial radionuclide contamination, as well as damage to urbanized areas and other human settlements. Networks of sensors, both installed on remote (e.g. airborne or spaceborne) platforms and in fixed/discrete and dynamic/quasi-ubiquitous locations will play increasingly importantly roles in the realm of homeland security.

In Chapter 8, Samuel Brody and Sammy Zahran discuss how specific characteristics of the built-up environment can affect the level of damage in a flooded community at the regional scale. Their research examines the relationship between the built-up environment and flood impact using data from Texas. They calculate property damage resulting from 423 flood events over a five-year period (1997 to 2001) at the county level. Their results suggest that naturally occurring wetlands play a particularly important role in mitigating flood damage. Their findings can potentially provide guidance to homeland security experts and flood managers on how to most effectively mitigate the economic impact of floods at the community level.

Chapter 9 by F. Benjamin Zhan and Xuwei Chen presents an agent-based modeling approach for evacuation planning. The complexity of evacuation planning in an urban environment requires a modeling framework that can incorporate multiple factors into the modeling process. Zhan and Chen illustrate the application of agent-based modeling and simulation in estimating the evacuation time from the Florida Keys, and report some preliminary results in planning a hypothetical route for evacuating the elderly from a nursing home on Galveston Island, Texas, based on network dynamics.

In Chapter 10, Christian Castle and Paul Longley start by assessing the reasons for the current surge of interest in developing building evacuation analysis using models focusing explicitly upon human individual behaviors. Next they present an extended overview of available software for modeling and simulating pedestrian evacuation from enclosed spaces. Based upon their thorough review, Castle and Longley propose guidelines for evaluating all the pedestrian evacuation software currently available. They also discuss a sequential conceptual framework of software and model specific conditions with a hypothetical building evacuation application as an illustration.

Mark Monmonier's Chapter 11 discusses the potential for using topographic LIDAR, capable of providing a half-foot vertical resolution, as a promising solution to the pressing need for better coastal elevation data. Monmonier shows how topographic LIDAR, when integrated with bathymetric LIDAR, can yield a seamless topographic/bathymetric dataset useful in high-resolution modeling of storm surge. This high-resolution coastal elevation data is the key to planning for storm surge and seal level rise—currently a clear and imminent danger to most coastal cities, and of great concern to the homeland security community.

Chapter 12 by Yanbo Huang, Yubin Lan, John Westbrook, and Wesley Hoffmann discusses the application of geospatial technologies for precision area-wide pest management and its implications for homeland security after 9/11.

Area-wide pest management essentially represents coordinated adoption of integrated pest management to the preventive suppression of a pest species throughout its geographic distribution. According to Huang and his colleagues, precision area-wide pest management systems were originally developed to reduce the country's daily risk of natural pest introduction. Now these systems are being developed to reduce the risk of every pest introduction, both natural and intentional. This is the only chapter that deals with potential agro-terrorism in this book.

In Chapter 13, Michael Ward reviews the development of spatial epidemiology over the past 150 years. Combined with analytical details and case studies, Ward presents a quite comprehensive review of the current approaches in spatial epidemiology, most of which are potentially valuable for public health and homeland security applications. The chapter also discusses some of the major obstacles to the consolidation of spatial analysis as one of the foundations of modern epidemiology, including the availability and quality of spatial disease data, information on the distribution of populations at-risk, and integration of methods seamlessly into epidemiologic software packages.

Chapter 14 by Jeremy Crampton examines the role of geospatial technologies in the production of the politics of fear. While Crampton acknowledges the utility of geospatial technologies in our dealing with terrorism, crime, or natural disaster, he also argues that geospatial technologies have contributed to the use of fear for political exploitation. Crampton further discusses the multiple factors sustaining the politics of fear. Using two case studies from nineteenth century mapping and contemporary crime mapping, Crampton suggests that if geospatial technologies continue to produce knowledge on populations at risk, then a politics of fear can be exploited to justify mass geosurveillance.

Chapter 15 by Matthew Hannah argues that one of the chief dangers of using geospatial technologies in the post-9/11 world lies not in the knowledge they produce but rather in the ways they tend to transform the lack of knowledge into grounds for the withdrawal of rights from disadvantaged groups. Drawing on Foucault's ideas on 'race war discourses' within the context of stigmatizations of groups deemed 'inscrutable' or 'subversive' in US history, Hannah suggests that we should regard the 'underscrutinized' as an emerging 'race.' He expands his argument by discussing the 'techno-political' context through an account of the nation-wide census boycott movement in West Germany in 1987. Hannah concludes that the uneven geographical coverage of a geospatial data set can lead to stigmatization and discrimination against the unregistered, even in the absence of any intent on the part of experts and state authorities.

Chapter 16 by Robert Bednarz and Sarah Bednarz reminds us that we live in uncertain times in the context of war against terrorism. The authors contend that we cannot eliminate uncertainty and its effects, but we can minimize the disruption and loss that result from it if citizens receive appropriate training in spatial thinking. Using results from their classroom-based research, Bednarz and Bednarz discover that simply teaching students how to use geospatial technologies is insufficient for them to deal with uncertainty unless they are also trained how to think spatially. It is further argued that spatial thinking can be learned and should be taught.

They also provide recommendations on how to develop effective ways to integrate spatial thinking and geospatial technologies in education.

Chapter 17 by David McIntyre and Andrew Klein continues the discussion on topics related to GIS and homeland security education. According to the authors, we must now improve the integration of GIS with homeland security in our classrooms today in order to create a better tomorrow. To meet homeland security challenges, it is necessary to develop integrated solutions that cross discipline boundaries and incorporate new technologies like the geospatial technologies discussed in this book. Developing a curriculum that successfully incorporates the full potential of geospatial solutions requires effort on the part of GIS experts to acquaint themselves with the challenges of homeland security and the components of the solutions.

Chapter 18 by Michael Goodchild serves to wrap up the book. According to Goodchild, the acronym GIS can be decoded in three distinct ways—GISystems, GIScience, and GIStudies. This framework is used to provide an overarching synthesis of the diverse topics discussed in this book, and to ask whether the chapters of this book provide a complete picture of geospatial issues and applications to homeland security. Goodchild identifies three characteristics that distinguish homeland security applications from other domains—the need for speed, the difficult environments in which technology must operate, and the impossibility of anticipating many relevant kinds of events in either space or time. The author argues that one of the strongest factors impeding the effective use of geospatial technologies is the lack of collaboration between institutions and the various cultures of emergency response. Goodchild also identifies four themes (dynamics, geo-collaboration, sensor networks, and volunteered geographic information) that he believes are largely missing in the book, and elaborates on how they represent opportunities and challenges for future research.

1.3 Geospatial Technologies and Homeland Security: Next Steps

Continuing the research themes as outlined or discussed in Greene (2002), Cutter et al. (2003), van Oosterom et al. (2005), and Committee for Planning Catastrophe (2007), the chapters in this volume fill in many gaps in the current literature on geospatial technologies and homeland security under the broader context of the geography of war and peace (Flint 2005). Nonetheless, it should be recognized that these two fields are very dynamic, and exciting new developments are being constantly reported. I hope this volume will serve as a springboard to launch new research explorations into the complex issues in homeland security using geospatial technologies.

In the near future, I believe the tight coupling of geospatial technologies and homeland security will keep increasing along the technical, scientific, as well as legal/ethical fronts. Close interactions of these two dynamic fields will be mutually beneficial to both GIS and homeland security studies.

Along the technical front, the growing maturity of Web 2.0, coupled with accelerated development in ubiquitous, open source, grid computing, and sensor networks will provide unprecedented information technology infrastructure, which will lead us to revolutionary new ways of collecting, storing, accessing, analyzing, visualizing, and disseminating information on multiple spatial (from genetic to global) levels and temporal (from seconds to decades) scales. The rapid development of volunteered (user created) geographic information and the concept of 'humans as sensors' both deserve particular attention (Goodchild 2007). Although the US Congress did not approve the total information awareness (TIA) project in 2004 due to privacy concerns, the concept of TIA has been resurrected in many interesting disguises. With the development of wikimapia and intellipedia, both geospatial technologies and homeland security are converging towards a new wikification paradigm, which raises a series of new research questions related to interoperability, metadata standards, system control, and access, among others (Sui 2008).

Many of the challenging issues in both geospatial and homeland security cannot simply be confined to the technical domain. In fact, finding the root causes of terrorism, and eventually reducing, if not eliminating, them require basic scientific research related to human behavior and perception, the global economic system, and human interaction across ideological, political, social, cultural, and religious boundaries (Alexander 2000; Flint 2003; de Blij 2005; National Research Council 2006b; Ronczkowski 2006). One particular area deserving more attention is the recent interdisciplinary research on network science (Newman et al. 2006), popularly known as the small world phenomena (Sui 2006). Preliminary studies have shown that the behavior of terrorist cells or groups often exhibits small world characteristics, not fundamentally different from other events observed within social or natural systems (Arquilla et al. 1999). As I discussed earlier, emerging network science in general, and the literature on small world phenomena in particular, can also provide a new way of representing the world, very different from the views embedded in the current generation of GISystems. Thus, basic research in homeland security can also contribute significantly to expanding GIScience, which in turn may help develop better tools for handling complex homeland security tasks.

This book does not espouse the uncritical view that science and technology are the panacea that can help us solve all the problems of homeland security. On the contrary, I would like to remind the readers that throughout human history, there never was a technology that solved some problems without creating new ones (Tenner 1996). In other words, technologies are not only solutions; sometimes they are also part of the problems. In the area of homeland security, most experts in the field would agree that part of the field's daunting challenges are due to the accessibility of many advanced technologies (chemical, biological, nuclear) not only to big brothers (i.e. nation-states) but also to many little brothers, even possibly terrorist groups or individuals. In the case of geospatial technologies, they are certainly like double-edged swords—a theme elaborated by Crampton (Chapter 14) and Hannah (Chapter 15) in this book. RAND's influential report evaluates the homeland security implications of the publicly available geospatial data (Baker et al. 2004). For example, Google maps/earth have provided millions of peace-loving people in the world with easy access to a vast volume of geospatial information at their finger tips, but they have also enabled less peaceful people to use on-line geospatial information to better plan attacks on sensitive targets with more precision and accuracy (Sui 2007). The technologies are in many ways ahead of the legal and ethical framework that we currently have for dealing with the complex situation regarding geospatial technologies and homeland security.

In summary, there is a general consensus among policy makers and researchers that successfully addressing complex issues in homeland security requires further seamless integration of state-of-the-art geospatial technologies into every step of a typical emergency response cycle in order to accurately delineate the spatial distribution of populations at risk (MacFarlane 2005; National Research Council 2007; Thomas et al. 2007). Many unresolved issues along the technical, scientific, and legal/ethical fronts demand concerted interdisciplinary collaboration between experts working in diverse fields. It is also imperative that we infuse the new exciting research findings into our classrooms, both K-12 and higher education (National Research Council 2006c). There is also a need to educate citizens how to think spatially. Interdisciplinary cooperation and better prepared citizens are more important than ever in this age of uncertainty.

References

Alexander, D. (2000). *Confronting catastrophe*. (Oxford, UK: Oxford University Press)

Arquilla, J., Ronfeldt, D. & Zanini, M. (1999). Networks, netwar, and information-age terrorism. (In I.O. Lesser, B. Hoffman, J. Arquilla, D. Ronfeldt, & M. Zanini (Eds.), *Countering the new terrorism* (pp. 39–84). Santa Monica, CA: RAND)

Baker, J.C., Lachman, B.E., Frelinger, D.R., O'Conneil, K.M., Hou, A.C., Tseng, M.S., Orletsky, D. & Yost, C. (2004). *Mapping the risks: Assessing the homeland security implications of publicly available geospatial information*. (Santa Monica, CA: RAND/National Defense Institute)

Beck, R.A. (2003). Remote sensing and GIS as counterterrorism tools in the Afghanistan war: A case study of the Zhawar Kili region. *The Professional Geographer*, 55, 170–179

Briggs, D.J., Forer, P., Järup, L. & R. Stern (Eds.) (2002). *GIS for emergency preparedness and health risk reduction*. (Berlin: Springer)

Bullock, J.A., Haddow, G.D., Coppola, D., Ergin, E., Westerman, L. & Yeletaysi, S. (2006). *Introduction to homeland security* (2nd edition). (Amsterdam, The Netherlands: Elsevier)

Committee on Planning for Catastrophe, National Research Council (2007). *Successful response starts with a map: Improving geospatial support for disaster management*. (Washington, DC: The National Academies Press)

Cutter, S.L. (2003). GIScience, disasters, and emergency management. *Transactions in GIS*, 7(4), 439–445

Cutter, S.L., Richardson, D.B. & Wilbanks, T.J. (Eds.) (2003). *The geographical dimensions of terrorism* (New York: Routledge)

de Blij, H. (2005). *Why geography matters: Three challenges facing America: Climate change, the rise of China, and global terrorism*. (New York: Oxford University Press)

ESRI (Environmental Systems Research Institute) (2001). *GIS for homeland security: An ESRI White Paper*. (Redlands, CA) Retrieved October 25, 2006 from http://www.esri.com/library/whitepapers/pdfs/homeland_security_wp.pdf

FGDC (Federal Geographic Data Committee) (2001). *Homeland security and geographic information systems*. (Washington, DC: Department of the Interior) Retrieved October 25, 2006 from http://www.fgdc.gov/library/whitepapers-reports/white-papers/homeland-security-gis

Flint, C. (2003). Terrorism and counterterrorism: Geographic research questions and agendas. *The Professional Geographer*, 55(2), 161–169

Flint, C. (Ed.) (2005). *The geography of war and peace*. (New York: Oxford University Press)

Goodchild, M.F. (2007). Citizens as sensors: The world of volunteered geography. *GeoJournal*, 69(4), 211–221

Greene, R.W. (2002). *Confronting catastrophe: A GIS handbook*. (Redlands, CA: ESRI Press)

Homeland Security Council (2007). *National strategy for homeland security*. (Washington, DC: The White House)

Katoaka, M. (2007). *GIS for homeland security*. (Redlands, CA: ESRI Press)

MacFarlane, R. (2005). *A guide to GIS applications in integrated emergency management*. (London: Emergency Planning College, Cabinet Office) Retrieved October 25, 2006 from http://www.ukresilience.info/publications/gis-guide_acro6.pdf

National Commission on Terrorist Attacks Upon the United States (2004). *The 9/11 Commission Report: Final report of the national commission on terrorist attacks upon the United States* [Electronic version]. (New York: W.W. Norton). Retrieved October 25, 2006 from http://www.gpoaccess.gov/911/

National Governors' Association (2006). *State strategies for using IT for an all-hazards approach to homeland security* [Electronic version]. Retrieved October 25, 2006 from http://www.nga.org/Files/pdf/0607HOMELANDIT.PDF

National Research Council (2005). *Frameworks for higher education in homeland security*. (Washington, DC: The National Academies Press)

National Research Council (2006a). *Beyond mapping: Meeting national needs through enhanced geographic information science*. (Washington, DC: The National Academies Press)

National Research Council (2006b). *Facing hazards and disasters: Understanding human dimensions*. (Washington, DC: The National Academies Press)

National Research Council (2006c). *Learning to think spatially: GIS as a support system in the K-12 curriculum*. (Washington, DC: The National Academies Press)

National Research Council (2007). *Tools and methods for estimating populations at risk from natural disasters and complex humanitarian crises*. (Washington, DC: The National Academies Press)

Newman, M., Barabasi, A.L. & Watts, D.J. (Eds.) (2006). *The structure and dynamics of networks*. (Princeton, NJ: Princeton University Press)

Patterson, S.A. & Apostolakis, G.E. (2007). Identification of critical locations across multiple infrastructures for terrorist actions. *Reliability Engineering & System Safety*, 92(9), 1183–1203

Ronczkowski, M.R. (2006). *Terrorism and organized hate crime: Intelligence gathering, analysis and investigations* (2nd Edition). (Boca Raton, FL: CRC)

Sui, D.Z. (2006). Geography and the small world: Challenges for GIScience. *GeoWorld*, June Issue, 24–26

Sui, D.Z. (2007). Is GIS a liability in the war against terrorism? *GeoWorld*, March Issue, 20–22

Sui, D.Z. (2008). The wikification of GIS and its consequences: Or Angelina Jolie's new tattoo and the future of GIS. *Computers, Environment, and Urban Systems*, 32(1), 1–5

Tenner, E. (1996). *Why things bite back: Technology and the revenge of unintended consequences*. (New York: Knopf).

Thomas, D.S.K., Ertugay, K. & Kemec, S. (2007). The role of geographic information systems/remote sensing in disaster management. (In H. Rodriguez, E.L. Quarantelli, & R. Dynes (Eds.), *Handbook of disaster research* (pp. 83–96). New York: Springer)

van Oosterom, P., Zlatanova, S. & Fendel, E.M. (2005). *Geo-information for disaster management*. (Berlin: Springer)

Chapter 2
Remote Sensing and GIS as Counterterrorism Tools for Homeland Security: The case of Afghanistan

John (Jack) Shroder

Abstract The classic buffer state of Afghanistan, created to avoid confrontation in the 19th century, has in the 21st century become a focal point for armed conflict that can only be alleviated by attention to the country's problems, many of which can be studied or alleviated through geospatial technologies. Instead of the present weakness and ineptitude, a Marshall-Plan style of post-war rebuilding could have led to a robust Islamic democracy instead of re-energization of the Taliban and the creation of other problems hindering the country's reemergence from its long national nightmare. Winning every battle could still result in losing the overall war for Afghanistan unless stronger redevelopment measures are undertaken promptly.

Keywords Afghanistan, Afghanistan Information Management Service (AIMS), Landsat & ASTER satellite imagery analysis, Osama bin Laden, Soviet-Afghan war

2.1 Introduction

2.1.1 Background

The problems in Afghanistan over the past few decades have been legion; unfortunately, the solutions attempted have been minimal, incorrect, or too expensive to maintain. This chapter is a discussion of some of the chief problems that might be acted upon to beneficial effect for homeland security in Afghanistan, as well as most importantly, for the USA. Most of these factors have a direct geospatial technology element. Part of the future of the War on Terrorism is likely to depend fairly heavily on solutions to some of these problems (Shroder 2002a). Data bases, web reports, and geographic information system (GIS) assessments of many of these

University of Nebraska at Omaha

D.Z. Sui (ed.) *Geospatial Technologies and Homeland Security*,

issues leading to terrorism coming from Afghanistan are especially important. Certainly if the country of Afghanistan is allowed to continue to drift back into the failed state it once was and could become again (Rubin 1995, 2007), its future prospects would not bode well for the region, or for the USA's long term prospects to achieve some victory in its struggle with the Islamists and the more dangerous Jihadis.

Some of the background to geospatial assessment of Afghanistan began in the middle of the past century. As a result of the ongoing considerable interest in Afghanistan throughout the Cold War era, largely because of its geostrategic location on the southern border of the USSR, a group of American scholars on Afghanistan were searching in the late 1960s for a financially committed, academic home base. Christian Jung, not yet finished with his doctorate on the Hazara migration to Kabul, asked the author to help him, and together in 1972 we convinced the administration of the University of Nebraska at Omaha (UNO) to establish the Afghanistan Studies Center (ASC). Mr. Jung died in 1973 so I went with administrators to Kabul for a few months to help hire Thomas Gouttiere, the new director of the nascent program on Afghanistan, as well as to initiate mapping of the country in a project to produce a National Atlas of Afghanistan in collaboration with Kabul University (Shroder 1975). An initial geospatial assessment of Afghanistan in the form of a large (1 × 2 m) mosaic of the country made from early Landsat images (Fig. 2.1) was presented to Afghanistan President Daoud to introduce the concept of remote sensing to the nation (Shroder 1977, 1978). In 1977–78, with money from USAID through the National Science Foundation, the author undertook the National Atlas Project full time in Afghanistan, which led to extensive travels and mapping throughout the country, as well as low-altitude over-flights and aerial photography in a small airplane provided by US Ambassador T. Eliot, even though such photography was not permitted by the Government of Afghanistan (GOA). Following the Communist coup d'etat of May 1978, the author was placed under house arrest for some months and then deported from the country prior to the full Soviet invasion in 1979. In spite of this temporary setback, many atlas maps were saved and sent out of Afghanistan by diplomatic pouch so that they became the nucleus of an important research collection at UNO thereafter. The Atlas Project was then taken over, in name only, by the Geokart organization from the Communist Eastern Block nation of Poland, and a variety of maps were published by them in 1987.

The Soviet invasion of Afghanistan was soon met with armed resistance by the people of Afghanistan who had been building up profound resentment to the incompetence and brutal actions of their own Communist factions (Bradsher 1983). The Soviet–Afghanistan War unfolded throughout the 1980s (Crile 2003) with the USA providing over $600 million in humanitarian aid to the refugees, plentiful maps to covert operatives (Figs. 2.2, 2.3), and over $3 billion in covert aid to the *mujahideen* resistance, channeled through the Government of Pakistan (Goodson 2001; Coll 2004). An additional $5 billion went to Pakistan at the same time. In part because the more fundamentalist Moslems of Afghanistan were leading the charge against the 'godless Communists,' the more religiously fanatic of the *mujahideen*

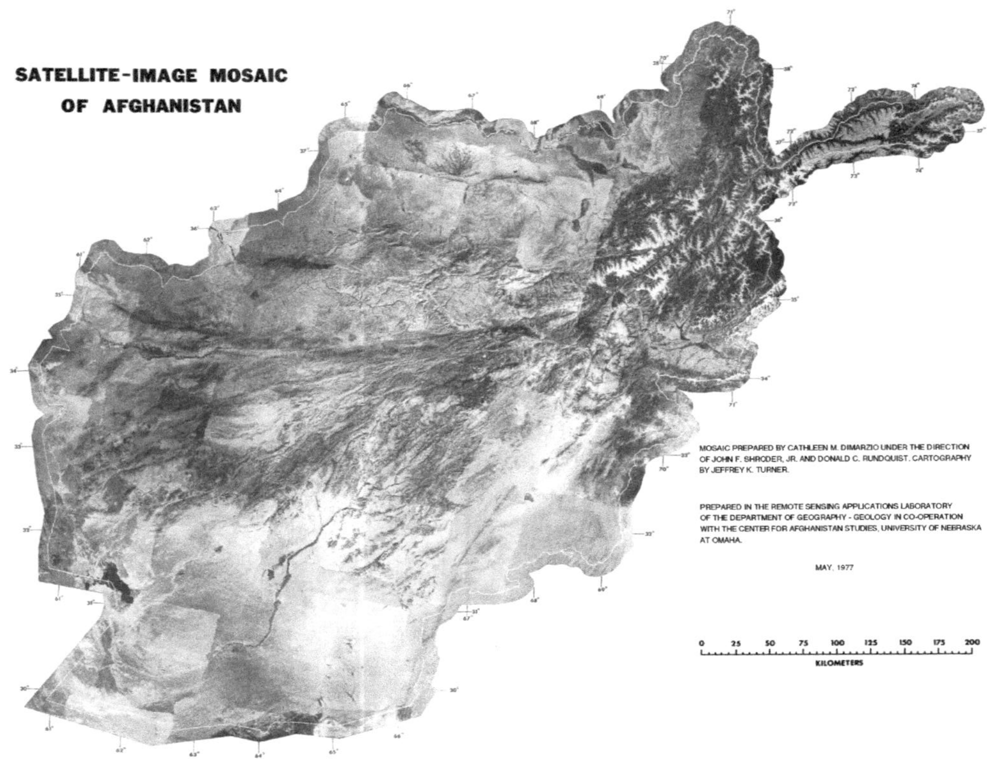

Fig. 2.1 Black and white Landsat image mosaic of Afghanistan (Shroder, 1977)

Fig. 2.2 Part of the Kabul and Pansher landsat mosaic (1:500,000 scale) printed by the CIA during the Soviet–Afghan War showing the area between Salang Pass, Baghram Air Base, and Kabul

resistance (Edwards 2002) were unfortunately supported more than the less fanatic, thus inadvertently helping give rise later to anti-West, pro-Jihadi factions that are now such a threat to our homeland security.

As a result of the dramatic educational needs described by the Afghan resistance leaders to the author in Peshawar, Pakistan in 1983–84, UNO was shortly thereafter able to obtain the funding to help run an Afghanistan Ministry of Education in exile there, which was responsible for some 1300 schools inside the country (Coll 2004). As well as fairly basic education in reading, writing, and arithmetic, education in

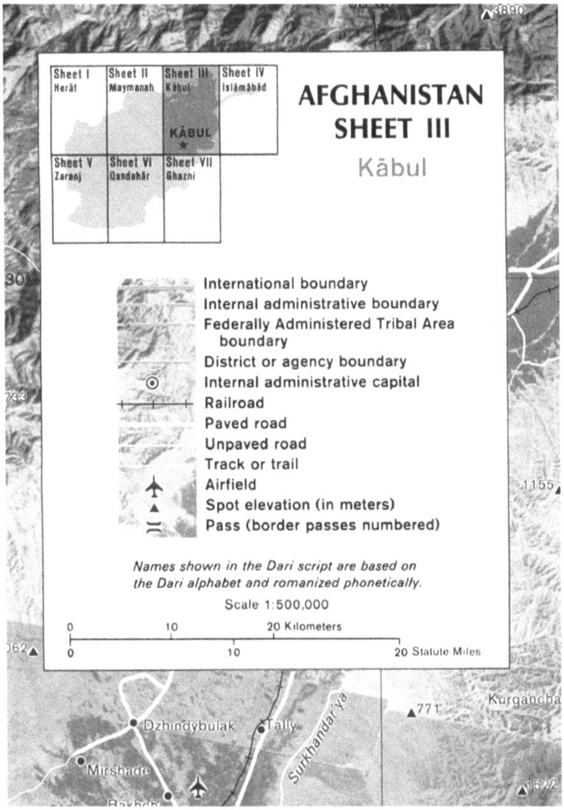

Fig. 2.3 Example of the legend of part of the Kabul III sheet at 1:500,000 scale of the Landsat mosaic in false color made by the CIA in the early 1980s, with gazetteer information on the back

the avoidance of land mines was a special feature of the schoolbooks that were written in Dari and Pushto by the Afghans in our project. Whereas a few pages at the height of the war presented arithmetic problems involving guns and dead Russians (Fig. 2.4), nothing like the supposed rampant militarism and 'blatant Islamist propaganda' was ever produced. This is in contradistinction to the rather overwrought and misleading opinions of Dreyfuss (2005) who apparently seeks to blame UNO in part for leading to, or supporting later terrorism.

As the Soviet–Afghanistan War played out in the 1980s, the author published much on the role of natural resources in the Soviet interest in Afghanistan (Shroder 1980a, b, 1981, 1982, 1983, 1987, 1989a; Shroder and Tawab Assifi 1987). The Soviets had produced a series of excellent new geology maps and resource reports about the country that the author had obtained. Following withdrawal of the defeated Soviet Army in 1988–89 (Borovik 1990; Grau and Gress 2002), the possible role of the relatively rich natural resource base in the rebuilding of war-torn Afghanistan became a focal point of interest to a number of people, and attention

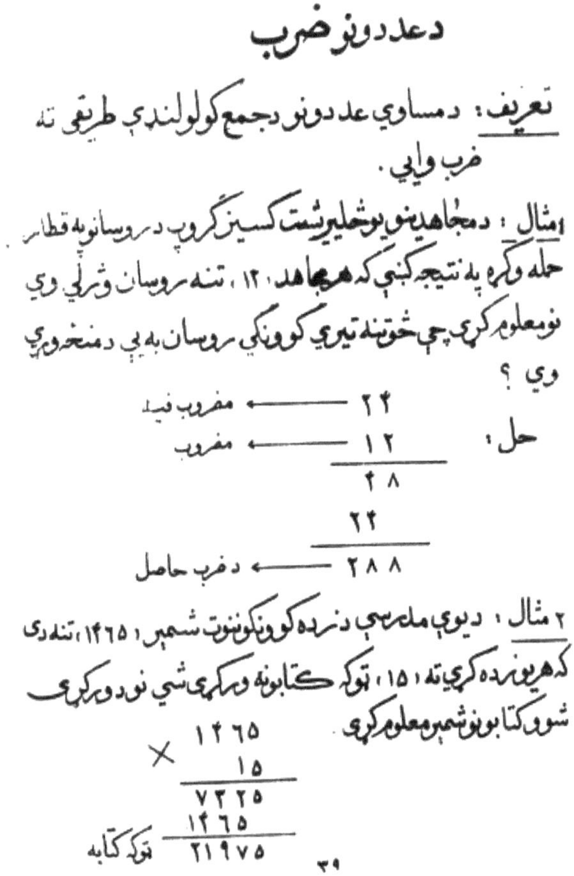

د عددونو ضرب

تعریف: د مساوي عددونو د جمع کولو لنډې طریقې ته ضرب وايي.

مثال: د مجاهدینو یو څلیرویشت کسیز ګروپ د روسانو په قطار حمله وکړه په نتیجه کښې که هر مجاهد ۱۲، تنه روسان وژلي وي نو معلوم کړئ چې څو تنه تیري کوونکي روسان به یې د منځه وړي وي؟

حل:
۲۴ ← مضروب فیه
۱۲ ← مضروب
۴۸
۲۴
۲۸۸ ← د ضرب حاصل

۲ مثال: د یوې مدرسې د زده کوونکو شمېر ۱۴۶۵، تنه دی که هر یو زده کړي ته ۱۵، توکه کتابونه ورکړې شي نو د ورکړې شوو کتابونو شمېر معلوم کړئ.

۱۴۶۵ ×
۱۵
۷۳۲۵
۱۴۶۵
۲۱۹۷۵ توکه کتابه

۲۹

Numbers multiplication

- Description: The shortest way of adding equal numbers is called multiplication.

Examples:
A group of 24 mujahid attacked the Russians. As a result each mujahid killed 12 Russian. Find out how many aggressive Russians were eliminated?
24 x 12 = 288

Example 2
A madrasa (RELIGIOUS SCHOOL) has 1465 students. Each student was given 15 books. How many books were distributed?
1465 x 15 =

Fig. 2.4 Facsimile page (in Pushto, with translation) from one of the textbooks produced by the Afghanistan Ministry of Education in Exile and printed by the UNO team in Peshawar, Pakistan, in the 1980s. The text was designed to teach arithmetic skills and was commensurate with the daily lives of the Afghanistan *mujahideen* resistance to the Soviet invasion and their refugee children was focused upon this aspect (Shroder and Watrel 1992; Shroder 2003, 2004). At the time of the Soviet departure, the Nathan–Berger Company also approached the author about producing a major report for USAID on the mineral resources of Afghanistan in order to prepare for reconstruction of the country (Nathan–Berger 1991).

But in spite of this optimistic report, and the protests of wiser heads in the US Department of State (P. Tomsen, 2003, oral communication), in the face of the increasing chaos and civil war erupting in Afghanistan, the USA also left Afghanistan after the Soviet departure and left the failed state to its own devices (Coll 2004; Goodson 2001). In 1993 the World Trade Center in New York was bombed by the Jihadis who had received some of their training in Afghanistan fighting against the Soviets. By 1994, the Pakistani-supported Taliban were becoming a potent force in Afghanistan, and by 1996 they were established in Kabul as the rulers of the country (Rashid 2001). In addition, Osama bin Laden reestablished himself in Afghanistan that same year (Bergen 2006).

Also at that time, because of the breakup of the Soviet Union and the need for communication and export to the outside world by the economically crumbling Central Asian republics, pressure was building up to export their abundant hydrocarbons by pipeline to the West. This led to the involvement of the Unocal Oil Co. in a project to bring a pipeline across Afghanistan to sell gas to Pakistan (Rashid 2001; Griffin 2001). UNO was hired to train some 400 Afghan teachers, electricians, carpenters, and pipe fitters to help lay the pipeline. Taliban leaders were excited about the prospects for revenue that such transshipments would produce. Mounting opposition to their brutal suppression of women, among other things, and most especially after bin Laden's terrorist training camps were bombed by the USA following the attack by his operatives on the American embassies in Africa, finally led Unocal to scrap the project in 1998.

As most people know, the rise of the al Qaeda terrorists under the tutelage of Osama bin Laden and others in Afghanistan (Bergen 2004, 2006), led to the subsequent horrors of the 9/11 airplane hijackings and destruction of the World Trade buildings in New York and damage to the Pentagon in Washington. There were no Afghan terrorists aboard those airplanes; however, the Taliban had allowed the terrorist training camps to exist in their country as a part of the traditional Afghan cultural hospitality. Almost immediately thereafter, Osama bin Laden and al Qaeda were identified as the prime suspects (Berntsen and Pezzullo 2005). Within a month after the attacks in New York and Washington the invasion of Afghanistan by the USA and its allies was underway.

2.2 Hunt for Bin Laden

Sometime soon after 11 September 2001, Osama bin Laden and his second in command, Ayman al-Zawahiri, sat with their two cohorts in a seemingly indistinguishable ravine somewhere and praised the 9/11 attacks before the Al Jazeera cameraman. The video camera was apparently set up on a tripod and was far enough back to take in the high walls, surface topography, and rock types of the area. For all but the last few frames, however, the tight focus and restricted field of view down in the ravine showed only the highly weathered rock and sediment directly behind the terrorists, as well as a probable shear or fault zone (Shroder 2005a; Schollmeyer 2006). Most of the media presentations of the time focused only on the tight close-up on the men, rather than the full scene. As the taping was finishing, it is thought that the video camera was still running when it was tipped upward to take it off the tripod. This produced a few quick frames of the rocky *tor* landforms on the horizon. The unusual migmatitic gneiss rocks of the tors were clearly visible, but because they went by so quickly, apparently few recognized them, and most people focused on the regolith sediments directly behind bin Laden so clearly visible in the background (Beck 2003). When the author first saw the video, the tors and shear zone were obvious and seemed similar to other rocks and landforms observed by the author when traveling in Nangahar Province near there.

With the geology maps of Afghanistan (Chmyriov and Mirzad 1972; Wittekindt and Weippert 1973) and the US Department of Defense (DOD) 1:100,000-scale topographic maps in our collections for comparison, the western part of the Spin Ghar Range appeared to be the most likely area of the video. In the course of conversations at the ASC, there seemed little harm in mentioning this to colleagues, because wherever the video had been made, the participants were not likely to be there afterwards; no one doubted that bin Laden was indeed somewhere in Afghanistan, and most likely to the east in the Pushtun tribal areas. Shortly thereafter, as various media people were asking questions around the ASC, a colleague chanced to mention that the author knew the approximate location of the video. One reporter asked about the location. Not realizing the implications, an approximate location was provided. The next day the media blitz began in earnest. Photographs of the author and various allegations appeared worldwide, which apparently caught the attention of many people (Beck 2003). Neither the tor landforms nor the shear zone had been mentioned, and instead the focus was on the apparent sedimentary rocks that were misleadingly noted as most likely in Paktia or Paktika Provinces in southern Afghanistan along the border with Pakistan.

Shortly after the video had been broadcast, in fact, several of the national intelligence agencies contacted the author. They advised on how to behave with the media, to remove a number of items from the Atlas of Afghanistan web site that could be of use to the terrorists, and to be careful about not revealing too much publicly while they attempted to zero in on bin Laden's whereabouts. On 25 October 2001 a Workshop on Earth Science Issues for Afghanistan was held at the main offices of the US Geological Survey (USGS) in Reston, VA, including a discussion about the interpretation of terrain by map and satellite imagery for locating terrorist camps or underground facilities. Four of the attendees among us had field experience in Afghanistan and there were another eleven USGS employees assigned to actively study or coordinate efforts on Afghanistan geology, hydrology, image analysis, geography, and GIS.

A number of us with experience in Afghanistan worked out the most probable location of bin Laden in the western Spin Ghar Range, near a prominent fault zone, and we discussed the wide variety of caves and bunkers possible in the region. Ideas for cave detection were presented (Hall 2001), and a classification of the caves and bunkers was constructed for use by the US Military that did not include all the hyperbole about fantastic equipment in many of them that the media were touting at the time. Transmission of this cave information out to the field was apparently not straightforward, however, perhaps because of the speed with which events were unfolding, perhaps because of military lethargy, and the tendency of the intelligence services to maintain excessive secrecy within their own units.

Shortly after the media onslaught, e-mail traffic into UNO increased, including death threats from supporters of al Qaeda. Strong criticisms also arrived from those who thought that information about bin Laden's possible whereabouts should be suppressed so as not to enable his escape. What the critics did not know, however, was that the public exposure established an address for e-mails from bounty hunters. The US government analysts already knew some location details from news reporters who had interviewed bin Laden in Afghanistan in the past, which enabled cross checking for veracity. We had our greatest success when photographs arrived of one of bin Laden's houses on the southwest side of the Spin Ghar, as well as pictures of his engineered bunkers. The concrete- and iron-reinforced, double doors of the bunkers excavated in crystalline rock had been previously observed, so we knew that the e-mail information was likely authentic, and the unusual oak and juniper vegetation behind the house was further biogeographic information that helped pinpoint the location. The names of the areas provided by bounty hunters were located in the Gazetteer of Afghanistan, as well as on the large-scale DOD maps.

In addition to the photographs, the e-mail traffic also reported the movement of the al- Qaeda leadership close to the south side of the Spin Ghar in Paktia and also near Tora Bora on the north side of the range in Nangarhar Province (Fig. 2.4). Osama bin Laden was reported as moving there mostly at night, but always close to the Pakistani border, near the Kurram Agency of Pakistan. The source also reported that those areas offered the best caves, easiest defense, and kept open the options of his crossing into Pakistan.

In December 2001, the name 'Tora Bora' first emerged in the media reports about the fighting by US-led, coalition troops who moved into the Spin Ghar region in pursuit of bin Laden. Gary Berntsen, the CIA's key field commander in Afghanistan at that time, requested a full battalion (~800 men) of US Army Rangers to trap bin Laden, but his repeated requests were denied by US Armed Forces Central Command (Berntsen and Pezzullo 2005).

By this time, however, the full flow of information within the US Government and the military was apparently not going as well as many had hoped, as those leading the assault apparently lacked adequate information about the caves and bunkers, or even about the most likely behavior of the always mercurial Afghan troops from the Northern Alliance and others who had apparently let bin Laden slip out of the Tora Bora area and over the undefended border into Pakistan (Weaver 2005). This was not surprising to cultural geographers knowledgeable of the region (oral communication with Nigel Allan), but many Americans seemed shocked at what appeared to be the embarrassing consequences of incompetence.

Richard Beck, who like the author, had experience of mapping geography and geology along the border of Afghanistan and Pakistan, had also been giving the US Military information about underground caves along the border (Beck 2003; Shroder 2005a). Ultimately, in spite of several other missed opportunities and a few mistakes, it took only a little longer than two months for 110 CIA officers, 350 Special Forces soldiers, ~60 British SAS troops, allied with ~15,000 Northern and eastern Alliance Afghans, with as many as 100 combat sorties per day, to defeat a Taliban army estimated to be 50,000–60,000 strong, along with several thousand al-Qaeda fighters (Berntsen and Pezzullo 2005; Moore 2003). By early 2002, the remnants of the Taliban had fled over the border into Pakistan, or hidden away their guns in Afghanistan and retired from active combat to await another perhaps more auspicious day for them when the enemy guard was down, as Afghans have been so famous for doing in many prior episodes of foreign invasion (Bearden 2001).

2.3 Geospatial Technologies in Afghanistan and Homeland Security

Modern GIS technologies are used in a tremendous number of applications (Bishop and Shroder 2004), and their utilization in Afghanistan has been carried forward to a new art by the necessity for keeping track of the tremendous quantity of information required to help that beleaguered nation arise from its bed of ashes. In effect the use of geospatial technologies to study and reconstruct Afghanistan is viewed by many as an essential element for ensuring that the failed state has increasing security and will never again become a threat to the West, thereby increasing our homeland security.

Mapping of data and the establishment of high quality, accurate, and useful GIS maps and data sets are required for the redevelopment of Afghanistan and have been developed by the Afghanistan Information Management Service, the highly efficient and modern data collection and mapping agency administered by UNDP and funded by the European Community (EC) and USAID in Kabul. In a similar vein, the Afghanistan Research and Evaluation Unit (AREU) was established as an independent research institution to inform policy, improve practice, and increase the impact of humanitarian and development programs. Among their many tasks, keeping track of the host of agencies and programs operating in Afghanistan has been paramount (Wakefield and Wilder 2003).

The creation of the AIMS organization in Kabul was a necessity born of the past Cold War confrontations between Western Powers and Soviet-backed agencies to whom mapped data were anathema. At the present time AIMS provides a host of detailed map and data products online (Table 2.1) that are not treated as classified products, even despite the objections of many senior Afghan civil servants to whom

Table 2.1 Maps available on-line from the Afghanistan Information Management Service (AIMS) (www.aims.org.af/)

Standard maps (554 documents)
National maps (50 documents)
Afghanistan urban areas (8 documents)
Afghanistan provincial maps (32 documents)
Afghanistan district maps (328 documents)
District vulnerability maps (6 documents)
Afghanistan topographic maps with background (65 documents)
Afghanistan topographic maps without background (65 documents)
Regional maps (19 documents)
Hirat maps
Jalalabad maps
Kabul maps
Kandahar maps
Kunduz maps
Mazar maps
Afghanistan regional maps
United Nations regions
Custom maps (82 documents)
Water/sanitation
Security
Governance and civil society
Natural disaster management
Health
Emergency assistance
Environment
Education
Agriculture
Floods
Refugees
Mine action
IDPs (internally displaced people)
UNDP opium poppy cultivation survey by hectares
External resources maps (67 documents)
Refugees
Provincial health maps in English
Major overland route map into Afghanistan

mapped information at the time of the Cold War was regarded as subversive or dangerous because someone on the other side would try to take advantage of the mapped knowledge. The website (www.aims.org.af/) is sometimes hard to get into, perhaps because of continual power outages in Kabul. The AIMS project introduces GIS technology to UN agencies, attempts to build information capacity in government and the NGOs, and delivers a variety of information management services, training, database services, and maps to organizations all across Afghanistan. It has offices in Kabul, Jalalabad, Kadahar, Kundiz, Mazar-i-Sharif, and Herat. AIMS maps generally use the WGS 84 Datum convention for their projection. This is also what the satellite GPS system is based upon, and this enables ease of location of new data points gathered in the field.

Interestingly, in the 1960s large scale (1:50,000-scale) topographic maps of Afghanistan were freely available in Kabul (Shroder 1983), but as the Cold War drew to a close in the final catastrophic implosion of the Soviet Empire following their failed machinations in Afghanistan, most maps became a major sticking point to the National Atlas of Afghanistan Project mentioned above. Both the Soviet military and the US Department of Defense (DOD) produced a parallel series of large-scale topographic maps of Afghanistan in the 1960s. Many of these large scale DOD topographic maps became basic references in our National Atlas Project hosted at Kabul University and UNO.

Remote sensing of Afghanistan in the early days, mainly by university academics, was rather crude; but at the time was still viewed as an important geospatial technology for useful applications in the country. The large (1 × 2 m) satellite image mosaic (Fig. 2.1) of the whole country (Shroder 1977), and publication of a number of papers on remote sensing of Afghanistan, were seen as a way to enable better understanding and mapping of bedrock geology, vegetation distribution, desertification, water resources, glacierization, and development (Shroder 1978). During the Soviet–Afghanistan War in the 1980s, some of the unclassified remote sensing of Afghanistan involved assessment of irrigated agriculture to better determine where refugees claimed to be from and who needed the most humanitarian assistance (Shroder 1989a, b, c, d; Langran and Shroder 1990). At that time, the covert support of the Afghan resistance by the USA resulted in the production of many new satellite image mosaics of different parts of Afghanistan by the US Central Intelligence Agency (undated a, b; Figs. 2.2–2.5). These 1:500,000-scale map

Fig. 2.5 Part of the Kabul III sheet showing Kabul in the upper left, the Spin Ghar (Safed Koh) range, the Parachinar ('parrot's beak') border reentrant of the Kurram Agency part of the Northwest Frontier Province of Pakistan into which bin Laden escaped, and the many enumerated passes along the border that are used for smuggling people, drugs, and weapons

111	Tora Tigga Kandao	توره تیکه کندو	111
112	Kot Mohmand Kandao	کوت محمد کندو	112
113	Oghaz Kandao (Owghaz Kandaw, Afghanistan)	اوغاز کندو اوغاز کندو	113
114	Papin Kandao	پاپین کندو	114
115	Agam Pass	آگام کوتل	115
116	______	______	116
117	Shah Hussaini Kandao	شاه حسینی کندو	117
118	______	______	118
119	Rishak Kandao	ریشک کندو	119
120	Spinkai Raghai	سپینکی رغائی	120
121	______	______	121
122	Zarai Khande Kandao	زراعی خنده کندو	122
123	Gawai Tangai	گوائی تنگی	123
124	Kirkai Tangai	کیرکی تنگی	124
125	______	______	125
126	Bargawi Kandao	برگاوی کندو	126
127	______	______	127
128	Peiwar Kotal (Peywâr Kandaw, Afghanistan)	پیوار کوتل (پیوار کندو)	128
129	______	______	129
130	Sursurang Kandao	سورسورنگ کندو	130
131	Shautalo Chhapri Kandao	شاوتلو چهاپری کندو	131
132	Zhâwar Kandao (Zhâwar Kandaw, Afghanistan)	ژاور کندو (ژاور کندو)	132
133	Margho Kandao	مرغو کندو	133
134	Darwezi Kandao	دروزی کندو	134
135	Mandatt Kandao	مندت کندو	135
136	Dre Drang Kandao	دره درنگ کندو	136
137	Swaro Gawai	سوارو گاوائی	137

Border passes have been numbered for ease of reference. Blank spaces have been left for adding the English and Dari names as they become known. Unless otherwise indicated, pass names are those used on the Pakistan side of the border.

Fig. 2.6 Small part of the back of the Kabul III sheet with passes enumerated, whether their name is known or not, with a space left to fill in the name of the pass if discovered

materials were annotated with major and minor political boundaries, bilingual place names (English and Dari) located by latitude and longitude on the map posterior, transportation routes down to tracks and trails, and spot elevations. They were unusual in also listing on their backs the names or numbers of as many passes as were known or suspected that were used for smuggling over the border between Pakistan and Afghanistan, with blank spaces for names to be filled in later by covert operatives (Fig. 2.6). Following the withdrawal of the Soviets, most, if not all, of these maps apparently became unimportant to the US Government and the duplicates were sent to UNO for the storage in the ongoing atlas project.

Nothing much happened with civilian remote sensing in Afghanistan in the 1990s, but subsequently in the new millennium, a major push began with the new, high-resolution (15 m) ASTER imagery in the GLIMS (Global Land Ice Measurements from Space) Project at UNO, which is providing a host of new environmental information in our laboratories (Bishop et al. 2000, 2001; Kargel et al. 2005; Shroder 2005b; Shroder and Bishop 2004, in press 2007; Shroder et al. 2007). The idea is to systematically map the snow and ice resources of the region, and to thereby keep track of changes or diminution of these vital melt-water sources. In the late 1990s the US Geological Survey had approached both governments of Afghanistan and

Pakistan about GLIMS, but neither government expressed any interest at the time. Consequently, a GLIMS Regional Center for Southwest Asia was established at UNO, in which a staff of five and numerous graduate and undergraduate students are digitizing glaciers and assessing change over time. The advent of Google Earth© has brought a new, open-source dimension to our ongoing mapping of glaciers in Afghanistan, because our analysts use multiple-screen presentations for digitizing glaciers, while simultaneously running multiple oblique views, or 'virtual helicopter' investigations of the terrain in order to facilitate the glacier mapping.

The Soviet economic collapse in the 1990s led to the sale of all of their relatively high quality topographic maps to the West, which have also become an essential comparison tool in our interpretations of ASTER satellite imagery. In the US invasion of Afghanistan in late 2001, NIMA (National Intelligence Mapping Agency) converted and updated, at very short notice after 9/11, all of the old DOD contour maps into digital elevation models. These digital topographic sources are now reclassified by the USA, so the old Soviet maps are the ones that have been scanned and made available for development projects. The continuation of the old Cold War idea of secret geospatial information has put a special pressure on AIMS in Kabul, but fortunately they have risen to the challenge and are still in the business of providing useful unclassified maps (Table 2.1).

The old National Atlas of Afghanistan Project (Shroder 1975) was originally designed as a pedagogical device for the schools in Afghanistan, as well as to help development agencies in the country. Unfortunately, because of political events, that concept lost out in the escalating violence as the Cold War played out its final battles in Afghanistan, and the Polish Geokart (1987) effort became little more than a historical curiosity. But the concept of using geospatial technologies to rejuvenate a new atlas for the same purposes of education and redevelopment has reemerged and may become a reality if its value is recognized to help in the effort to bring Afghanistan back into the fold of nations emergent from failure. If honest assessments and useful mapped analyses are made available more than they were in the Cold War, then a new geospatial atlas might be feasible; only time and new effort will tell.

2.4 Win every Battle and Lose the War?

In late 2006 and 2007 the news media was replete with stories and information about a resurgent Taliban, the record opium poppy production, and failures by the Washington and Kabul governments to really effect much necessary change in Afghanistan in such a way as to bring it back from the brink of failed states (Sinha and Schaffer 2006; Ignatieff 2002; Jennings 2003; Moseley 2006; Rohde and Sanger 2007, and many other sources). To be sure the rapid defeat of the Taliban and al Qaeda fighters in late 2001 was seen as a major victory by the United States and its coalition allies, including most recently a strong NATO force (www.nato.int/issues/afghanistan/040628-factsheet.htm) (www.nato.int/issues/afghanistan/index.html), but Afghans are renowned for biding their time after an invasion until the time is right for counterattack. The coalition alliance is trying to succeed where many previous armies have suffered alarming defeats, including the British in the 19th century and the Soviets in the 20th. Conquest of Afghanistan has always been relatively easy for invasionary forces, but later withdrawals have been more problematic (Bearden 2001). The overwhelming military superiority of the present NATO and International Security Assistance Force (ISAF) coalitions may allow them to win every battle by brute force, but in the long run the war may still be lost by too little positive rebuilding of the country, too late to avoid a possible ultimately overwhelming return of the Taliban in the 'forever war' (Danner 2005). The necessity of a well designed and efficient neo-Marshall plan at the outset of 2002 could have turned the beleaguered state into an example of American beneficence and good will (Shroder 2003), but that has not happened and the situation has continued to deteriorate (Anonymous 2006a, b).

The Taliban insurgency has regrouped since their retreat in 2001 and by 2007 has escalated into a wave of attacks by improvised explosives and suicide bombers, particularly after NATO took on the fight in pitched battles against the insurgents since spring 2005. Operation Medusa near Kandahar in southwestern Afghanistan, and Operation Mountain Fury in eastern Afghanistan are the latest attempts to control the escalating violence, but as the Taliban insurgency gains in strength and sophistication, not only is Pakistan implicated (Walsh 2006a; Gall 2007), but the Taliban can keep the pressure on until the Afghan public gets so frustrated that eventual defeat of the central government will be inevitable.

Part of the ongoing problem has been the necessity of redevelopment teams requiring strong security around them. This has resulted in the formation of military-based, Provincial Reconstruction Teams (PRTs) to accomplish redevelopment, which is not how traditional armies are trained (www.army.mil/professionalwriting/volumes/volume4/march_2006/3_06_4.html). In order ever to have a hope of defeating the illusive insurgents requires very different skills, such as the knowledge of foreign languages and cultures, policing, intelligence, information operations, and civil affairs (Boot 2006a, b).

At the present time there are some 24 multinational (USA, UK, Canada, Denmark, Germany, Sweden, Hungry, Italy, Lithuania, Netherlands, Norway, South Korea, Spain, and other nationalities) PRTs in Afghanistan operating out of the main provincial centers.

The character and priorities of the PRTs vary regionally and between the US coalition forces and the NATO/ISAF teams (BAAG 2006). Coalition PRTs are primarily geared towards winning "hearts and minds" of people in support of intelligence gathering in counter-insurgency operations. The NATO/ISAF teams try to stabilize the country, assist the government, and help with security sector reform.

In general, a high-ranking, field-experienced military officer and about 60 personnel drawn from Special Forces, Civil Affairs, USAID, and the State Department, as well as other coalition representations command each US coalition PRT. The PRTs are backed up militarily by some 10 Forward Observation Bases, 16 military airfields scattered across the country, 17 separate military camps, and two main compounds (www. globalsecurity.org/military/facility/afghanistan.htm) to keep the peace while the somewhat limited redevelopment is underway.

The notion of using military-based PRTs to rebuild Afghanistan has received criticism from the non-governmental organizations (NGOs) who provide other humanitarian relief services. The NGOs worry about being lumped together in the eyes of the Afghans with the coalition belligerents, that the PRTs will work with the wrong people, and that they will have security or other agendas that do not necessarily match the pressing priorities of Afghans (Jennings 2003).

Reconstruction progress has also been much slower than originally envisioned. The international community initially pledged $4.5 billion over five years for reconstruction. Of the $1.8 billion promised for 2002 only $1 billion was committed, with even less implemented as assistance in the field. Furthermore, 80 percent of the total was disbursed to fund relief programs rather than actual reconstruction initiatives (Jennings 2003). USAID, the major donor in Afghanistan, has even recently decreased its contribution by 29.2 percent (Table 2.2), apparently despite US Government recognition that things have not been going so well together with awareness of the return of the Taliban insurgency. The overwhelming disparity of total military expenditures of $82.5 billion against the paltry $7.3 billion of development aid has led many Afghans to conclude that neither their pitiful economic situation nor their extreme poverty will ever be relieved (Senlis Council 2006).

Table 2.2 USAID contributions to Afghanistan reconstruction for the past five years. The financial assistance promised by USAID for 2007, however, has declined by 29.2 percent (www.usaid.gov/policy/budget/cbj2007/ane/af.html), although a new contract for $1.4 billion through 2011 was awarded recently for infrastructure development (www.usaid.gov/pressreleases/2006/pr060922.html)

Other donors include Japan, UK, Germany, India, Canada, the Netherlands, Italy, Iran, Norway, Denmark, Saudi Arabia, Sweden, the Russian Federation, Spain, Pakistan, France, China, United Arab Emirates, Switzerland, Finland, Australia, the Republic of Korea, Kuwait, Belgium, Qatar, Ireland, Turkey, Austria, Luxembourg, Oman, Greece, New Zealand, Portugal, and Poland. Major multilateral donors in Afghanistan include the European Community, World Bank, Asian Development Bank, Organization of the Islamic Conference, and the United Nations. US assistance exceeds the total of all other multilateral aid combined.

Year	Amount
2003	$428,492,000
2004	$1,133,280,000
2005	$1,568,750,000
2006	$617,742,000
2007	$802,800,000
Percent of FY budget for each area in 2005 (USAID Country Profile 2005)	
Economic growth	46%
Democracy & governance	19%
Agriculture & environment	17%
Education	9%
Health	9%

Clearly, ignoring extreme poverty can only destroy nation-building efforts, making the ground even more fertile for a revival of the Taliban. In addition, the apparently futile counter-narcotics policies only intensify the security and poverty crises.

Illicit opium is now one of the largest sectors in the national economy, accounting for nearly a third of the Afghan GDP. The opium GDP is estimated at ~$2.6–2.7 billion, equivalent to ~36 percent of the licit GDP in 2005/06 (Byrd and Buddenberg 2006). This sheer size means that it seriously affects Afghanistan's economy, state, society, and politics, and that it is a massive source of corruption and gravely undermines the credibility of the government and its local representatives. The massive and detailed World Bank report edited by Byrd and Buddenberg (2006) on the opium trade in Afghanistan shows the overall dimensions of the pressing problem that may cause state failure once again, with obvious implications for homeland security there as well as in the USA and Europe.

Reports from the Helmand area from former USAID opium poppy specialist, Richard Scott, who has worked in the region for over three decades up through 2004, are not optimistic (Scott 20/12/2006, 25/09/2007, written communication). As he reports:

> The 2007 opium poppy crop for central Helmand has been planted and germinated, thanks to some early unexpected rains. The local buyers, the smugglers, and the anti-central government political groups, including the Taliban, are all happy about this event. It fits with their plans to undercut respect for the central government, as well as the reliability of the foreign donor community that in the past promised so much and delivered so little.

The solutions to the opium problem have long been fairly simple and rather obvious (Table 2.3); but instead, more drastic and ultimately short-sighted and environmentally and politically dangerous interdiction techniques such as aerial spraying (Walsh 2006b) are too often used, to the detriment of redevelopment success in Afghanistan. Geospatial technologies using satellite images to detect poppies (Sader 1992: Chuinsiri et al. 1997; UNODC 2007; Srinivas et al. 2004) are also fairly simple and have been used effectively to find out where the problem

Table 2.3 Recommendations to do the obvious that the farmers have repeatedly requested in reducing opium production in Afghanistan, particularly in the Helmand area (in part after R. Scott, 20/12/06, written communication)

- **Water** – Continue to improve the irrigation system… providing more water and increasing the efficiency with which it is distributed and used.
- **Jobs** – Use hand labor for virtually all development work, putting as many people to work as possible at reasonable wage levels.
- **Transport** – Improve and develop all farm and main roads throughout the region, keeping in mind that the region gets only ~4+ inches (~8.2+cm) of rain a year… but needs good drainage.
- **Credit** – Introduce at least a limited agricultural credit system to compete with the informal opium-based credit system.
- **Cotton** – Increase the price paid for cotton, making it the convenient, reliable, and economically viable market it was in the 1970s.
- **Legalization** – Investigate advisability of producing some opium legally under tight controls for conversion to pharmaceuticals.

crops are located in Afghanistan. A strategy of heavy-handed crop eradication, however, where chemical sprays also affect limited water supplies, and the prioritization of counter-narcotics above the survival of Afghans is unfathomable to the country's rural communities; such tactics only undermine the international community's more beneficial efforts (Senlis Council 2006). American military personnel have even complained to the author that simply destroying the poppies as they are sometimes ordered to do, with no crop replacement strategies or other help for the farmers, is hardly likely to win hearts and minds. Furthermore in the ongoing global climate change with recurrent water shortages, drought-resistant opium poppies are often a last resort crop for an already desperately poor people. Furthermore, the obvious licensing of opium production in Afghanistan for inexpensive pharmaceutical purposes would alleviate the lack of opiate-based medicines in Afghanistan and other developing countries but it does not seem to have been considered.

2.5 Conclusion

Problems *in* Afghanistan seem to result in problems *from* Afghanistan. Failed states allow terrorists to prosper, and we most definitely do not want anything like 9/11 to happen again. Geospatial technologies and the preservation of our homeland security related to problems in Afghanistan are obvious elements in most aspects of our military and agency deployment there; the threat assessment that they must do, the PRT work, opium-poppy mapping and alternative crop production, water and mineral resource exploitation, transportation-corridor assessments and enhancements, environmental databases, and a host of other uses and applications are evident. Much of the technology is classified, but there is enough unclassified so that AIMS, AREU, and the many other NGOs are still able to accomplish useful geospatial tasks with good maps and GIS applications.

At 5th from the bottom, Afghanistan remains one of the world's poorest countries, with an estimated per capita GDP of but US$315, little changed for the past half century. Social indicators are also among the world's worst, with high rates of infant and maternal mortality (one quarter of all children <5 die), 44-year life expectancy overall, 3rd highest illiteracy rate in the world, malnutrition, gender disparity (World Bank 2007a, b; BAAG 2006), rampant criminality, and rising jihadism once again. In 1996, during Taliban times, Afghanistan ranked 169th out of 174 countries in the UNDP Human Development Index, but at the present time it has declined even more to be second-worst in the world after Sierra Leone (BAAG 2004).

Optimistic plans for rebuilding the country in the post 9/11 world have been slow for many reasons, not the least of which has been lack of attention and only limited financing from the US Government that has apparently been too preoccupied with other things, especially the failing invasion of Iraq which is undermining the War on Terror (Fallows 2004). Brute force can certainly win every battle but perhaps not the whole war. The rejuvenation of the Taliban in Afghanistan under these circumstances of questionable competence (Chayes 2006) and weak reconstruction was predicted repeatedly by many analysts (Bearden 2001; Shroder 2002b; Anonymous 2004; Scheuer 2006; Moreau et al. 2006; United Nations 2006); the reemergence of a strong Taliban will now dramatically increase costs and may lead to overall failure in this part of the War on Terror.

Even the US Secretary of State, Condoleeza Rice, has said that the West must not fail in Afghanistan, so it seems clear that the administration has recognized the problem (news.bbc.co.uk/1/hi/world/south_asia/5340892.stm). Homeland security in the US however, seems somewhat further jeopardized by ongoing political and intelligence failures in Afghanistan, but perhaps recovery can occur with the new Secretary of Defense, Robert Gates. The many elements of geospatial technologies will figure in the task, especially if used wisely to benefit the people of Afghanistan. This can only be done, however, by considerable attention to: (1) robust security with good intelligence; (2) strong cultural sensitivity; (3) full engagement of academic communities having expertise and teaching capabilities; (4) the reputation for American good will and honesty; (5) more robust, Marshall Plan-style redevelopment; and (6) continuing to build a democracy while it may be possible. Continued attention to these details may improve domestic homeland security, perhaps by helping better ensure the security of the people of Afghanistan better than has been done up to now. Afghans are their own worst enemy, as they well recognize, but they always liked Americans in the 1970s and 1980s, and if a more subtle and sensitive approach than mere brute force is used to help, they might just do so again. We can only hope.

References

Anonymous (M. Scheuer) (2004). *Imperial hubris: Why the west is losing the war on terrorism.* (Dulles, VA: Potomac Books)

Anonymous (2006a). The test in Afghanistan. *The Economist*, November 25, pp. 12

Anonymous (2006b). Predictions of its death were premature. *The Economist*, November 25, pp. 24–26

BAAG (British Agencies Afghanistan Group) (2004). On-line reports, September, 2004; Retrieved September 28, 2007 from http://www.baag.org.uk/downloads/monthly%20review%2004/52%20-September%202004.pdf

BAAG (British Agencies Afghanistan Group) (2006). *Afghanistan briefing pack: An introduction to working in Afghanistan*, on-line report, 52 pp. Retrieved September 28, 2007 from http://www.baag.org.uk/downloads/reports/BAAG%20Briefing%20Pack%202006.pdf

Bearden, M. (2001). Afghanistan, graveyard of empires. *Foreign Affairs* (November/December), pp. 17–34

Beck, R.A. (2003). Remote sensing and GIS as counterterrorism tools in the Afghanistan War: A case study of the Zhawar Kili region. *The Professional Geographer* 55, 170–179

Bergen, P.L. (2004). The long, long hunt for bin Laden. *The Atlantic*, October, 294(3), 88–100

Bergen, P.L. (2006). *The Osama bin Laden I know*. (New York: Free Press/Simon & Schuster)

Berntsen, G. & Pezzullo, R. (2005). *Jawbreaker: The attack on bin Laden and al-Qaeda.* (New York: Three Rivers Press/Crown Publishing Group)

Bishop, M.P., Kargel, J.S., Kieffer, H.H., MacKinnon, D.J., Raup, B.H. & Shroder, J.F., Jr. (2000). Remote-sensing science and technology for studying glacier processes in high Asia. *Annals of Glaciology*, 31, 164–170
Bishop, M.P., Bonk, R., Kamp, U., Jr. & Shroder, J.F., Jr. (2001). Terrain analysis and data modeling for alpine glacier mapping. *Polar Geography*, 24(4), 257–276
Bishop, M.P. & Shroder, J.F., Jr. (Editors) (2004). *Geographic information science and mountain geomorphology*. (London: Praxis)
Boot, M. (2006a). *War made new: Technology, warfare, and the course of history*. (New York: Gotham Books)
Boot, M. (2006b). From Siagon to Desert Storm: How the US military reinvented itself after Viet Nam. *American Heritage*, November/December, 28–37
Borovik, A. (1990). *The hidden war: A Russian journalist's account of the Soviet war in Afghanistan*. (New York: Atlantic Monthly Press)
Bradsher, H.S. (1983). *Afghanistan and the Soviet Union*. (Durham, NC: Duke University Press)
Byrd, W.A. & Buddenberg, D. (Editors) (2006). *Afghanistan: Drug industry and counter-narcotics policy*. World Bank, 214 pp. [Electronic version] Retrieved September 28, 2007 from http://web.worldbank.org/WBSITE/EXTERNAL/COUNTRIES/SOUTHASIAEXT/0,,contentMDK:21133060~pagePK:146736~piPK:146830~theSitePK:223547,00.html
Central Intelligence Agency (not identified on map) (undated a; but early 1980s). *Afghanistan: Kabul and Panjshār*. Landsat imagery Sept. 1972–June 1979; 505317 (547059F) 9–82; 1:500,000 scale
Central Intelligence Agency (not identified on maps) (undated b; but early 1980s). *Afghanistan*:
Sheet I Herāt; 505081 (547059-H) 2–83; 1:500,000 scale.
Sheet II Maymanah; 505082 (547059-G) 8–82; 1:500,000 scale.
Sheet III Kābul; 505083 (547059 F) 2–83; 1:500,000 scale.
Sheet IV Islamābad; 505084 (547059 E) 10–82; 1:500,000 scale.
Sheet V Zaranj; 505085 (547059 D) 5–84; 1:500,000 scale.
Sheet VI Qandahār; 505086 (547059-C) 6–82; 1:500,000 scale.
Sheet VII Ghazni; 505087 (547059 B) 1–83; 1:500,000 scale.
Chayes, S. (2006). *The punishment of virtue: Inside Afghanistan after the Taliban*. (New York: Penguin Press)
Chmyriov, V.M. & Mirzad, S.H. (1972). *Geological map of Afghanistan*, 1:1,000,000. Royal Government of Afghanistan
Chuinsiri, S., Blasco, F., Bellan, M.F. & Kergoat, L. (1997). A poppy survey using high resolution remote sensing data. *International Journal of Remote Sensing*, 18(2), 393–407
Coll, S. (2004). *Ghost wars*. (New York: Penguin Press)
Crile, G. (2003). *Charlie Wilson's war*. (New York: Atlantic Monthly Press)
Danner, M. (2005). Taking stock of the forever war. *The New York Times Magazine*, September 11, 44–53, 68–71
Dreyfuss, R. (2005). *Devil's game: How the United States helped unleash fundamentalist Islam*. (New York: Owl Books/Henry Holt & Co.)
Edwards, D.B. (2002). *Before Taliban: Genealogies of the Afghan Jihad*. (Berkeley, CA: University of California Press)
Fallows, J. (2004). Bush's lost year. *The Atlantic*, October, 294(3), 68–84
Gall, C. (2007). At border, signs of Pakistan role in Taliban surge. *The New York Times*, Sunday, January 21, CLVI (53831), 1 & 12
Grau, L.W. & Gress, M.A. (2002). *The Soviet-Afghan war: How a superpower fought and lost*. (Lawrence, KS: University Press of Kansas)
Geokart (1987). *National atlas of the democratic republic of Afghanistan*. (Warsaw, Poland: Geokart)
Goodson, L.P. (2001). *Afghanistan's endless war: State failure, regional politics, and the rise of the Taliban*. (Seattle, WA: University of Washington Press)
Ignatieff, M. (2002). How to keep Afghanistan from falling apart. *The New York Times Magazine*, July 28, Section 6, 26–31, 54–59

Griffin, M. (2001). *Reaping the whirlwind.* (London: Pluto Press)
Hall, S.T. (2001). *Mapping limestone caverns of southeastern Afghanistan using structural geology, aerial photography and geophysical surveys.* Unpublished report from Geo-Environmental, Inc., Reston, VA
Jennings, R.S. (2003). *The road ahead: Lessons in nation building from Japan, Germany, and Afghanistan for postwar Iraq.* United States Institute of Peace, Peaceworks No. 49, Washington, DC, 41 pp. [Electronic version]. Retrieved September 28, 2007 from http://www.usip.org/pubs/peaceworks/pwks49.html
Kargel, J.S., Abrams, M.J., Bishop, M.P., Bush, A., Hamilton, G., Jiskoot, H., Kääb, Keiffer, H.H., Lee, E.M., Paul, F., Rau, F., Raup, B., Shroder, J.F., Soltesz, D., Stainforth, D., Stearns, L. & Wessels, R. (2005). Multispectral imaging contributions to global land ice measurements from space. *Remote Sensing of the Environment; Special Issue on Terra/ASTER Science*, 99, 187–219
Langran, K.J. & Shroder, J.F., Jr. (1990). Using remote sensing to monitor the decline of karez agriculture in Zamin Dawar, Afghanistan. *American Association of Geographers Program and Abstracts*, Toronto, Canada, pp. 133
Moore, R. (2003). *The hunt for bin Laden: Task Force Dagger.* (New York: Random House)
Moreau, R., Yousafzai, S. & Hirsh, M. (2006). Losing Afghanistan: The rise of Jihadistan. *Newsweek International Edition*, Monday 02 October 2006. Retrieved September 28, 2007 from http://www.truthout.org/cgi-bin/artman/exec/view.cgi/64/22749
Moseley, W.G. (2006). America's lost vision: The demise of development. Association of American Geographers, *AAG Newsletter*, 41(10), 17
Nathan-Berger (Nathan Associates & Louis Berger International, Inc.) (1991). *Afghanistan mineral resources study.* Nathan Associates Inc. Unpublished report for USAID
Rashid, A. (2001). *Taliban: Militant Islam, oil and fundamentalism in Central Asia.* (New Haven, CN: Yale University Press)
Rohde, D. & Sanger, D.E. (2007). *How the 'Good War' in Afghanistan went bad.* The New York Times, August 12, CLVI (54,034, 1, 12–13)
Rubin, B.R. (1995). *The search for peace in Afghanistan: From buffer state to failed state.* (New Haven, CN: Yale University Press)
Rubin, B.R. (2007). Saving Afghanistan. *Foreign Affairs*, 86(1), 57–78
Sader, S. (1992). Remote sensing of illicit narcotic crops from aircraft and commercial satellite platforms. *International Symposium on Remote Sensing of Environment*, 24th, Rio de Janeiro, Brazil, USA; 27–31 May 1991, 849–857
Scheuer, M. (2006). Clueless into Kabul. *The American Interest*, 2(1), 111–119
Schollmeyer, J. (2006). Terrorism: A shifting landscape. *Bulletin of the Atomic Scientists*, March/April, 8–9.
Scott, R. (2006). *Helmand follow up XX: The way backward.* Unpublished report to author on record opium poppy production in southwest Afghanistan
Senlis Council (2006). *Nation-building priorities in the wrong sequence.* (12/2/2006). Retrieved September 28, 2007 from http://www.senliscouncil.net/modules/publications/014_publication/chapter_05
Shroder, J.F., Jr. (1975). National Atlas of Afghanistan: A call for contribution. *Afghanistan Journal*, 2, 108–111
Shroder, J.F., Jr. (1977). *Satellite-image mosaic of Afghanistan.* Prepared by C.M. Dimarzio under direction of J.F. Shroder, Jr. and D.C. Rundquist with cartography by J.K. Turner; Remote Sensing Applications Laboratory, University of Nebraska at Omaha. Republished as frontispiece in: Debon, F., H. Afzali, P. LeFort, J. Sonet & J.L. Zimmerman, 1987. Plutonic rocks and associations in Afghanistan: Typology, age and geodynamic setting. Mémoires Sciences de la Terre, Mémoire No. 49
Shroder, J.F., Jr. (1978). Remote sensing of Afghanistan. *Afghanistan Journal*, 5, 123–128
Shroder, J.F., Jr. (1980a). *Mineral wealth of Afghanistan.* United Press International report (24 January 1980); published widely, including newspapers in New York, Albany, Boston, Baton Rouge and Omaha

Shroder, J.F., Jr. (1980b). Afghanistan's minerals. *Business Week*, 27 October 1980, 8
Shroder, J.F., Jr. (1980c). Special problems of glacial inventory in Afghanistan. *World Glacier Inventory Proceedings Reideralp Workshop, September 1978 (IAHS-AISH)* Publication No. 126, Hydrological Sciences Bulletin, 142–147
Shroder, J.F., Jr. (1981a). Physical resources and the development of Afghanistan. *Studies in Comparative International Development*, XVI(3–4), 36–53
Shroder, J.F., Jr. (1981b). Russians grab Afghan resources. *Omaha World Herald*, Thursday, 19 March, 9
Shroder, J.F., Jr. (1982). Afghanistan's unsung riches. *The Christian Science Monitor*, 74(54), Thursday, 11 February
Shroder, J.F., Jr. (1983). The USSR. and Afghanistan mineral resources. (In A. Agnew (Ed.), *International mineral resources a national perspective* (pp. 115–153). American Association for Advancement of Science)
Shroder, J.F., Jr. (1987). Afghanistan resources and the Soviet Union. *Geotimes*, 32, March, 4–5
Shroder, J.F., Jr. (1989a). Afghanistan resources and Soviet Policy in Central and South Asia. (In M. Hauner & R. Canfield (Eds.), *Afghanistan and the Soviet Union: Collision and transformation* (pp. 101–119). Westview Press)
Shroder, J.F., Jr. (1989b). *The multi-temporal and spatial characteristics of agriculture in the Helmand and Ghazni Provinces of Afghanistan.* Unpublished report for Volunteers in Technical Assistance (VITA) and US AID
Shroder, J.F., Jr. (1989c). *Remote sensing and mapping of Afghanistan.* Unpublished annex D of report by RONCO and US AID Afghanistan Agricultural Project redesign team, November
Shroder, J.F., Jr. (1989d). *Preliminary report #2 on agriculture of Helmand area.* Unpublished report for Volunteers in Technical Assistance (VITA) and US AID
Shroder, J.F., Jr. (2002a). Resource assessment for Afghanistan and alleviation of terrorism. *EOS, Transactions, American Geophysical Union*, 83(19): S2–S3, AGU Spring meeting, Washington, DC, 07/05/2002
Shroder, J.F., Jr. (2002b). *Afghanistan press release.* American Geophysical Union, AGU Spring meeting, Washington, DC, 07/05/2002
Shroder, J.F., Jr. (2003). Reconstructing Afghanistan: Nation building or nation failure? *Geotimes*, October, 5
Shroder, J. (2004). Afghanistan redux: Better late than never? *Geotimes*, October, 34–38
Shroder, J. (2005a). Remote sensing and GIS as counterterrorism tools in the Afghanistan War: Reality plus the results of media hyperbole. *Professional Geographer*, 57, 592–597
Shroder, J.F., Jr. (2005b). Global land ice monitoring from space (GLIMS) Project Regional Center for southwest Asia (Afghanistan and Pakistan). (In: *Mountains, Witnesses of Global Changes: Research in the Himalaya and Karakoram*: **SHARE**-Asia Project (**S**tations at **H**igh **A**ltitude for **R**esearch on the **E**nvironment), Ministry of Regional Affairs, Office of the Prime Minister, Government of Italy, Abstract Book, 55)
Shroder, J.F., Jr. & Tawab Assifi, A. (1987). Afghan mineral resources and Soviet exploitation. (In R. Klass (Ed.), *Afghanistan and the Soviet Union: The great game revisited* (pp. 97–134). New York: Freedom House)
Shroder, J.F., Jr. & Watrel, R.H. (1992). Mineral resources in Afghanistan (summary of report for USAID). *AACAR Bulletin* (*Association for the Advancement of Central Asian Research*), 5(2) 9 pp
Shroder, J.F., Jr. & Bishop, M.P. (2004). Glacier change detection in the Hindu Kush of Afghanistan. *Eos Transactions, American Geophysical Union*, 85(47), F778
Shroder, J.F., Jr., Bishop, M.P., Bulley, H.N.N., Haritashya, UK & Olsenholler, J.A. (2007). Global land ice monitoring from space (GLIMS) project regional center for Southwest Asia (Afghanistan and Pakistan). (In R. Baudo, G. Tartari & E. Vuillermoz (Eds.) (J.F. Shroder, Jr. Series Editor; Developments in Earth Surface Processes, 10) *Mountains witnesses of global changes in the Himalaya and Karakoram: SHARE Asia project* (pp. 187–208). Amsterdam: Elsevier)

Shroder, J.F., Jr. & Bishop, M.P. (in press, 2008), Satellite glacier inventory of Afghanistan. (In R.S. Williams, Jr. & J.G. Ferrigno (Eds.), *Satellite Image Atlas of Glaciers*. U.S. Geological Survey Professional Paper 1386-F)

Sinha, F. & Schaffer, T. (2006). The reconstruction of Afghanistan: A fight for survival. *South Asia Monitor*, Center for Strategic and International Studies, Number 97, August 1, 2006, Washington, DC, 3 pp

Srinivas, P., Das, B.K., Saibaba, J. & Krishnan, R. (2004). Application of distance based vegetation index for agricultural crops discrimination. *International Society for Photogrammetry and Remote Sensing XXth Congress, Geo-Imagery Bridging Continents*, 12–23 July 2004, Istanbul, Turkey, Commission VII papers, Vol. XXXV, part B7. Retrieved September 28, 2007 from http://www.isprs.org/istanbul2004/comm7/papers/215.pdf

United Nations (2006). *Afghanistan could return to being a failed state*. Press release, Sunday 26 November 2006 4:00AM. Retrieved September 28, 2007 from http://www.scoop.co.nz/stories/WO0611/S00395.htm

UNODC (United Nations Office on Drugs Crime) (2007). *Afghanistan opium survey 2007*. Retrieved September 28, 2007 from http://www.unodc.org/pdf/research/AFG07_ExSum_web.pdf

Wakefield, S. & Wilder, A. (2003). *The A to Z guide to Afghanistan assistance*. Afghanistan Research and Evaluation Unit. 2nd Edition (Kabul: AREA)

Walsh, D. (2006a). As Taliban insurgency gains strength and sophistication, suspicion falls on Pakistan. *The Guardian*, Monday, 13 November

Walsh, D. (2006b). Afghanistan's opium poppies will be sprayed, says US drugs tsar. *The Guardian*, Monday, 11 December

Weaver, M.A. (2005). Lost at Tora Bora. *The New York Times Magazine*, September 11, Section 6, 54–58

Wittekindt, H. & Weippert, D. (1973). *Geological Map of Central and Southern Afghanistan*, 1:500,000. Bundesanstalt für Bodenforschung, (Geological Survey of the Federal Republic of Germany), Hannover

World Bank (2007a). Afghanistan data Profile. Retrieved March 8, 2008 from http://devdata.worldbank.org/external/CPProfile.asp?PTYPE=CP&CCODE=AFG

World Bank (2007b). Afghanistan at a glance. Retrieved March 9, 2008 from http://devdata.worldbank.org/AAG/afg_aag.pdf

Chapter 3
Economic Impacts of Terrorist Attacks and Natural Disasters: Case Studies of Los Angeles and Houston

Qisheng Pan[1], Peter Gordon[2], James E. Moore[2], II, and Harry W. Richardson[2]

Abstract Large metropolitan regions are vulnerable to terrorist attacks and natural disasters. Ports and downtown business districts could be targets of terrorist attacks and are also prone to substantial losses from natural disasters like earthquakes or hurricanes. It is important for stakeholders and decision makers to be aware of the spatial distribution of these losses and recognize the potential economic losses from various hypothetical terrorist attacks and natural disasters on these crucial facilities and core sites. The Southern California Planning Model (SCPM), a GIS-based regional planning model developed initially for the five-county Los Angeles metropolitan area, is capable of endogenizing freight and passenger flows and allocating impacts spatially via unexpected impedances to trips and shipments through the regional highway network. This chapter presents the SCPM model and describes several applications via three case studies of hypothetical events: (1) A radiological bomb or so-called "dirty bomb" attack and conventional bomb attacks on the twin ports of Los Angeles and Long Beach; (2) A radiological bomb attack on a large office building in Downtown Los Angeles Financial District; and (3) A hurricane striking the Houston-Galveston-Brazoria (HGB) region. The results show that the model can allocate the losses to various types of impact analysis zones or political jurisdictions. The methods used in this study are adaptable to almost any kind of terrorist and natural disaster attacks and also transferable to other large metropolitan areas.

Keywords Economic impacts, terrorist attacks, natural disasters, GIS, Los Angeles, Houston

[1] Texas Southern University, Houston; pan_qs@tsu.edu

[2] University of Southern California, Los Angeles; pgordon@usc.edu, jmoore@usc.edu, hrichard@usc.edu

D.Z. Sui (ed.) *Geospatial Technologies and Homeland Security*,

3.1 Introduction

The 9/11 attacks on the New York World Trade Center in 2001 have raised concerns over the safety of downtowns and similar sites and the socio-economic effects of terrorist attacks on such facilities. The Government Accounting Office (2002) reports that the September 11 attacks on the two World Trade Center buildings cost about $83 billion. Similarly, hurricanes Katrina and Rita of 2005 caused significant economic losses in the Gulf Coast states and indirectly throughout much of the United States. The total number of fatalities directly and indirectly related to Hurricane Katrina is 1,833 and a preliminary estimate of total damage is about $81.2 billion (Knabb et al. 2006a). Though Hurricane Rita spared the heavily populated Houston-Galveston Area and made landfall on the Texas-Louisiana border, it claimed 120 direct and indirect deaths and total damages of about $10 billion (Knabb et al. 2006b). Catastrophes such as these illustrate why it is necessary to investigate the potential losses of earthquakes in California, tornados in the mid-east region, and other types of natural disasters in the rest of America, to consider mitigation steps, arrange emergency response efforts, and allocate manpower and resources for disaster planning more cautiously and efficiently.

This chapter summarizes our economic impact research in three case studies. Two of them are from studies we completed for the Economic Modeling Group at the Center for Risk and Economic Analysis for Terrorism Events (CREATE) at the University of Southern California. One explores possible radiological bomb attacks at the Ports of Los Angles and Long Beach combined with conventional bomb attacks on access bridges. The other examines a radiological bomb attack on a major downtown office building in Los Angeles. The third case study estimates the economic losses of a hypothetical hurricane event in the Houston-Galveston Area.

We apply a well-established model to measure economic impacts, called the Southern California Planning Model (SCPM). This is a metropolitan input–output model in the Garin-Lowry tradition that is spatially disaggregated (with more than 3,000 Traffic Analysis Zones (TAZs) in the 2005 version of the model), to which we add a highway network with endogenously determined loadings, including freight and passenger flows. SCPM enables us to estimate spatially detailed output and job impacts of a variety of exogenous shocks, including policies, projects, and plans. In this case, the exogenous shock is a terrorist attack or a natural disaster.

3.2 Methodology

It is important for federal, state, and local policy makers to address disaster preparation and response questions by utilizing plausible loss estimates of terrorist attacks and natural disasters at fine levels of spatial detail. The Southern California Planning Model (SCPM) provides useful tools and functions to estimate total economic impacts and allocate the impacts spatially over a large metropolitan area.

It is capable of allocating the losses to various types of impact analysis zones or political jurisdictions. Multiple versions of SCPM offer flexibility to determine the geographical location of indirect and induced economic losses through a simple Garin-Lowry style approach or a complex integrated spatial interaction model with an endogenized transportation network.

3.2.1 The Southern California Planning Model Version 1 (SCPM1)

The Southern California Planning Model version 1 (SCPM1) was developed for the five-county Los Angeles metropolitan region, and has the unique capability to allocate all impacts, in terms of jobs or the dollar value of output, to 308 sub-regional zones, mainly individual municipalities defined by the Census Bureau. This is the result of an integrated modeling approach that incorporates two fundamental components: input–output and spatial allocation. The approach allows the representation of estimated spatial and sectoral impacts corresponding to any vector of changes in final demand. Exogenous shocks treated as changes in final demand are fed through an input–output model to generate sectoral impacts that are then introduced into the spatial allocation model to determine their spatial distribution.

The first model component is built upon the well-known IMPLAN input–output model, which has a high degree of sectoral disaggregation (509 sectors). The second basic model component is used for allocating sectoral impacts across the 308 geographic zones in Southern California. The key is to adapt a Garin-Lowry style model for spatially allocating the induced impacts generated by the input-output model. The building blocks of the SCPM1 are the metropolitan input–output model, a journey-to-work matrix, and a journey-to-nonwork-destinations matrix. This is a journey-to-services matrix that is more restrictively described as a 'journey-to-shop' matrix in the Garin-Lowry model.

The journey-to-services matrix includes any trip associated with a home-based transaction other than the sale of labor to an employer. This includes retail trips and other transaction trips, but excludes non-transaction-based trips such as trips to visit friends and relatives. Data for the journey-to-services matrix include all trips classified by the Southern California Association of Governments (SCAG) as home-to-shop trips, and a subset of the trips classified as home-to-other and other-to-other trips.

The key innovation associated with SCPM1 is to incorporate the full range of multipliers obtained via input–output techniques to obtain detailed economic impacts by sector and by submetropolitan zone. The SCPM1 follows the principles of the Garin-Lowry model by allocating sectoral output (or employment) to zones via a loop that relies on the trip matrices. Induced consumption expenditures are traced back from the workplace to the residential site via a journey-to-work matrix and from the residential site to the place of purchase and/or consumption via a journey-to-services matrix. See Richardson et al. (1993) for a further summary of SCPM1.

3.2.2 The Southern California Planning Model Version 2 (SCPM2)

Incorporating the Garin-Lowry approach into spatial allocation makes the transportation flows in SCPM1 exogenous. These flows are also relatively aggregated compared with transportation models, defined primarily at the level of political jurisdictions. Most transportation models use TAZs, which are much smaller. However, with no explicit representation of the transportation network, SCPM1 has no means to account for the economic impact of changes in transportation capacity. Terrorist attacks and natural disasters, especially against the transportation system, may induce such changes, including capacity losses that will contribute to reductions in network level service and increases in travel delays. SCPM1 does not account for such changes in transportation costs, underestimating the costs of any exogenous shock. Treating the transportation network explicitly endogenizes otherwise exogenous Garin-Lowry style matrices describing the travel behavior of households, achieving consistency across network costs and origin-destination requirements. SCPM2 makes distance decay and congestion functions explicit. This allows us to endogenize the spatial allocation of indirect and induced economic losses by endogenizing choices of route and destination. This better allocates indirect and induced economic losses over zones in response to direct losses in trade, employment, and transportation accessibility (Fig. 3.1). See Cho et al. (2001) for a more detailed summary of SCPM2.

3.2.3 The Southern California Planning Model Version 2005 (SCPM 2005)

The most recent version of the model is more sectorally disaggregated with 47 industrial sectors. We call these the USC (University of Southern California) Sectors because they have been constructed to reconcile various databases and to integrate SCPM with a national model, NIEMO (National Interstate Economic Model), See Park et al. (2007) for a detailed description. The updated transportation network follows the definition in the SCAG's 2001 regional transportation model with 3,217 TAZs and 89,356 network links. This disaggregation is an important feature of a major update of SCPM, called SCPM05.

SCPM05 includes more up-to-date data and other refinements beyond SCPM2. Also, the model makes use of 2005 Freight Model estimates. In general, freight flows are more difficult to estimate than passenger flows, so it is quite important to obtain external validation for the accuracy of these estimates. To test this, our 2005 estimates were compared with the SCAG 2003 Annual Average Weekday Truck Traffic Counts (SCAG/LAMTA 2004). Under a variety of assumptions about PCEs (Passenger Car Equivalents), we plotted estimated against actual freight flows, and obtained R^2s in the 0.67–0.80 range.

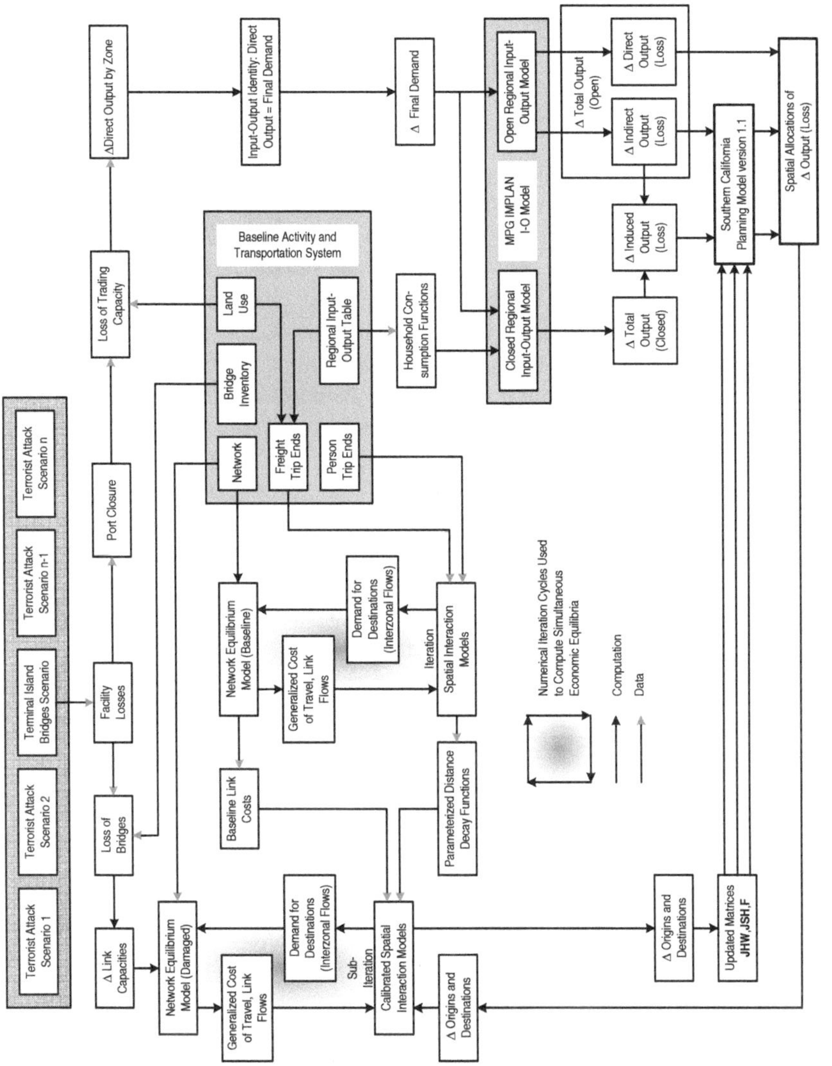

Fig. 3.1 SCPM2 data flows and calculations

3.3 Case study I: The Cost of a Terrorist Attack on Terminal Island in Los Angeles[1]

The twin ports of Los Angeles and Long Beach have played a significant role in the local and national economy. They are the third largest port complex in the world. They rank first and second nationally in terms of containerized traffic and accounted for 135 million tons of seaborne trade in 2005 (USACE 2005). There are half a million workers, about seven percent of the region's labor force, working directly or indirectly for freight traffic that goes through these two ports. The combined import and export trade flows of more than $218 billion through these two ports in 2003 (USDOT 2004) is equivalent to about 26 percent of the greater Los Angeles gross regional product. Reflecting trends in the national economy, imports are much more important than exports ($184 billion compared with $34 billion). About one-half of the imports and two-thirds of the exports originate outside the region. In other words, the ports fulfill a national function, even more than a regional function.

In this case study, we hypothesized a possible radiological bomb attack on Terminal Island and proposed three scenarios to examine the impacts of port closure on regional economics and transportation system performance: 15-day closure with no bridge damage, 120-day closure with bridge damage, and one-year closure with bridge damage called Terminal Island scenario. One of our previous research (Gordon et al. 2006) estimated that the closure of the Los Angeles and Long Beach Ports for anywhere from 15–120 days would cost the US economy in a range of $4 to $34 billion—or 26,000 to 212,000 person-years of employment (see Table 3.1

Table 3.1 Output and employment losses of a 15-day, 120-day and one-year closure of Terminal Island (Authors' calculations in Gordon et al. 2006a)

	Output ($1,000s)			Jobs (Person-Years)		
	15-Day	120-Day	One-Year	15-Day	120-Day	One-Year
City of Los Angeles	423,152	3,385,384	4,537,531	2,639	21,111	28,503
City of Long Beach	87,787	700,310	815,517	657	5,249	5,787
County of Los Angeles	1,034,070	8,271,386	10,914,814	6,513	52,097	68,535
County of Orange	262,434	2,100,029	2,796,342	1,669	13,352	17,791
County of Ventura	72,609	580,860	774,623	435	3,482	4,641
County of Riverside	64,052	512,697	680,566	421	3,371	4,475
County of San Bernardino	89,270	714,515	949,071	568	4,548	6,044
Sum of Five Counties	1,522,436	12,179,488	16,115,416	9,606	76,850	101,485
Regional Leakages	2,736,487	21,891,893	28,754,513	16,914	135,316	178,482
Total	4,258,923	34,071,381	44,869,929	26,521	212,165	279,967

[1] This part of the paper draws upon research results originally reported in the 'The Costs of a Terrorist Attack on Terminal Island at the Twin Ports of Los Angeles and Long Beach' by Peter Gordon, James E. Moore II, Harry W. Richardson and Qisheng Pan, Chapter 3 in John D. Haveman and Howard J. Shatz (eds.) *Protecting the Nation's Seaports: Balancing Security and Cost*. San Francisco: Public Policy Institute, 2006.

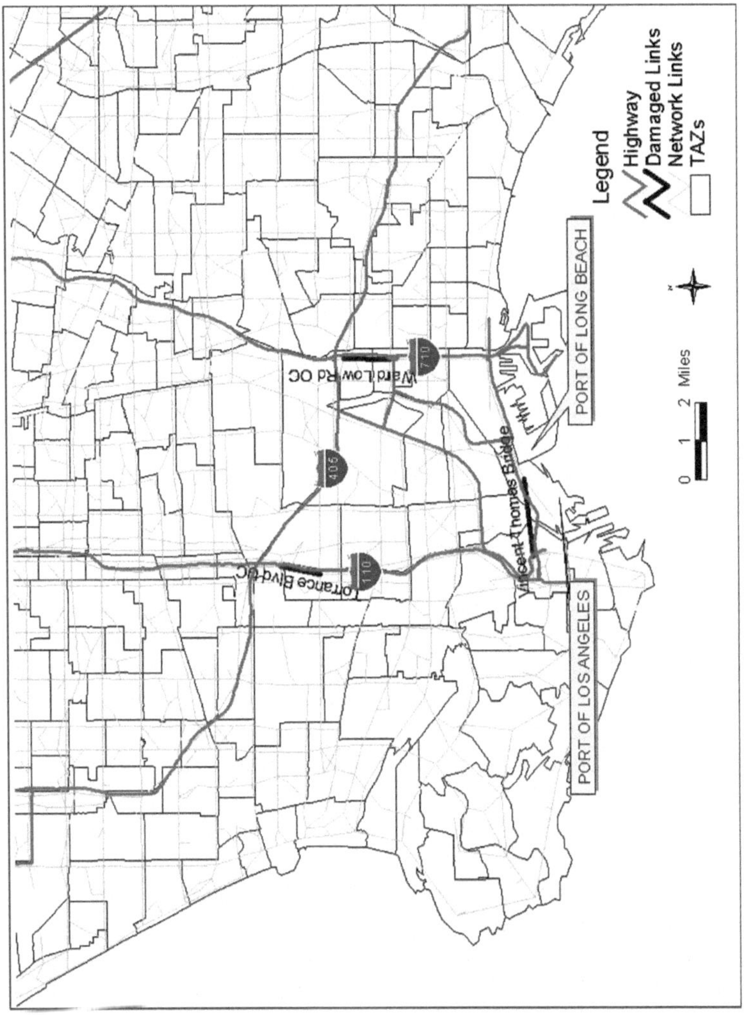

Fig. 3.2 Geographic locations of the network links where the damaged bridges are located

for aggregate results and county-level details). SCPM2 provided economic results in much greater spatial detail, to the level of counties, cities, census tracts, or TAZs if required. This preliminary work established the ground for the 15-day closure and 120-day closure scenarios.

This study also determined that many of the ports' vulnerabilities arise from restricted highway access to most of the docks. Therefore, we decided to further study the implications of bridge attacks intended to isolate all or part of the ports complex. In particular, freight going to and from Terminal Island accounts for a significant portion of combined port activities. The best estimate is 55 percent of total trade dollars although Port authorities were unable to provide exact figures.

The Terminal Island docks are accessed by four major bridges: the Vincent Thomas Bridge, the Gerald Desmond Bridge and the Commodore Schuyler F. Heim Lift Bridge, and a rail bridge (Badger Bridge) parallel to the Heim Bridge, which handles 21 percent of Terminal Island trade. These bridges are elevated facilities that permit ship traffic in the waters between the coast and Terminal Island. The Desmond Bridge, for example, is 250 feet above the water. In addition to these bridges, we also identify the critical facilities on the Interstate Highway 110 and 710 (two major highways connecting the Ports of Los Angeles and Long Beach), including the Torrance Boulevard Underpass and Willow Street Overpass, as the most vulnerable locations (Fig. 3.2). After the bridges break down, the network links where the bridges are located become inaccessible.

The 15-day scenario assumes no bridge damage so that Terminal Island is still accessible through the bridges. However, the 120-day estimates are based on scenarios that involve destruction of various access bridges, which significantly multiplies the downtime of the ports. The ports could reopen and shippers could resort to congested surface streets, but at a substantial efficiency cost. Thus, an additional $90 million dollars in transportation network delay costs are incurred in the 120-day scenario. This scenario includes a loss of network capacity in this period because of bridge damage and a reduction in transportation demand because of port closures. The model estimates the associated changes in network flows and costs.

SCPM2 simulations also revealed that an attack that incapacitates these bridges for twelve months would create economic losses of almost $45 billion per year, accounting for job losses of nearly 280,000 person-years. Thirty-five percent of these losses impact the local five-county area, and the remaining 65 percent are spread throughout the US (see Table 3.1 for county-level results and Fig. 3.3 for detailed spatial results for the region).

Tables 3.2a, b compare changes in transportation costs across all three scenarios. In the case of the Terminal Island scenario, network costs increase by $58 million per year. This value is lower than the increase in delay costs associated with the 120-day scenario in the radiological bomb plus bridge access study because the Terminal Island scenario represents only a partial elimination of port capacity. It is important to note that these are delay costs; we have not made precise estimates of bridge repair costs.

We do not know how quickly access to Terminal Island could or would be restored. If under normal circumstances bridge repairs on this scale take up to two years, then our approach can be used to approximate the benefits of speedier

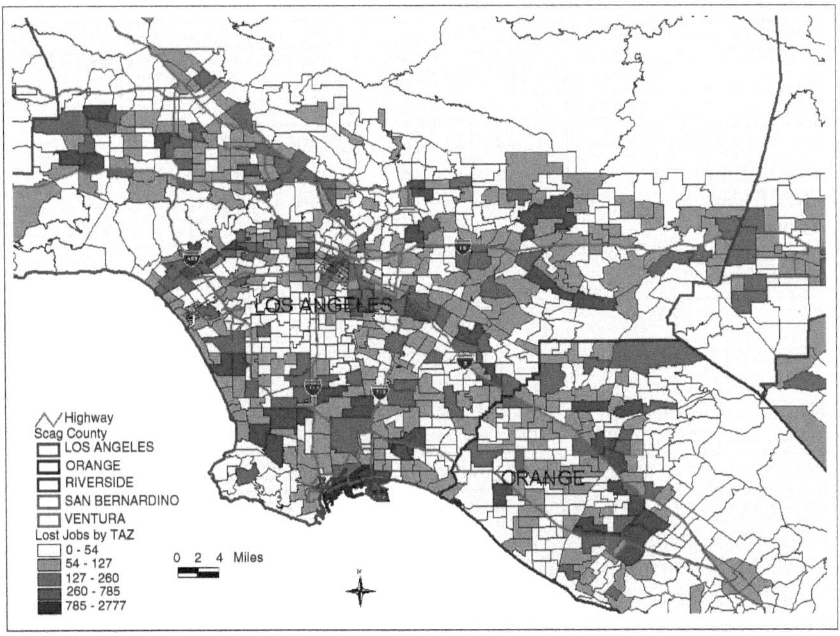

Fig. 3.3 Spatial distribution of job losses, one year closure of Terminal Island

Table 3.2a Network performance for multiple impact scenarios (total travel costs in $million)

	Freight travel costs[a]	Personal travel costs[b]	Total costs[d]	Personal travel cost[c]	Total costs
Baseline–15 day	1,031.88	2,507.11	3,539.00	5,014.23	6,046.11
Baseline–120 day	8,255.07	20,056.91	28,311.98	40,113.82	48,368.89
Baseline–one year	25,109.18	61,006.43	86,115.62	122,012.87	147,122.05
15 Day Radiological Scenario	1,007.18	2,494.93	3,502.11	4,989.85	5,997.03
120-Day Radiological Scenario	8,137.61	20,160.48	28,298.09	40,320.95	48,458.56
Terminal Island Scenario	24,771.72	61,204.11	85,975.82	122,408.21	147,179.93

[a]Freight trip cost is assumed to be $35.00 per PCE per hour

[b]Personal trip cost is assumed as $6.50 per PCE per hour

[c]Personal trip cost is assumed as $13.00 per PCE per hour

[d]Total trip cost is the sum of freight trip cost and personal trip cost

It should be noted that the model does not yet explicitly include the rail network. The highway network cost estimates do not accommodate the truck traffic accessing the rail

repairs, including the installation of temporary facilities. Temporary high capacity bridges might be constructed relatively quickly, but these low-design facilities would block ship traffic in the channels separating Terminal Island from the remainder to the port complex.

Table 3.2b Change in total travel costs for multiple impact scenarios ($million) (Authors' calculations)

	Freight Travel Costs	Personal Travel Costs	Total Costs	Personal Travel Cost	Total Costs
15 Day Radiological Scenario	−24.70	−12.18	−36.89	−24.38	−49.08
120-Day Radiological Scenario	−117.46	103.57	−13.89	207.13	89.67
Terminal Island Scenario	−337.46	197.68	−139.80	395.34	57.88

Table 3.3 Total output and highway network losses for alternate bridge reconstruction periods (Authors' calculations)

	Output Loss	Network Loss	Total Loss
24 Month Bridge Loss	$89.74 billion	$115.8 million	$89.856 billion
18 Month Bridge Loss	$67.31 billion	$86.9 million	$67.392 billion
12 Month Bridge Loss	$44.87 billion	$57.9 million	$44.928 billion
6 Month Bridge Loss	$22.44 billion	$29.9 million	$22.464 billion

Table 3.3 shows the total losses for the Terminal Island scenario by six-month increments up to two years. The estimated benefits of accelerated repairs are approximated by the differences between Row 1 of Table 3.3 and the row corresponding to the actual repair period. The differences are significant, and the implications are obvious: it is highly cost-effective to analyze emergency bridge reconstruction options and formulate plans for the protection of the Terminal Island access routes or their speedy replacement.

Recent estimates from the California Department of Transportation are that the costs of the replacement span for the Oakland Bay Bridge are in the range of $5.1 billion. This span carries 275,000 passenger-car equivalents each day, approximating the scale of the Vincent Thomas Bridge. The other two bridges now serving Terminal Island are comparatively smaller, and would be cheaper to replace. A $12 billion total reconstruction cost for all bridges is a reasonable estimate, but it is unknown to what extent these costs might arise if construction is accelerated. Accepting the linearity assumptions associated with our alternative loss estimates, accelerating access to all the bridges would have an economic benefit of $3.75 billion per month.

Planning now to protect these facilities or for reconstruction or rapid temporary replacement of these critical bridges is easily justified. The costs of accelerated repairs to the Santa Monica Freeway bridges following the Northridge Earthquake were easily justified. Our modeling approach makes it possible to be specific *ex ante* about the efficiency gains of accelerated repairs.

3.4 Case study II: The Economic Impacts of a Terrorist Attack on Downtown Los Angeles Financial District

This case examines a representative terrorist attack on the financial district in downtown Los Angeles, aimed at an unspecified major office building. It mimics the 9/11 World Trade Center attack in some respects in that it is an attack on a major downtown office building in Los Angeles. The mechanism, however, is different by incorporating a radiological bomb rather than airplanes. We choose to simulate a radiological rather than a conventional bomb attack because we are interested in examining non-localized attacks. This chapter focuses on business interruption effects only. We do not attempt to estimate accurately the number of deaths and injuries and their costs. Also, we do not estimate the costs of physical damage to the attacked building and other nearby buildings.

Hypothetically, we assumed the explosion of a 50-lb bomb, the maximum portable without requiring a vehicle as the delivery instrument. Blast damage would be quite limited, with deaths and serious injuries within a range of perhaps 50 meters and with moderate damage to physical infrastructure, except at ground zero. The outer evacuation zone would include all areas with exposure >1 REM (Roentgen Equivalent Man), a unit of radiation dose. We assume a hypothetical radiation plume, a long narrow rectangle four kilometers long and more than 200 meters wide with an inner and more contaminated zone of about 100 meters radius (an area of 0.03 km^2), an oversimplification of plume representations that are not open source. The critical early phase of exposure lasts about four days (EPA guidelines); the time frame for intermediate and later phases is variable and subjective (weeks, months, even years). We assume a one-year evacuation in the Inner Zone, and a one-month evacuation in the Outer Zone. With respect to the Outer Zone, this may be conservative because some firms may trickle back with a lag after being given permission to return. Health factors will dictate an immediate evacuation, but because the health effects are long-term, the decision to allow a return will be determined by political rather than scientific considerations.

By dividing the Inner and Outer Zone in the plume area in terms of the variation of damage levels and evacuation times, we examine three limiting cases: first, an exit scenario where firms disappear (either close down or move out of town); second, a relocation scenario where all the evacuating firms relocate to other subcenters within the five-county metropolitan region; and third, a hybrid scenario where the Inner Zone firms exit and the Outer Zone firms relocate. These are just three of an almost limitless set of scenarios, and they are all based on the assumptions of a one-year evacuation of the Inner Zone and a one-month evacuation of the Outer Zone. These time periods are based on discussions with experts on radiological contamination, but alternative time periods are easily substitutable.

Our field of research does not focus on deaths and/or injuries but on business interruption. The health costs of a radiological attack stretch out over a long time, but a blind guess of the immediate toll might be 20 deaths and 200 hospital-related injuries. The duration of the disruption determines the length of time for which firms throughout the region will be non-operational or operating below normal levels of service. This allows the calculation of exogenously prompted reductions in demand by these businesses. These are introduced into the interindustry model as declines in final demand. The I/O model translates this production shock into direct, indirect, and induced costs. The indirect and induced costs are spatially allocated over the 3,000-plus zones in terms consistent with the endogenous transportation behavior of firms and household.

Although our study takes transportation networks into account, the transportation repercussions of a downtown closure are relatively modest. First, there are no freeways in the inner zone. Second, as a major service center, downtown attracts fewer deliveries and pick-ups than the rest of the metropolitan region. Our data show only a two percent PCE (passenger cars equivalent) truck flow rather than the seven percent region wide. Third, only 8,620 jobs are affected, a drop in the bucket compared with the nine million jobs in the region. Fourth, and more important, most trips to the downtown area are through traffic rather than traffic with origins and/or destinations in the downtown area. Our analysis assumes that if motorists roll up their windows and keep the air conditioning off that they can pass through the plume area in relative safety. If the authorities mandate a different and more coercive procedure, the transportation impacts would be magnified. As it is, the network effects in this particular case study are so small that they are not worth reporting.

We utilized the latest version of the Southern California Planning Model, called SCPM 2005, to measure economic impacts. Similar to its previous versions, it is also a metropolitan input–output model in the Garin-Lowry tradition, but it utilizes more industrial sectors (47 USC Sectors) and is more spatially disaggregated (with more than 3,000 TAZs), to which we add an up-to-date highway network with endogenously determined loadings, including freight and passenger flows in 2001. The SCPM 2005 includes 2005 Freight Model estimates, which has better validation results in comparison to the Southern California Association of Governments 2003 Annual Average Weekday Truck Traffic Counts. It enables us to estimate the geographical output and job impacts of a variety of recent exogenous shocks, including policies, projects and plans. In this case, of course, the exogenous shock was a terrorist attack.

The estimates for jobs and households impacted by the radioactive plume after the bomb attack are based on the 1997 SCAG Employment Data and the Census 2000 Summary File. There are 7,843 jobs and no households affected in the Inner Zone for a one-year evacuation period. There are more than 8,500 jobs and 60,000 people in the Outer Zone, but because the evacuation period is only one month and the model is run as an annual model, the model input is 710 jobs and 2,424 households. Despite the absence of households, the economic impacts of evacuation in the Inner Zone are much larger than those in the Outer Zone (the economic impact from a lost job is greater than that from a lost household, by a factor of more than three).

The Exit Scenario assumes that firms and households exit from the region for the evacuation periods. The true regional economic impact is the discounted value of the stream of future output and job losses because the firms and households may be gone forever. Thus, the Exit Scenario is merely a measure of the losses during the evacuation period under the assumption that there are no regional offsets in the form of positive relocation impacts.

Table 3.4 shows the results of the Exit Scenario. Although impacts that distinguish between the Inner and the Outer Zones have been estimated for all scenarios, we report such results only for the Exit Scenario. The Inner Zone impacts were much larger ($5.624 billion of output and 38,000 jobs) than those in the Outer Zone ($0.278 billion of output and 2,391 jobs) for a total of $5.901 billion of output and 40,391 jobs (Table 3.4). As a generalization, one-half of the overall impacts (indirect and induced as well as direct) occur in the City of Los Angeles (of course, all the direct impacts are in the City), and about two-thirds occur in Los Angeles County. Regional leakages (i.e. spillovers in the indirect and induced effects) are small ($0.726 billion of output and 5,408 jobs); this reflects the fact that the local component of the financial and office sectors is very high, with minimal reliance on imports from outside the region (such as computing, other information technology equipment, materials and supplies). Overall, however, the indirect and induced effects are larger than the direct effects, implying a sizeable output multiplier (2.25) and an even larger employment multiplier (4.82), reflecting highly paid workers in the Financial District that generate above-average consumption (and induced jobs in the retail and service sectors).

Table 3.4 Economic impact of a terrorist attack on downtown Los Angeles, exit scenario, for all businesses and households moving out of the Inner and Outer Zones (Authors' calculations)

	Output ($1,000s)				Jobs			
	Direct	Indirect	Induced	Total	Direct	Indirect	Induced	Total
City of Los Angeles	2,304,493	136,563	499,580	2,940,636	7,257	961	5,171	13,389
County of Los Angeles	2,304,493	362,308	1,301,283	3,968,084	7,257	2,412	13,472	23,141
County of Orange	0	116,622	459,508	576,130	0	863	4,758	5,622
County of Ventura	0	24,227	107,274	131,501	0	172	1,110	1,282
County of Riverside	0	36,558	202,202	238,761	0	276	2,101	2,376
County of San Bernardino	0	44,422	216,455	260,877	0	313	2,248	2,562
Sum of Five Counties	2,304,493	584,137	2,286,722	5,175,352	7,257	4,037	23,689	34,983
Regional Leakages	312,733	80,190	332,954	725,878	1,363	623	3,425	5,408
Total	2,617,225	664,327	2,619,676	5,901,229	8,620	4,660	27,114	40,391

The Relocation Scenario is more complex than the Exit Scenario because it requires a procedure for relocating both firms and households out of downtown and the outer zone, which stretches northeast of downtown. Households were relocated using an empirically estimated distance-decay function with a negative exponential formula. According to a study by Clark et al. (2002) on the association between residential changes and commuting behavior in the Greater Seattle area, the mean move distance was 6.28 miles. Using this as a template, the probability distribution function (PDF) for household relocation is as follows:

$$F_X(x)=1-e^{-x/6.28}, x\geq 0. \tag{1}$$

where x is moving distance.

This function is used to randomly generate moving distances for the households. The final destinations of a relocating household are locations close to the estimated moving distance with similar median housing rents/prices to the origin locations. The Census 2000 (Summary File) data on household income and house price data (combined with the relocation distance assumptions) are employed to identify the probable destinations.

There were no households living in the inner zone in the year when the database was constructed based on the Census 2000 data at block level, but there were 2,424 households in the outer areas. All these 2,424 households are relocated over the Five-County Los Angeles region. Of these, all but 31 relocated within Los Angeles County, with most of the rest in Orange County. Finally, Census 2000 blocks with the moving-in households are further aggregated into the SCAG 1999 TAZs for modeling purposes. Household consumption at the new locations is calculated by using the average propensity to consume from the Consumer Expenditure Survey for Los Angeles.

Businesses moving out from the inner and outer impact zones of downtown Los Angeles are relocated in the region based on the job vacancy and job distribution by sector in the business submarkets. Based on the second quarter office vacancy report in MarketBeat Mid-Year 2005 by Cushman and Wakefield (2005), there are over 50 submarkets with an average 14 percent of vacancy rate in Los Angeles North, Southern, Central, West, and the Tri-Cities Offices sub-regions. After the development of a correspondence table between submarkets and TAZs, the office vacancy rates are recalculated from submarkets to TAZs.

SCAG 1999 TAZs with the 3191 internal zones are used as a base for business relocation. SCAG 1997 employment by business establishment by SIC code is translated into employment by USC Sector and finally aggregated into SCAG TAZs. The jobs moving out of the inner and outer impact areas are relocated into these TAZs based on the vacancy rate and the job distribution by sector in the TAZs. Most, but not all, of the jobs relocate within Los Angeles County. The move-in jobs are converted into dollar values of output by applying the dollars per job ratio obtained from the regional input–output model.

Overall, the Relocation Scenario is a 'wash' with minimal changes at the County level (a decline of $77 million of output and 217 jobs in Los Angeles County relocated to Ventura County; Tables 3.5a, b). The major impacts take place at the city level, especially in Los Angeles (a net loss of $1.567 billion with an outward movement of $2.941 billion and an inward movement of $1.373 billion and 5,099 jobs with 13,389 jobs out and 8,289 in). The major gainers were Torrance, Industry, El Monte, Glendale and Pasadena in terms of output and Torrance, El Segundo, Pasadena, Glendale, Beverly Hills, and Santa Monica in terms of jobs. All of the top 25 gainers (in terms of output and job gains) were in Los Angeles County with the exception of Thousand Oaks (in Ventura County) (Tables 3.6a, b). Figure 3.4 shows the spatial distribution of the relocated jobs throughout the region; it illustrates their wide geographical dispersion, with concentrations at subcenters derived from the submarket analysis.

The Exit Scenario does not seem plausible for activities in the Outer Zone if the evacuation lasted for only a few weeks. So we developed the Hybrid Scenario where Inner Zone firms exited while Outer Zone firms and households temporarily relocated. The numbers in the Hybrid Scenario are a modified version of the Exit Scenario, reflecting the dominance of Inner Zone impacts. They total $5.624 billion of output and 38,000 jobs (Table 3.7). One half of the output losses and almost three-fifths of the job losses occur in Los Angeles County. The City of Los Angeles experiences a modest output loss of $57.826 million and a job loss of 413 jobs. The main cities gaining from relocation are El Segundo, Torrance, Glendale, Pasadena, Beverly Hills, Commerce, and Santa Monica in that order.

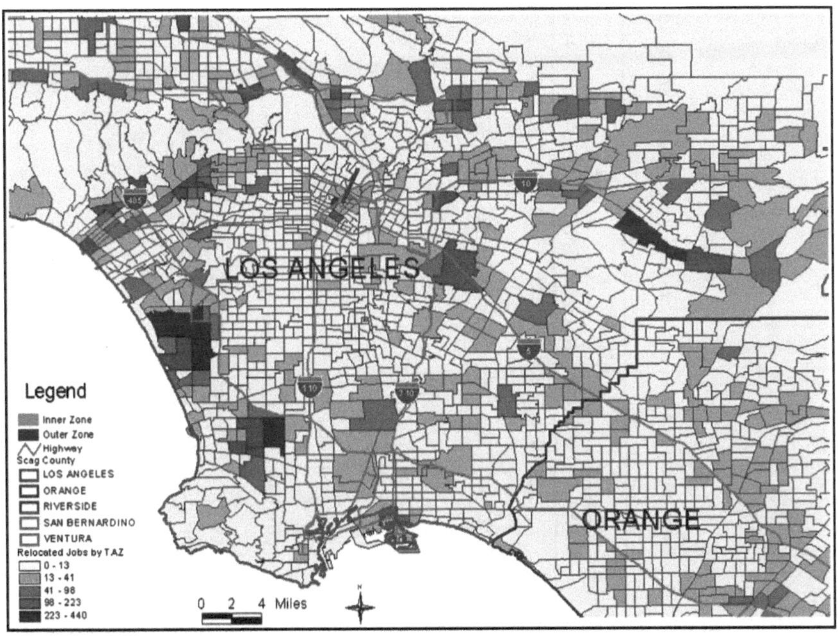

Fig. 3.4 Relocated jobs from a terrorist attack on downtown Los Angeles (relocation scenario) for all businesses and households moving out of the Inner and Outer Zones, five-county region, 2000

Table 3.5a Economic impacts of all businesses and households from the Inner and Outer Zones: Relocation scenario (output, $1,000s, 2001) (Authors' calculations)

	Positive				Negative				Net			
	Direct	Indirect	Induced	Total	Direct	Indirect	Induced	Total	Direct	Indirect	Induced	Total
City of Los Angeles	737,221	136,563	499,581	1,373,365	2,304,493	136,563	499,580	2,940,636	−1,567,272	0	1	−1,567,271
County of Los Angeles	2,227,433	362,308	1,301,284	3,891,024	2,304,493	362,308	1,301,283	3,968,084	−77,060	0	1	−77,059
County of Orange	56	116,622	459,507	576,185	0	116,622	459,508	576,130	56	0	−1	55
County of Ventura	76,996	24,227	107,274	208,496	0	24,227	107,274	131,501	76,996	0	0	76,995
County of Riverside	0	36,558	202,202	238,761	0	36,558	202,202	238,761	0	0	0	0
County of San Bernardino	9	44,422	216,455	260,886	0	44,422	216,455	260,877	9	0	0	9
Sum of Five Counties	2,304,493	584,137	2,286,722	5,175,352	2,304,493	584,137	2,286,722	5,175,352	0	0	0	0
Regional Leakages	312,733	80,190	332,954	725,878	312,733	80,190	332,954	725,878	0	0	0	0
Total	2,617,225	664,327	2,619,676	5,901,229	2,617,225	664,327	2,619,676	5,901,229	0	0	0	0

Table 3.5b Economic impacts of all businesses and households from the Inner and Outer Zones: Relocation scenario (jobs 2001) (Authors' calculations)

	Positive				Negative				Net			
	Direct	Indirect	Induced	Total[a]	Direct	Indirect	Induced	Total[a]	Direct	Indirect	Induced	Total[a]
City of Los Angeles	2,157	961	5,171	8,289	7,257	961	5,171	13,389	−5,099	0	0	−5,099
County of Los Angeles	7,039	2,412	13,472	22,924	7,257	2,412	13,472	23,141	−217	0	0	−217
County of Orange	1	863	4,758	5,622	0	863	4,758	5,622	1	0	0	1
County of Ventura	217	172	1,110	1,499	0	172	1,110	1,282	217	0	0	217
County of Riverside	0	276	2,101	2,376	0	276	2,101	2,376	0	0	0	0
County of San Bernardino	0	313	2,248	2,562	0	313	2,248	2,562	0	0	0	0
Sum of Five Counties	7,257	4,037	23,689	34,983	7,257	4,037	23,689	34,983	0	0	0	0
Regional Leakages	1,363	623	3,425	5,408	1,363	623	3,425	5,408	0	0	0	0
Total	8,620	4,660	27,114	40,391	8,620	4,660	27,114	40,391	0	0	0	0

[a] Total impact is the sum of direct, indirect, and induced effects

Table 3.6a Relocation scenario: Impacts of businesses and households from Inner and Outer Zones (2001, $1,000s)

			Positive[a]			Negative[a]			
Rank	County	Place	Households	Businesses	Total	Households	Businesses	Total	Net
1	LOS ANGELES	Los Angeles	5,577	1,367,788	1,373,365	6,732	2,933,904	2,940,636	−1,567,271
2	LOS ANGELES	Torrance	128	353,697	353,825	115	38,661	38,775	315,050
3	LOS ANGELES	Industry	59	166,563	166,622	56	21,550	21,606	145,016
4	LOS ANGELES	El Monte	60	124,233	124,293	40	15,124	15,163	109,129
5	LOS ANGELES	Glendale	134	135,522	135,657	101	37,034	37,135	98,521
6	LOS ANGELES	Pasadena	273	124,371	124,644	96	34,476	34,572	90,072
7	LOS ANGELES	Monterey Park	28	79,094	79,123	26	9,217	9,243	69,880
8	VENTURA	Thousand Oaks	79	96,838	96,917	79	28,269	28,348	68,570
9	LOS ANGELES	UNCOR-LOS ANGELES	182	113,487	113,669	131	53,103	53,234	60,434
10	LOS ANGELES	West Covina	55	71,297	71,351	37	13,573	13,609	57,742
11	LOS ANGELES	El Segundo	38	70,759	70,797	37	14,314	14,351	56,446
12	LOS ANGELES	Santa Clarita	48	71,765	71,812	46	19,154	19,200	52,613
13	LOS ANGELES	Commerce	40	65,517	65,557	39	15,541	15,580	49,976
14	LOS ANGELES	Santa Monica	128	78,251	78,379	77	28,432	28,509	49,869
15	LOS ANGELES	Avocado Heights	12	42,168	42,180	7	2,794	2,801	39,379
16	LOS ANGELES	Alhambra	89	47,496	47,584	37	13,944	13,981	33,603
17	LOS ANGELES	Beverly Hills	121	52,467	52,587	63	22,236	22,299	30,289
18	LOS ANGELES	Arcadia	31	36,536	36,566	29	10,682	10,712	25,854
19	LOS ANGELES	Cerritos	35	37,596	37,631	35	13,032	13,066	24,564
20	LOS ANGELES	West Hollywood	33	34,186	34,219	33	12,105	12,138	22,081
21	LOS ANGELES	Monrovia	23	30,022	30,045	22	8,624	8,646	21,399
22	LOS ANGELES	Manhattan Beach	40	25,878	25,918	20	7,291	7,311	18,608
23	LOS ANGELES	Agoura Hills	7	20,652	20,659	7	2,567	2,575	18,085
24	LOS ANGELES	Burbank	73	42,148	42,221	70	25,326	25,396	16,825
25	LOS ANGELES	Culver City	43	27,403	27,447	43	15,902	15,945	11,502

[a] Direct + Indirect + Induced

Table 3.6b Relocation scenario: Impacts of businesses and households from Inner and Outer Area (2001, jobs)

			Positive[a]			Negative[a]			
Rank	**County**	**Place**	Households	Businesses	Total	Households	Businesses	Total	**Net**
1	LOS ANGELES	Los Angeles	62	8,228	8,289	75	13,314	13,389	−5,099
2	LOS ANGELES	Torrance	1	1,037	1,038	1	356	357	681
3	LOS ANGELES	El Segundo	0	616	616	0	125	125	491
4	LOS ANGELES	Pasadena	3	830	833	1	345	346	487
5	LOS ANGELES	Glendale	1	765	766	1	361	362	404
6	LOS ANGELES	Beverly Hills	1	482	483	1	206	207	276
7	LOS ANGELES	Santa Monica	1	545	547	1	280	281	266
8	LOS ANGELES	Industry	1	430	430	1	203	204	226
9	VENTURA	Thousand Oaks	1	475	476	1	281	282	194
10	LOS ANGELES	El Monte	1	324	325	0	148	148	177
11	LOS ANGELES	Manhattan Beach	0	236	236	0	73	73	163
12	LOS ANGELES	UNCOR-LOS ANGELES	2	636	638	1	476	477	161
13	LOS ANGELES	Commerce	0	288	288	0	137	137	151
14	LOS ANGELES	West Covina	1	274	274	0	138	138	136
15	LOS ANGELES	Burbank	1	371	371	1	239	240	132
16	LOS ANGELES	Monterey Park	0	219	219	0	91	91	128
17	LOS ANGELES	Arcadia	0	231	231	0	106	107	125
18	LOS ANGELES	Alhambra	1	238	239	0	127	128	112
19	LOS ANGELES	Cerritos	0	217	218	0	125	126	92
20	LOS ANGELES	West Hollywood	0	206	207	0	122	123	84
21	LOS ANGELES	Monrovia	0	161	161	0	82	82	79
22	LOS ANGELES	Santa Clarita	0	248	249	0	171	172	77
23	LOS ANGELES	Culver City	0	230	231	0	158	159	72
24	LOS ANGELES	Avocado Heights	0	83	83	0	26	27	56
25	LOS ANGELES	Agoura Hills	0	72	72	0	24	24	48

[a] Direct + Indirect + Induced

Table 3.7 Economic impacts from businesses and firms of a terrorist attack on downtown Los Angeles: Hybrid scenario (Authors' calculations)

	Output ($1,000s)				Jobs			
	Direct	Indirect	Induced	Total	Direct	Indirect	Induced	Total
City of Los Angeles	2,162,218	130,270	478,036	2,770,525	6,643	907	4,953	12,503
County of Los Angeles	2,216,711	347,420	1,245,170	3,809,300	6,643	2,283	12,906	21,832
County of Orange	56	111,077	439,692	550,825	0	814	4,559	5,373
County of Ventura	3,269	23,138	102,648	129,054	0	162	1,063	1,226
County of Riverside	0	35,045	193,485	228,531	0	262	2,012	2,274
County of San Bernardino	9	42,746	207,124	249,878	0	298	2,154	2,452
Sum of Five Counties	2,220,044	559,426	2,188,119	4,967,589	6,643	3,820	22,694	33,157
Regional Leakages	284,059	70,813	301,237	656,109	1,200	544	3,101	4,843
Total	2,504,103	630,239	2,489,356	5,623,698	7,843	4,363	25,795	38,000

In summary, we examined three scenarios (Exit, Relocation, and Hybrid) in this case study. The aggregate impacts in the Exit Scenario are $5.901 billion of output and 40,391 jobs, somewhat less in the Hybrid Scenario ($5.624 billion and 38,000 jobs). The Relocation Scenario is neutral from a regional perspective, although direct losses in the impacted zones are 7,257 jobs and $2.617 billion of output. The City of Los Angeles is the main loser, with a net output loss of $1.567 billion and a net job loss of 5,099 jobs. The County-level changes are insignificant with a small loss in Los Angeles County balancing an equivalent increase in Ventura County (the City of Thousand Oaks). Otherwise, jobs decentralize to major subcenters in Los Angeles County.

3.5 Case study III: The business Interruption Losses of a Hypothetical Hurricane event in the Greater Houston Area

In this case, the exogenous shock was a natural disaster. The analysis utilizes an SCPM-style model to capture the business interruption losses of a hypothetical hurricane event in the Houston-Galveston Area. The hypothetical hurricane strike is designated by adopting the storm parameters of Hurricane Rita at its landfall hour from the HURREVAC 2000 database and shifting the storm track of Hurricane Rita to the southwest by about 85 miles, which causes the hurricane to make landfall at

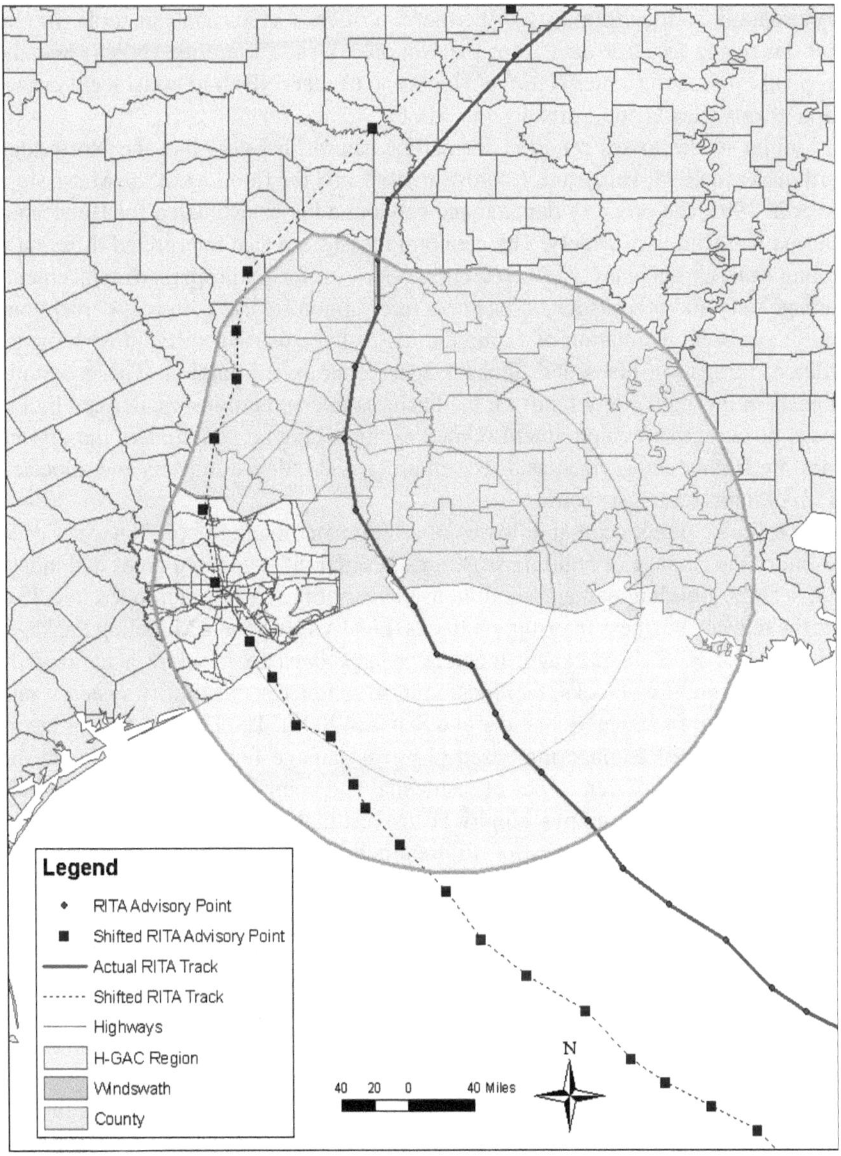

Fig. 3.5 Storm track and advisory points for the hypothetical hurricane

the coast of Galveston and then follow Interstate 45 north to downtown Houston. This scenario is considered the worst case scenario for a hypothetical hurricane event for the Houston-Galveston region (Fig. 3.5).

HURREVAC 2000 is a Windows-based hurricane decision assistance tool for government emergency managements developed by the Sea Island Software Incorporation for the FEMA and the Army Corps of Engineers. It provides historical storm data in various formats. Furthermore, it offers various analysis tools for real time hurricane forecast data from the National Weather Service (NWS) and the Tropical Prediction Center/National Hurricane Center (NHC) to assist local evacuation efforts (Sea Island Software Inc. 2006).

Similar to the losses reported for earlier natural disasters like the Northridge earthquake in 1994, Hurricane Andrew in 1992, and the flooding of the Mississippi River in 1993, the property damage and economic losses estimated for Hurricanes Katrina and Rita vary widely. The economic losses are also interpreted differently among various agencies and researchers. Some refer to property (replacement) damage only and others include business interruption losses. Different perceptions and inconsistent estimation of economic losses have made it difficult to compare efficient mitigation plans and allocate appropriate relief funding. The economic impacts in this case study focus on the business interruption losses, defined by the losses of output and employment related to the economic activities in the region. They are further categorized as direct, indirect, and induced impacts and allocated to small impact analysis zones.

To estimate direct economic losses of a hurricane in a region, it is necessary to inventory the damage of building structures in different residential types and industrial sectors. Building damages and their recovery time can be employed to calculate the relevant business interruption losses. FEMA's Hurricane Modeling Package, the so-called HAZUS package, includes a large database to store a nationwide inventory of building blocks, facilities, transportation systems, utility systems, and hazard material facilities (Schneider and Schauer 2006). The HAZUS model incorporates a series of engineering-based physical damage functions to estimate the damage states for different types of residential and commercial buildings in a hurricane event. It also employs empirical loss estimation techniques to determine repair costs and recovery times for different buildings.

The limitation of the HAZUS Hurricane Model is that there is no module to calculate indirect and induced impacts of hurricanes on industry purchases and household consumption (FEMA 2006a). Indirect impacts represent the ripple effects from direct final demand changes on related industrial purchases, while induced impacts reflect the effects on regional industries caused by the changes of household consumption due to the effects of direct final demand changes. Unlike the Earthquake Model and the Flood Model (FEMA 2006b), the Hurricane Model in the recently released HAZUS (HAZUS MH-MR2) does not have a module to estimate either indirect or induced economic losses related to business interruption.

In addition, the direct and total losses of output estimated by the HAZUS Hurricane Model for a high-intensity hurricane are usually less than 5 percent of structure damage losses or 0.5 percent of total regional output, which are far less than the ratios found in other studies. For instance, Burrus et al. (2002) found that total business interruption impacts, including direct, indirect, and induced impacts, in a low-intensity hurricane are equivalent to between 0.8 and 1.23 percent of total regional output.

In this hypothetical hurricane scenario, HAZUS reports a total of $48.5 billion building-related economic losses, which includes $40.9 billion of property damage losses and $7.6 billion of income-related business interruption losses (See Table 3.8). The HAZUS Hurricane Model also estimates output and employment losses at a total of $1.6 billion and 9,293 jobs. The total of $1.6 billion direct output losses is only 0.4 percent of $402.3 billion output in the region, which is much lower than the ratios for low-intensity hurricanes estimated by Burrus et al. in 2002.

This study proposes an approach to calculate direct impacts related to business interruptions systematically. First, it utilizes building damage state probabilities for each specific occupancy class in each census tract estimated by the HAZUS Hurricane Model, and building recovery times for different damage states for different occupancy classes (borrowed from the technical documents of the HAZUS Earthquake Model) to calculate the building and service interruption times by occupancy class by census tract. Second, it links the estimated commercial building and service interruption times to small area employment data to calculate the lost jobs in terms of business and service interruptions. Finally, it converts the job losses to dollar values of output losses using dollar per job ratios. The calculation of job losses in a census tract or a loss analysis zone is:

$$L_{z,s} = J_{z,s} * (\sum_{d} (DS_{z,s,d} * BRT_{s,d} * F_{s,d})), \tag{2}$$

where
$L_{z,s}$ = the total losses of job*day in industrial sector s in zone z,
$J_{z,s}$ = the jobs in industrial sector s in zone z,
$DS_{z,s,d}$ = the probabilities of damage states in damage state d, occupancy class (or sector) s, and zone z,
$BRT_{s,d}$ = the building recovery times (including both construction and clean-up time) measured in days for buildings in damage state d and occupancy class (or sector) s, and
$F_{s,d}$ = the building and service interruption time multiplier for buildings in damage state d and occupancy class (or sector) s.

Equation (2) associates employment data with the corresponding building damage status to estimate the total losses of jobs by time, which are further converted to annual losses of jobs to be consistent with the regular time span for data analysis in input–output models.

After we determine the direct impacts, it is standard procedure in SCPM to calculate total indirect and induced effects from the direct final demand changes. In this case study, we develop a simple version of SCPM that is an analogue of the SCPM 1.

This SCPM1-style model transfers the methodology initially developed for the Southern California to the Houston-Galveston area. First, it constructs an IMPLAN input–output model for the H-GAC region. Second, it calculates indirect and induced impacts from direct impacts from this IMPLAN input–output model and aggregates the impacts in hundreds of industrial sectors calculated to a small number of sectors, i.e. 17 sectors. Third, it allocates the indirect effects spatially to 886 census tracts in the H-GAC region using employment-weighted attractions and

Table 3.8 Building-related economic losses by county ($1000) (Authors' calculations using HAZUS)

County	Property Damage				Income-Related Business Interruption				Total
	Building	Content	Inventory	Subtotal	Relocation	Capital	Wage	Rental Income	
Brazoria	3,301,559	1,539,740	21,539	4,862,838	475,645	43,497	54,400	164,644	5,601,024
Chambers	17,399	3,248	76	20,723	1,737	251	391	641	23,742
Fort Bend	841,619	269,285	3,581	1,114,485	126,332	6,413	7,852	42,156	1,297,238
Galveston	6,695,523	3,064,312	24,897	9,784,732	928,544	138,856	153,676	357,382	11,363,190
Harris	17,928,184	6,907,206	128,498	24,963,888	2,934,520	370,001	454,083	1,304,565	30,027,055
Liberty	8,680	1,670	14	10,364	576	35	60	229	11,265
Montgomery	155,311	41,604	164	197,079	10,071	1,546	1,512	5,174	215,381
Waller	1,276	116	0	1,392	15	0	0	8	1,415
Total	28,949,551	11,827,181	178,768	40,955,500	4,477,440	560,598	671,973	1,874,798	48,540,310

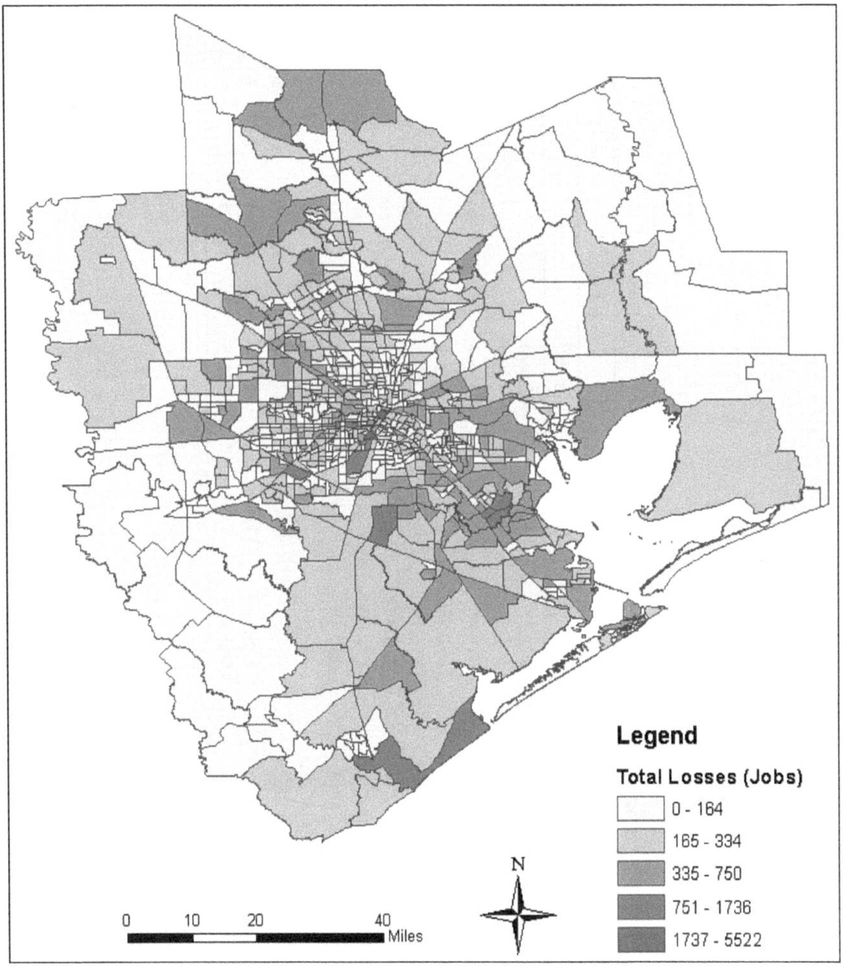

Fig. 3.6 Total losses of jobs from the hypothetical hurricane event

productions. Finally, it distributes induced impacts using a Garin-Lowry style model with two trip origin-destination (O-D) matrices, a journey home-to-work matrix, and a journey home-to-shop matrix. The journey home-to-work O-D matrix traces wages from workplace back to home and the journey home-to-shop matrix traces housing expenditures from the home to the retail stores or service establishments. The results of output and job losses in this hypothetical hurricane event are summarized in Table 3.9. Total impacts including direct, indirect, and induced impacts measured by job by census tract are shown in Fig. 3.6.

Table 3.9 Output and job losses from the hypothetical hurricane event (Authors' calculations)

	Output ($1,000s)				Jobs			
	Direct	Indirect	Induced	Total	Direct	Indirect	Induced	Total
City of Houston	5,188,013	2,377,868	2,926,709	10,492,589	40,201	15,490	30,918	86,610
Brazoria County	790,277	154,107	491,683	1,436,066	5,577	964	5,457	11,998
Chambers County	7,827	24,375	60,699	92,901	56	153	659	869
Fort Bend County	238,171	300,781	822,678	1,361,630	1,334	1,669	8,724	11,727
Galveston County	962,184	134,033	457,894	1,554,110	8,218	894	5,164	14,277
Harris County	7,209,271	3,340,298	5,707,274	16,256,843	54,365	21,629	60,380	136,375
Liberty County	456	28,450	128,895	157,800	3	164	1,436	1,603
Montgomery County	17,225	257,194	733,068	1,007,487	117	1,457	7,794	9,368
Waller County	44	17,479	63,023	80,546	0	108	676	784
Sum of Eight Counties	9,225,454	4,256,717	8,465,213	21,947,383	69,671	27,039	90,290	187,000
Regional Leakages	3,724,930	1,529,657	3,104,553	8,359,140	25,302	10,236	32,981	68,519
Regional Total	12,950,384	5,786,373	11,569,766	30,306,523	94,973	37,275	123,271	255,518

Table 3.10 The 20 cities with the highest losses of output in the hypothetical Hurricane Rita event (Authors' calculations)

		Output ($1,000s)				Jobs			
Index	City Name	Direct	Indirect	Induced	Total	Direct	Indirect	Induced	Total
1	Houston	5,188,013	2,377,868	2,926,709	10,492,589	40,201	15,490	30,918	86,610
2	Pasadena	448,527	99,425	230,836	778,788	3,200	550	2,479	6,229
3	Galveston	260,244	42,583	81,588	384,415	2,669	306	942	3,918
4	Texas City	265,844	32,478	81,645	379,967	1,888	194	918	3,000
5	Sugar Land	50,268	105,509	133,564	289,341	301	504	1,394	2,199
6	The Woodlands	9,493	109,493	127,498	246,485	54	484	1,314	1,852
7	Pearland	139,341	21,016	84,221	244,578	1,179	147	911	2,236
8	League City	123,143	16,469	99,276	238,887	1,515	142	1,121	2,778
9	Webster	175,369	13,813	18,326	207,508	1,994	108	206	2,308
10	Stafford	105,282	77,498	19,815	202,595	406	357	206	970
11	Missouri City	46,421	25,888	118,238	190,547	350	172	1,241	1,763
12	Deer Park	105,656	21,733	57,644	185,033	651	139	616	1,406
13	La Porte	95,386	21,874	61,067	178,327	559	141	653	1,352
14	Alvin	130,112	15,480	27,862	173,454	1,012	95	302	1,410
15	La Marque	93,233	12,011	29,635	134,879	374	59	330	763
16	Channelview	64,023	23,588	45,671	133,281	282	125	480	888
17	Baytown	27,240	23,779	81,164	132,184	231	173	883	1,287
18	Friendswood	68,162	7,825	49,104	125,092	794	63	546	1,403
19	Aldine	35,833	18,675	26,501	81,010	271	135	280	687
20	Conroe	871	31,299	48,720	80,890	8	210	531	749
Total		7,432,459	3,098,305	4,349,084	14,879,849	57,940	19,596	46,272	123,809

Table 3.9 shows that total output losses in the region would be over $30.3 billion, including about $13.0 billion direct losses, $5.8 indirect losses, and $11.6 billion induced losses. The region as whole also loses over 255,518 employment person-years, among which 94,973 person-years are direct losses, 37,275 are indirect losses, and 123,271 are induced losses.

There are also about $8.4 billion output losses and 68,519 job losses as regional leakages, which account for about 27 percent of total regional losses of output and employment. This high percent of losses in areas outside of the H-GAC region indicates the important role of the region in the national economy.

In the eight-county metropolitan area, Harris County accounts for about half of the total losses in output and employment. Galveston County ranks second by loss among all the counties. Brazoria County ranks third, mostly due to its geographic location along the Gulf Coast.

Table 3.10 lists the top 20 cities ranked by losses of output in dollars after the hypothetical Hurricane Rita event. The City of Houston, the one suffering the greatest impacts on the list, loses over 86,000 jobs, which accounts for about one-quarter of the total job losses in the region. The output losses in the City are $10 billion, which is about one-third of the regional losses. The total losses in the top 20 area cities are $16.9 billion output and 146,000 jobs, which is slightly higher than the losses in Harris County.

3.6 Conclusions

This chapter has examined three case studies on the economic impacts of terrorist attacks and natural disasters, including a terrorist attack on Terminal Island at the Los Angeles-Long Beach ports complex, a plausible radiological bomb attack on Los Angeles' Downtown Financial District, and a hypothetical hurricane event in the Houston-Galveston area.

The Terminal Island case study utilized a more detailed version of the Southern California Planning Model, i.e. SCPM2, to estimate spatial and sectoral impacts corresponding to the exogenous shocks of exploding a radiological bomb in the port area and the destruction of bridges accessing Terminal Island. The results demonstrated that a relatively simple terrorist attack could inflict massive damage not only to the Southern Californian local economy but also to the national economy. This study also suggests a high *ex ante* payoff to accelerate restoration in the event of a successful attack.

In the second case study, a hypothetical radiological bomb attack on Downtown Los Angeles Financial District, an up-to-date version of SCPM called SCPM 2005 was employed to simulate household and firm relocation in three scenarios; an exit scenario where households and firms in both the inner and the outer zone disappear; a relocation scenario where households and firms in the inner and the outer zone relocate to somewhere else within the five-county metropolitan region; and a hybrid scenario where the households and firms in the inner zone disappear while those in the outer zone relocate. This case study finds that the effects on the inner zone play a dominant role because of its long evacuation period. The exit scenario has an aggregate impact of $5.9 billion of output losses and 40,391 job losses. The relocation scenario has neutral effects from a regional perspective but direct losses in the impacted zones are 7,257 jobs and $2.617 billion of output. The hybrid scenario may be the most realistic, but its impacts are marginally lower than the exit scenario because of the dominance of the inner zone impacts.

In the hurricane case study, the SCPM 1 approach initially established for the Southern California region was transferred to the Houston-Galveston area. This study designs a hypothetical hurricane by adopting the storm parameters of Hurricane Rita but shifting the storm's tracks to make landfall at the Galveston Coast and then follow Interstate 45 to strike downtown Houston. First, it utilizes the Hurricane Model in FEMA's HAZUS-MH package to estimate damage states of buildings in the area. Second, it calculates direct impacts using building damage states estimated by the HAZUS Hurricane Model, building recovery time from HAZUS technical documents, and employment data from local planning agencies. Third, it utilizes a regional input–output model to calculate indirect and induced effects from direct final demand changes. Finally, it uses SCPM 1 to allocate these predicted effects to small impact analysis zones and highlight the most vulnerable geographic areas in the region. All three case studies show that the economic impacts of terrorist attacks and natural disasters on a large metropolitan area would extend to the whole nation rather than be limited to the local area. The large economic impact costs justify considerable resource expenditures on prevention.

The SCPM model is adaptable to almost any kind of terrorist attack or natural disaster on major urban infrastructures and attractions, such as ports, airports, downtowns, and theme parks, etc. The Houston case study demonstrates that the SCPM model initially constructed for Southern California is transferable to another large metropolitan area where suitable data are available.

A noticeable advantage of SCPM models is the scalable functionality and flexible data requirements. SCPM 1 has minimum data requirements on model construction. It uses the regional input–output table, small area employment data, a journey home-to-work matrix, and a journey home-to-shop matrix to develop the base model. In addition to the data used by SCPM 1, SCPM 2, and SCPM 2005 need network link files to develop sophisticated network routing functions to analyze the change of network performance under different situations. SCPM 2005 can handle a large number of economic sectors, TAZs, and network links. The scalable model structures of SCPM enhance adaptability of the model to different scenarios with different data availability, which is demonstrated in the three case studies in this chapter.

References

Burrus, R.T., Jr., Dumas, C., Farrell, C.H. & Hall, W.W., Jr. (2002). Impact of low-intensity hurricanes on regional economic activity. *Natural Hazards Review*, 3(3) 118–125

Cho, S., Gordon, P., Moore, J.E., II & Richardson, H.W. (2001). Integrating transportation network and regional economic models to estimate the costs of a large urban earthquake. *Journal of Regional Science*, 41(1), 39–65

Clark, W.A.V., Huang, Y. & Withers, S.D. (2002). Does commuting distance matter? Commuting tolerance and residential change. *Regional Science and Urban Economics*, 33(2), 199–221

Cushman & Wakefield (2005). *MarketBeat Midyear*. (Los Angeles, CA: Cushman and Wakefield of California)

Federal Emergency Management System (FEMA) (2006a). *HAZUS MH MR2 Hurricane Model Technical Manual*. (Washington, DC: FEMA)

Federal Emergency Management System (FEMA) (2006b). *HAZUS MH MR2 Earthquake Model Technical Manual*. (Washington, DC: FEMA)

Gordon, P., Moore, J.E., II, Richardson, H.W. & Pan, Q. (2006). The economic impact of a terrorist attack on the twin ports of Los Angeles – Long Beach. (In H.W. Richardson, P. Gordon, & J.E. Moore, II (Eds.), *The Economic Impacts of Terrorist Attack* (pp. 262–285). Northampton, MA: Edward Elgar)

Government Accounting Office (2002). *Review of Studies of the Economic Impact of the September11, 2001, Terrorist Attacks on the World Trade Center*. (Washington, DC)

Knabb, R.D., Rhome, J.R. & Brown, D.P. (2006a). *Tropical Cyclone Report, Hurricane Katrina, 23–30 August 2005*. (Miami, FL: National Hurricane Center)

Knabb, R.D., Rhome, J.R. & Brown, D.P. (2006b). *Tropical Cyclone Report, Hurricane Rita, 18–26 September 2005*. (Miami, FL: National Hurricane Center)

Park, J.Y., Gordon, P., Moore, J.E., II, Richardson, H.W. & Wang, L. (2007). Simulating the state-by-state effects of terrorist attacks on three major U.S. ports: Applying NIEMO (National Interstate Economic Model). (In H.W. Richardson, P. Gordon, & J.E. Moore II (Eds.), *The Economic Costs and Consequences of Terrorism*. Cheltenham: Edward Elgar)

Richardson, H.W., Gordon, P., Jun, M.J. & Kim, M.H. (1993). PRIDE and prejudice: The economic impacts of growth controls in Pasadena. *Environment and Planning A*, 25(7), 987–1002

Schneider, P.J. & Schauer, B.A. (2006). HAZUS-Its development and its future. *Natural Hazards Review*, 7(2), 40–44

Sea Island Software Inc. (2006). *HURREVAC 2000 Documentation and User's Manual: Hurrevac for 2006 Season, Version 5.0*, (Mount Pleasant, SC)

U.S. Army Corps of Engineers (USACE) (2005). *Waterborne Commerce of the United States, Calendar Year 2005, Part 4- Waterways and Harbors Pacific Coast, Alaska and Hawaii* (New Orleans, LA)

U.S. Department of Transportation (USDOT) (2004). *America's Freight Transportation Gateways, Connecting Our Nation to Places and Markets Abroad*. (Washington, DC: Bureau of Transportation Statistics (BTS))

Chapter 4
From Crime Analysis to Homeland Security: A Role for Neighborhood Profiling?

David I. Ashby[1], Spencer Chainey[2], and Paul A. Longley[2]

Abstract This chapter presents a review and interpretation of the use of small area neighborhood profiles (geodemographics) in community policing and, by extension, in homeland security applications. We discuss the merits of a local focus in policing, and the data and analytical frameworks that are necessary to support this activity. Next we use case study examples to illustrate how priorities for neighborhood policing may be developed, and suggest that available public sector data may be used to drive improved bespoke classifications of neighborhoods. It is argued that better measures of local social capital and community cohesion may be used to tailor interventions to local circumstances, and to maintain and enhance community stability. We conclude with an overview of some ethical impediments to development of such approaches in homeland security applications.

Keywords Community policing, geodemographics, neighborhood profiling

4.1 Introduction

In this chapter, we introduce and describe a number of approaches to crime analysis and neighborhood profiling which are becoming of increasing interest and relevance to the homeland security and wider public service/governmental sector. Specifically, we outline recent advances in geospatial crime analysis from the identification of priority neighborhoods to the descriptive profiling of different types of neighborhoods. Fundamentally, such technologies and classifications of neighborhood traits are required in security settings where intelligence from local populations is paramount to help target services effectively, to promote appropriate interventions, and to identify local area conditions, which may lead to a tipping point in community cohesion or even to civil unrest.

---

[1]Dr Foster (Research) Ltd.

[2]University College London (UCL)

D.Z. Sui (ed.) *Geospatial Technologies and Homeland Security*,

We discuss how classification and profiling of neighborhoods can be used to interpret a variety of databases, in order to assist the police and law enforcement agencies. Practices of classifying neighborhoods have been widely used in the commercial sector for decades to segment customers and target appropriate services, and to identify the most effective channels of communication with appropriate messages. Such analyses can help target resources to areas of greatest need. They may also point to the correlation between certain neighborhood types and needs for other public services, suggesting ways in which long-term solutions to deprivation may lie in multi-agency partnerships. These approaches enable service providers to empower local communities through activities designed to increase social capital, community cohesion, and collective efficacy. Neighborhood service delivery, and hence local area analysis, has become increasingly pivotal in UK public services; the approaches outlined here can provide a shared understanding that can facilitate networking between individuals and organisations to improve policing and, potentially, homeland security.

Crime analysis aims to deliver the right information to the right people at the right time, and can be applied across all levels of geography to identify patterns and correlations between crime data and other relevant information sources (Cope 2004; Gill 2000; Gottlieb et al. 1994). The notion of Homeland Security does though tend to start from the geographical context of a federal effort to prevent and reduce the vulnerability of terrorist attacks, and minimizing the damage and recovering as quickly as possible when attacks do occur (United States Office of Homeland Security 2002). Whilst the national effort then seeks to coordinate efforts across all levels of government—federal, state, and local—the analysis of local threats and its situational influences can often not undergo the same rigor that is more routinely applied to the local analysis and situational influences of crime. Additionally, in the UK and several other countries such as Australia and New Zealand, the coordination of crime analysis and public safety at the local level feeds up along a chain to inform regional and national threats, with this local aggregation being added to along the way by those threats that require the coordination of regional and national government agencies. It is up through these levels that the threat of terrorism is then assessed, and fed back through the tiers in order to identify and help direct strategic priorities that may not only be national, but which require the support of particular local service delivery.

This coordination of intelligence therefore makes it more possible in some countries to harness their existing activities in crime analysis and apply this approach to the local assessment of homeland security. Additionally, this helps to allow for local flexibility into the issues that are considered. In the United States, the fragmented nature of law enforcement makes this more difficult, but in both situations, the opportunity exists to draw from the experiences, techniques, and technologies that are increasingly being applied in crime analysis to support the neighborhood assessment of homeland security concerns.

4.2 Geodemographics and Neighborhood Classification

Geodemographics are small area measures of social and economic conditions, which are available for much smaller and more detailed areal units than those typically used in crime analysis. In the UK, the basic unit of analysis is the unit postcode—typically 17 addresses, equivalent to Zip +4 in the US system—whereas public policy is more usually constrained to the electoral ward level of analysis. The academic roots to the geodemographic tradition of human geography can be traced to the work of Burgess, Park, and the other Chicago 'urban ecologists' in the 1920s. Geodemographics emerged from the social area analysis and factorial ecologies of the 1950s and 1960s, and has recently enjoyed considerable rehabilitation under research council initiatives such as the UK Economic and Social Research Council Centre for Neighborhood Research (www.neighbourhoodcentre.org.uk). The approach has been a very successful example of technology and knowledge transfer to the private sector, to the extent that almost all customer facing organisations today employ geodemographics as tactical and strategic decision making tools in the delivery of private goods and services. In recent years, there has been a sense in which geodemographics is returning to its roots, in that techniques that have been honed and refined to predict the demand for *private* goods and services have been applied to the demand and need for *public* goods, in an area of activity that has become known as *social marketing* (National Social Marketing Centre for Excellence 2005). In the present context, this is also leading to a reappraisal of local approaches to crime analysis.

4.3 Advances in Crime Analysis—Going Local

Crime analysis involves a set of systematic processes that aim to identify patterns and correlations between crime data and other relevant information sources (Cope 2004; Gill 2000; Gottlieb et al. 1994). Crime analysis can be applied across all levels of geography, but is increasingly being applied to the neighborhood level to help identify, understand, explain, and respond to local crime issues. As a process, crime analysis has developed alongside the paradigm of intelligence-led policing—an approach that seeks for all actions and responses to be informed by facts and evidence (Chainey and Ratcliffe 2005). Crime analysis also has foundations in problem-oriented policing (Goldstein 1979). This is an approach whereby formal criminology theory and scientific research methods are applied to the analysis of crime issues to conduct an in-depth examination of the crime problem, develop informed responses, and assess their impact (Boba 2003). Crime analysis has therefore evolved to respect the need for a more intelligence-led, problem oriented approach to tackling crime.

Recent developments in crime analysis have resulted from the recognition that an effective approach to solving crime problems requires understanding the specifics of the issues (Clarke and Eck 2003). The nature of being crime specific most typically requires the analysis to be local, exploring particular neighborhood and situational features that identify the probable cause to the crime problem. For example, in a study on vehicle thefts in Central London (UK), the Metropolitan Police overcame a perception that the problem was related to resident's cars being stolen outside their properties at night by identifying that each area where vehicle theft was a problem had different unique and specific qualities—in one area, for example, vehicle thefts were predominantly of motorbikes, scooters, and mopeds, from non-residents of the area, most usually from motorbike parking bays during daylight hours (Home Office 2005).

Crime analysis has also developed to recognize that police information systems may offer only a narrow view of criminal activity, and that by working with local partners the understanding of the problem can be improved, including considering how a multi-agency response (rather than a single operational police response) can address the problem. These developments have occurred at a time when many other public services are becoming increasingly neighborhood oriented in their delivery. For example, the language of local service delivery in the UK places emphasis on Neighborhood Wardens, regeneration programs that focus on Neighborhood Renewal, local community engagement through Neighborhood Forums, and more recently in policing, via Neighborhood Policing.

As the pressures to adopt a more neighborhood focus have increased, so the demand has increased for ever richer, more geographically accurate data. Over time, accuracy has improved with crime data being geocoded with greater care than ever before. Census data have become routinely available for small (and more policy-relevant) areas and developments have taken place in the supply and provision of other neighborhood demographic, social, and economic datasets—relating, for example, to deprivation, welfare benefits, and fire incidents (for UK examples see the Neighbourhood Statistics Service at www.neighbourhood.gov.uk).

Data that can be extracted from recording systems, administrative databases, and other demographic information sources provide only 'hard' data materials (i.e. data that are recorded in response to reported incidents), calls for service, or a formal census. Hard data may help reveal many qualities of the local neighborhood but may miss many other issues and characteristics. The intelligence gaps arising from using hard data alone are increasingly recognized with many law enforcement agencies now seeking to draw on softer data sources. That includes data that are gathered by proactive interaction and engagement with the community (such as neighborhood surveys or visual audits), and identification of issues that either go unreported or are difficult to capture in formal incident reporting. An example of this is the signal crimes perspective, developed by Innes and Fielding (2002), where an incident (that may not necessarily be a crime, but can be an event that prompts the public to react and behave differently) may act as a form of 'warning signal' that is disproportionately influential in causing a person or persons to perceive themselves as being at some degree of risk. This type of community intelligence is acutely situational, and can depend on a range of factors, not least the person's life experience. Hence it is methodologically important to capture a detailed neighborhood view by engaging with key individuals from the local community, in order to ensure the accurate capture of signal perception. Incorporation of hard data alongside such more qualitative and situational considerations poses several unique challenges, but these must nevertheless be addressed as part of an intelligence-led approach that tackles neighborhood issues through improved performance (reflected in reductions in crime) and effective reassurance of and engagement with the public.

4.4 Applicable Technologies

We have detailed elsewhere much of the history and developments of crime mapping (e.g. Chainey and Ratcliffe 2005), the application of GIS across public and commercial sectors (e.g. Longley et al. 2005) and the drive to visualize and explore crime data in a spatial manner specific security domains (e.g. Ashby and Craglia 2007). Here we outline the principal concepts and their application within crime analysis, as well as some potential implications for homeland security. In general terms, these applications are founded upon intensive data management strategies set within a GIS environment. They each arise out of the recent confluence of information technology, geodemographic techniques, and improved data availability.

4.4.1 Identifying Priority Neighborhoods

In recent years, a number of crime analysis methods have been developed to help perform the first stage of identifying where to target resources. Most often, the geographic targeting of resources with which to tackle crime employs hotspot analysis. This approach seeks to identify where crime concentrations are highest. A number of techniques exist for hotspot mapping, with kernel density estimation emerging as one of the most popular techniques (Eck et al. 2006; Chainey and Ratcliffe 2005).

While hotspot mapping is useful for identifying high crime areas, in recent years there has been a growing demand to identify neighborhoods that not only suffer from crime but also from issues that contribute to other forms of community breakdown and fragmentation. One such measure is the Index of Multiple Deprivation (ODPM 2003), a composite and singular domain index (available at the Lower Super Output Area level) that has received widespread use in the UK for helping target regeneration and renewal schemes. While useful, many involved in policing and crime reduction have begun to seek even more geographically precise and more application relevant measures to help not only identify high crime areas but neighborhoods that would benefit from specific police attention and community engagement.

Rose (2005) developed a composite index to support West Midlands Police (UK) in identifying its 'priority places.' By combining 18 indicators that grouped to five core priorities from a variety of partner agencies across the city, a single composite core priority indicator was calculated. The outputs that this index generates have continued to prove instrumental in helping the West Midlands Police develop its strategic plans for improving and monitoring community safety across the city.

A composite index that has begun to see widespread use for identifying priority neighborhoods is the Vulnerable Localities Index (VLI) (Chainey 2004). This index was developed to help police forces identify where they should direct priority attention. The VLI, developed initially as a measure to pre-empt any escalations of civil disorder, has now received more widespread attention as a practical tool to assist in the prioritisation of neighborhoods for the UK's Neighbourhood Policing programme (ACPO Neighbourhood Policing Team 2006). It has also begun to help inform similar requirements in many countries across Europe (GMAC 2006a).

The VLI is designed to act as the initial suggestive guide for identifying and measuring areas that are potentially experiencing problems of community breakdown and fragmentation. It uses crime, deprivation, education, and demographic data that are available for each area in the UK. To identify areas at the local level, it uses Census Output Areas (OAs) as the aggregate unit for these statistics. Each of the statistics can be presented on its own, but in a standardized indexed form it makes it possible to aggregate a range of statistics to form a single composite vulnerable index score for each locality (Chainey 2004). For example, it makes little sense to aggregate the count of burglaries in an area with the population score of 15–24 year olds. Instead, both are indexed to allow aggregation (for more details on this approach see Chainey 2004). A value of 200 indicates twice the propensity of vulnerability, whereas a score of 50 would indicate that the observed rate is only half of that which would be expected, relative to the study area's average.

Figure 4.1 shows a completed Vulnerable Localities Index (VLI) for the Sandwell district. It suggests that a number of localities would be worthy of further attention. In discussion with Sandwell Police, these areas supported certain perceptions of problem neighborhoods, but also revealed other areas that they had not previously considered. This was mainly because of a previous policing focus that was dominated by 'high crime neighborhoods' without considering some of the environmental drivers (e.g. deprivation) that contribute to community problems. The VLI is now used in many areas of the UK, including Greater Manchester (GMAC 2006b) and Lancashire (Dallison 2005).

4.4.2 Profiling Neighborhoods of Interest

> *What does effective engagement look like for the Police Service?*
>
> *A police service which is engaging effectively with the community will ... have a detailed, neighborhood level understanding of the demographics of the community it serves.*
>
> (Home Office 2004: 67)

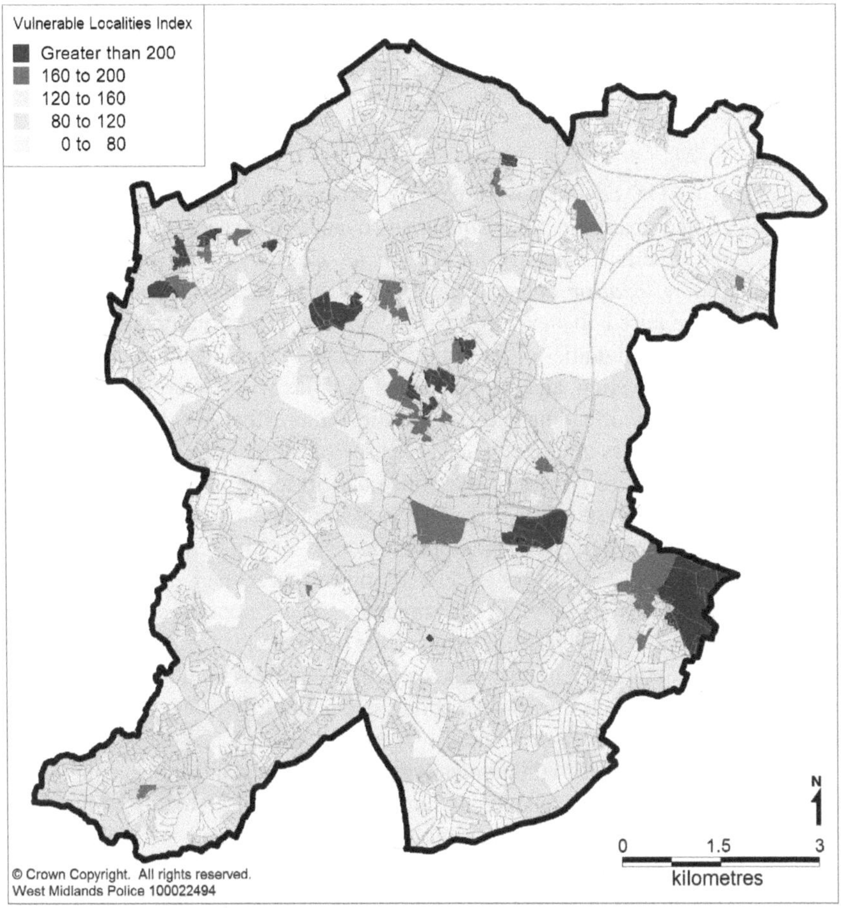

Fig. 4.1 The Vulnerable Localities Index for Sandwell in the West Midlands region of England. A score of 100 is representative of the study area's average measure. Values greater than 100 indicate higher measures of vulnerability. Areas with the highest VLI scores are those that should receive focused neighborhood policing attention (Chainey 2004)

The Home Office 2004 White Paper, *Building Communities, Beating Crime: A better police service for the 21st century*, called for increased neighborhood intelligence and actively promoted the widespread use of geodemographic techniques applauding one Constabulary's use of 'market analysis tools to reduce crime' (Home Office 2004: 56). This is an accurate reflection of the current climate and political direction for crime analysis/policing in the UK, yet there remains relatively little expertise in developing these techniques for crime and security purposes. The Statistics Commission recognized the progress made and advances offered by the likes of the Audit Commission national pilot study into neighborhood profiling of crime and anti-social behavior (see www.audit-commission.gov.uk/neighbourhoodcrime and Ashby and Webber 2006) and outlined:

> the problems posed for the local government's evolving local delivery framework by the lack of granularity of crime data at neighborhood-level. A variety of techniques and approaches are suggested based upon the techniques of *Geo-demographic Information* and the technologies of *Geographic Information Systems*.
>
> (Statistics Commission 2006: 71)

These moves by UK central government towards characteristic profiling of small areas marks a step-change in policy and great advances in the adoption of geospatial techniques and expertise. One-size-fits-all policies are now often rejected in favor of the so-called *new localism* agenda, where the characteristics and composition of local areas is required to develop bespoke, tailored solutions to neighborhood problems. Local communities differ not just in terms of their incomes, age distributions, levels of deprivation, and proportions of families with children, etc., but they also differ in levels of offending and victimization among their residents. They differ with respect to the types of crime which are perpetrated, in terms of whether or how they communicate these crimes to the police and in terms of the speed of police response and clear up rates. They likewise vary in terms of their attitudes towards the police, and the effectiveness and appropriateness of different crime prevention strategies. For example, whilst postcode marking and target hardening may prove a very effective policing strategy in predominantly student populated neighborhoods near to universities, campaigns to alert residents to the danger of rogue callers may be a much more appropriate use of police resources in areas with a high proportion of elderly residents. Such intricacies cannot be captured by a few crude summary performance statistics for large and heterogeneous areal units, and policing strategies and assessment measures should ultimately reflect the needs of the local community. Geodemographics offer an analytical framework and toolset for assessing these characteristics and directing strategy, prior to the capture and analysis of those 'softer' data sources (recall the signal crimes perspective above).

Three broad themes of geodemographic analyses have been identified as potential subjects that may inform policing strategy (Ashby 2005) and may similarly apply to homeland security. Primarily, the basic geodemographic profiling of areas such as Police Beats or city/divisional boundaries may provide an intelligence-based resource for strategic policing and performance review. Secondly, the profiling of operational crime data within these policing units may assist in the identification of local trends in crime incidents. Finally, geodemographic profiling of survey data may provide indications of the likely variations in attitudes and fear across different neighborhood types; data which may otherwise prove prohibitively expensive to acquire. Significantly, given the key unique identifier of a postcode, associated geodemographic profiles can be extrapolated and mapped across all unit postcode 'neighborhoods.' This approach therefore imparts significant predictive power whereby trend data can be extrapolated to those local areas for which insufficient data exist. This integration and extrapolation of varied data sources can indicate where certain issues have an enhanced probability of being located. Ashby and Webber (2006) provide detailed examples of each of these applications in ten study areas across the UK. Here we summarize some of their findings, specifically focusing upon the extrapolation of national survey data to a local level.

Table 4.1 presents a profile of a number of questions taken from the national British Crime Survey, extrapolated to local areas using a geodemographic approach. These profiles summarized for such questions may help to identify key challenges and issues within different types of neighborhoods. The numbers given here are standardized index scores (calculated in a similar manner to the indexing of the variables that collectively make up the Vulnerable Localities Index as explained in section 3) whereby a score of 100 represents the national average. A value of 200 would indicate twice the propensity for the given variable, whereas a score of 50 would indicate that the observed rate is only half of what would be expected.

From Table 4.1 one can identify the most likely problems, attitudes, or fears within an area (analysis by column) or one can compare the relative propensity of any one of the variables across the three study areas (analysis by row). This table concisely

Table 4.1 Weighted-average index scores from the British Crime Survey profiles for three example wards, taken from the Audit Commission pilot study (Ashby and Webber 2006)

Question, prompt or theme	Response	Area 1 (Inner-city, Liverpool)	Area 2 (Rural, Wales)	Area 3 (Coastal, Cornwall)
How common are burnt out cars?	Fairly common	118	142	54
How common is people using or dealing in drugs?	Fairly common	131	144	140
How common is rubbish?	Very big problem	159	184	118
How common is vandalism and graffiti	Very big problem	165	224	131
Have bad effect on your life?	Teenagers hanging around on the streets	153	177	99
Feel safe walking alone after dark?	Very unsafe	145	155	131
How worried about burglary?	Very worried	128	143	109
How worried about having car stolen?	Very worried	112	122	77
How worried about being insulted or pestered?	Very worried	125	139	82
How worried about mugging?	Very worried	127	144	109
Interest shown by police	Inadequate	114	120	114
Nice place to live?	Very good place to live	57	51	80
Social capital	Neighbours help each other	81	75	73
Rating of police	Very good	82	81	108

summarizes modelled output for a range of 'fear of crime' drivers which are rarely disentangled, or compared at such local scales. For illustrative purposes, one might take the example of Area 1 (an inner city area of Liverpool in the North West of England) and conclude that rubbish, litter, and graffiti are likely to be common problems high on the priorities of local residents. However, these issues may conceivably be of greater concern in Area 2 (a materially deprived area in the Rhonnda, Wales). Alternatively, one may highlight the fact that the modelled propensities suggest that abandoned cars seem to be unlikely to pose a problem in Area 3 (a deprived, former seaside resort near Margate, with a transient population), and that feeling safe walking alone after dark is a much greater concern for local residents.

Of course, much of the value of geodemographics is derived from the spatial reference key which enables the mapping and analysis of spatial trends. Index scores, of the sort detailed in Table 4.1 can be mapped to local areas and correlated with traditional hotspot maps of crime prevalence. The illustrative examples given in Figs. 4.2 and 4.3 provide local level estimates of perceived problems within the community (e.g. teenagers loitering on the street) and some form of proxy for social capital (i.e. the likelihood of neighbours to help one another in a time of need). Different colour ramps are used to represent the differing nature of these variables; the first is often used as a proxy indicator for levels of crime or fear of crime in a local area. The second is an indicator of potential levels of cohesion within an area.

4.4.3 *Neighborhood Classifications: Looking Forward*

Geodemographics remains fundamentally a successful inductive technology in generalization and data reduction, although there are four important trends that govern their current use. First, the pressures to improve responsiveness of the public sector have shifted the locus of cutting edge applications away from retailing and commerce. Second, the data economy of the public sector is changing profoundly as more and more government datasets come online for free download by the public. Third, profiling techniques now accommodate a far greater range of rich and frequently updated data sources. And fourth, there is recognition that world city status is difficult to accommodate in nationwide classifications, giving rise to bespoke classifications of particular cities, or special weighting of their unique neighborhood characteristics within nationwide classifications.

Nowhere are these developments of more profound importance than in crime analysis and reassurance policing, although their consequences are at first sight perplexing. Recent years have seen the innovation of 'market analysis tools' and 'social marketing' approaches to reassurance policing. Social Marketing has been defined by the National Social Marketing Centre for Excellence (NSMC 2005) as 'the systematic application of marketing concepts and techniques to achieve specific behavioral goals relevant to a social good.' Leading private sector geodemographics providers have attempted to respond to these developments by rebadging or even

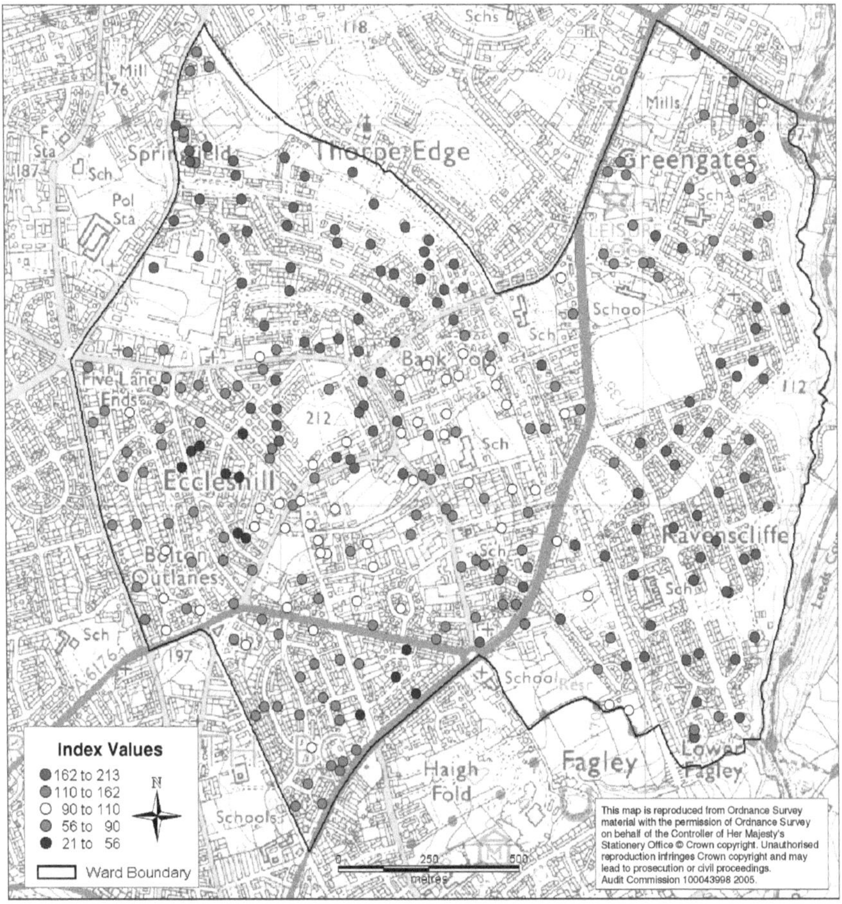

Fig. 4.2 Modelled propensities: The relative likelihood of the resident population perceiving teenagers hanging around as a very big problem in Eccleshill, Bradford (Ashby and Webber 2006)

re-engineering their classifications of existing datasets, while continuing to assert the value of frequent updates and non-Census data sources. To date, however, the research literature remains bereft of studies that demonstrate that classifications fundamentally designed to predict consumption of private goods and services should intrinsically be suitable for social marketing of public goods such as reassurance policing and law enforcement. Moreover, while it is intuitively plausible that more, and frequently updated, data sources may lead to improved classification, it is ironic that the sources most relevant to the security sector are not the surveys assembled by the private sector, but rather the under-utilized local data that are collected by the law enforcement sector itself. Finally, the 'black' (or at least 'grey') box nature of commercial solutions is less than wholly acceptable in crime science applications, where the consequences of discrimination in resource allocation to all stakeholders in law enforcement (e.g. tax-payers) need to be transparent.

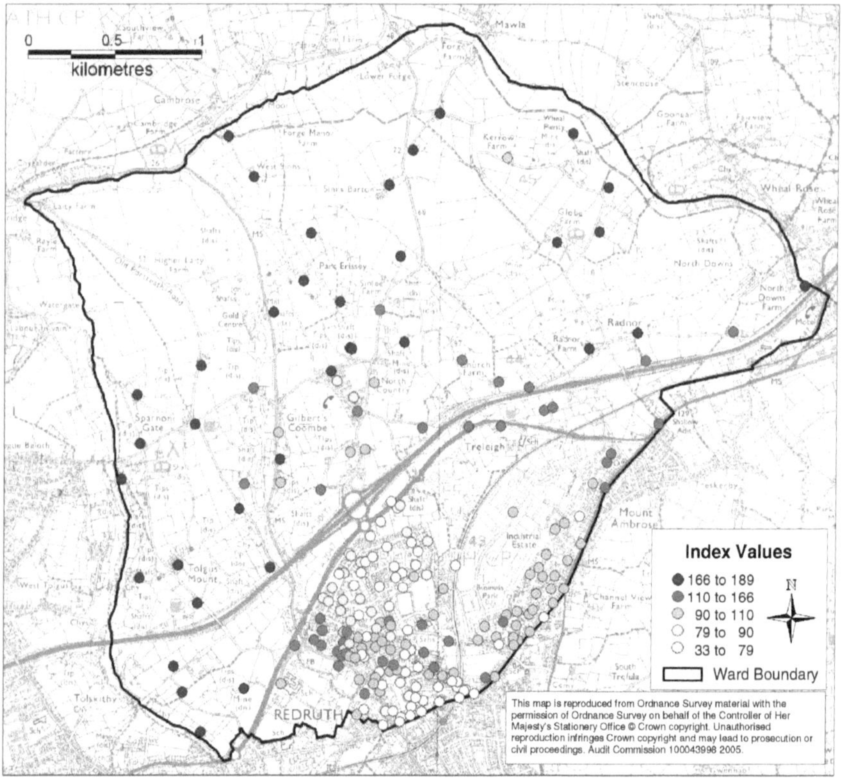

Fig. 4.3 Modelled propensities: The relative likelihood of the resident population to help their neighbours, within Redruth North, Kerrier (Ashby and Webber 2006)

With gathering momentum towards a more customer focused approach to public service delivery (even in the law enforcement sector where 'the customer is always wrong': Squires 1998: 169), we are currently experiencing a renaissance of academic interest in the use of geodemographics to improve public service delivery (Harris et al. 2005; Longley 2005). At the same time, developments in the dissemination of public sector information have empowered researchers in academia and elsewhere to develop free-to-access classifications—most obviously apparent in the UK in the creation of the 2001 Census of Population Output Area classification (OAC) under a collaboration between the Office for National Statistics and the University of Leeds (Vickers and Rees 2007). At the most fundamental level, this offers the prospect of augmenting the data and building blocks of general purpose classifications (typically built around a core of census data) with domain specific crime data collected at the local level. Even where domain specific data are available only for coarser aggregations (e.g. national crime surveys), there has been recent interest in appending geodemographic codes to national surveys in order to segment the attitudes and expectations of the population with regard to law enforcement issues.

Such procedures may be utilised to target community reinforcement initiatives such as 'neighborhood watch' schemes, for example.

In conceptual terms, the work has far-reaching implications for the social marketing agenda, not only in crime science but in relation to our understanding of individual decisions and our attempts to shape them. Within the domain of delivering public health, this is already evident in the stated (NCSM 2005: 4) need for well-developed audience 'segmentation' in the social marketing of primary health care. Here, there is recognition that a focus upon clearly defined and reasonably homogeneous groups leads to deeper understanding of the audience and consequently a better tailored marketing mix. Geodemographics is recognized as providing the intelligence to facilitate the physical targeting of marketing strategies (in terms of person, place, and time). The UK government is already actively promoting health-related social marketing, and recommended that a national strategy for health should include a social marketing component. As part of the associated delivery plans and Public Service Agreements, the UK Department of Health is implementing a number of social marketing campaigns and initiatives. There is growing international evidence, particularly from North America and Australia, that social marketing interventions are effective in modifying health-related behaviors, and it remains for researchers in crime science to consider the transfer of these ideas to the law enforcement domain.

4.5 Social Capital and Community Cohesion

Much has been written on the topics of social capital, community cohesion, and collective efficacy, and we do not attempt to address this broad theoretical literature here. Yet whilst the need to develop socially cohesive communities is apparent for a safe and secure homeland environment, it is evident that there has been relatively little empirical, quantitative research relating to the estimation, analysis, and evaluation of relative levels of social capital in different neighborhoods. In collaboration with colleagues, we have elsewhere developed heuristic models of neighborhood development, summarizing likely crime profiles, social capital neighborhood traits, and proposing some appropriate policing options (Ashby 2005). An extract is given in Table 4.2.

There is considerable scope for developing heuristic models (such as Table 4.2) alongside other frameworks within the neighborhood policing and community cohesion doctrines, which are founded on the fundamental premise that policing strategies need to be based upon neighborhood context. For if Sampson and Raudenbush (1999) are correct in their discussion of collective efficacy and the associated community vulnerability to crime and disorder, then it remains of paramount importance for the Police to diagnose relative levels of this social capital derivative and hence identify appropriate neighborhood-orientated policing strategies (Ashby 2005). This is also true for the homeland security sector. Geodemographics as a strategic diagnostic tool can provide the intelligence base to identify differential levels of social capital and community cohesion in local neighborhoods. Furthermore, the spatially referenced framework enables the efficient

Table 4.2 A segmented approach to Neighborhood Policing; neighborhood groups, crime profiles, fear profiles and social capital, with tentative examples of appropriate neighborhood policing styles (Ashby 2005)

	CRIME PROFILE				SOCIAL CAPITAL			
Geodemographic group	Common types of crime and disorder	Level of crime and disorder	Level of fear	Level of trust	Informal Contacts	Formal Association	Summary	Appropriate 'neighborhood policing' strategies
'Symbols of Success'	Fraud, Traffic Offenses	Low	Low	Fairly high (excluding "Global Connections")	Low	High	Networks are often instrumental and not locally based	Engage with local representatives. Leaflet drops to communicate information and promote campaigns.
'Ties of Community'	Alcohol related; domestic violence	Average	Moderate – low	High among local residents: apprehensive of outsiders	Quite high	Quite high	Tends towards 'self policing'.	Identify representatives and attempt to recognize parallel communities. Police Community Support Officers. Rapid response to environmental disorders such as abandoned cars
'Urban Intelligence'	Snatching; mugging; credit card theft	High	Considered fact of life	Low	Informal contacts are not local	Patchy: high for those who are not transient	Networks are often not local. Local networks are often transient and one dimensional	Reliance on communications programs; target hardening; intelligence led Policing; Police Community Support Officers.
'Welfare Borderline'	Drug dealing; child abuse; car crime	Very high	High	Low	Low	Very low	Low levels of social cohesion	Partnership work with housing department and social services; Community Support Officers.
'Grey Perspectives'	Few	Low	Underlying but unfocussed need for reassurance	High among known people/low among outsiders	High	Above average	Highly responsible for local conditions. Are at home much of the day. Natural 'wardens'	Reassurance; high visibility policing; anti fraud campaigns; neighborhood watch.
'Rural Isolation'	Theft of equipment; planned, high value burglaries	Low	Anxiety about quality of police response	High among known people	High	High	High levels of responsibility to other community members	Reassurance on response times. More intensive communications with community leaders

delivery of appropriately targeted strategies to small areas. Geodemographics therefore may provide the initial strategic review of neighborhoods and communities prior to the incorporation of additional local knowledge and appreciation of community level (and individual) nuances. A diagnostic of this nature would present the police and their partners with a detailed appraisal of those areas where informal social control may be activated within the community, and conversely where levels of collective efficacy are likely to be wholly absent and such strategies rendered ineffective.

Analyses are ongoing, and initial findings indicate that geodemographic analysis of health and education data profiles demonstrate differences at a neighborhood level that are at least as significant as those which we have observed for crime and policing. We anticipate similar avenues for exploratory research within the homeland security sector. By refocusing on the heterogeneity of neighborhoods at the finest levels of spatial granularity, it should be possible to customise the management of crime, security, and other public services, in order to better meet local needs.

Increasingly, demands are being placed on crime analysis to consider not only data and intelligence that is reported and recorded on police information systems, but also to consider data from softer intelligence sources such as surveys, visual audits, and other forms of community engagement. This type of community intelligence can often reveal many of those issues that go unreported or highlight certain 'signal events'—incidents that have a disproportionate influence on communities' perceptions and concerns about crime in their neighborhoods (Innes and Fielding 2002). Additionally, gathering community intelligence through this type of engagement may also have the benefit of improving community relations, identifying key individuals in the community who could help ensure the police are kept informed of representative opinion and who could support the promulgation of messages during times of any community tension, or issues of national importance (e.g. such as the identification of potential terrorist cells) that require local neighborhood support.

Handheld computing devices are increasingly being used by patrol officers to support the capture of this community intelligence in the field (Chainey and Ratcliffe 2005; Home Office 2005). Many of these devices incorporate streamlined GIS-based reporting tools, GPS, a gazetteer, and a camera, allowing field officers to quickly record details about incidents. In other examples they have proved useful for helping capture qualitative survey responses (Rose 2006). The potential also exists to use similar devices to actively collect intelligence from routine neighborhood engagement, which in turn could provide critical information that could be significant to a national threat (e.g. a bomb making factory in a residential property).

4.6 Conclusions

This review and interpretation has presented an upbeat assessment of the applicability of ideas from geodemographics to neighborhood profiling of attitudes to, and expectations of, law enforcement agencies. However, the approach offers very much less than a panacea for homeland security applications. Although collected for small area aggregations, geodemographics nevertheless still subsumes individual data within areal aggregations. The literature on geodemographics is replete with cautions about the dangers of 'ecological fallacy', that is confounding the characteristics of areas with the characteristics of individuals resident within them. There is (to our knowledge) no documented instance of this being proven to have affected the results of a real world application, although Openshaw (1984) provides persuasive evidence that changes in the scale of spatial analysis, or in the ways that aggregate zones are composed from elemental spatial units, can have profound effects upon the results. These problems are all the more crucial in homeland security applications where 'events' or 'targets' are likely to represent a tiny fraction of the population at large, and the implications of false inference arising from ecological fallacy have grave implications for civil liberties. It must be remembered that in a typical geodemographics application in the private sector, a target of doubling the response rate to a mail shot from 2 percent to 4 percent might be set: whilst this does not sound ambitious, it will if achieved double the effectiveness of the marketing budget. The consequent 'failure' rate of 98 percent or even 96 percent is unacceptable for almost all issues of homeland security, but may be acceptable within the remit of reassurance policing.

Fundamentally, this strand of research has built upon a foundation of classifying communities on the basis of *social similarity* rather than merely *locational proximity*; communities and localities have been examined using a neighborhood typology which groups on a basis of the underlying social conditions, rather than merely crude areal divisions. From this basis, one may begin to develop a framework for evidence-based policing strategies which accounts for the geodemographic composition and social capital found within different local communities. Such a foundation, at a fine spatial granularity, appears to be one clear pathway to achieve knowledge-based delivery of policing, and potentially related homeland security services, in line with much of the UK rhetoric regarding the new localism.

References

ACPO Neighbourhood Policing Team (2006). *Briefing paper on neighbourhood policing and the NIM* [Electronic version]. Association of Chief Police Officers & Centrex. Retrieved from www.neighbourhoodpolicing.co.uk

Ashby, D.I. (2005). Policing neighbourhoods: Exploring the geographies of crime, policing and performance assessment. *Policing and Society*, 15(4): 413–447

Ashby, D. & Craglia, M. (2007). Profiling places: geodemographics and GIS. (In T. Newburn, T. Williamson, & A. Wright (Eds.), *Handbook of criminal investigation* (pp. 517–546). Cullopmton: Willan)

Ashby, D.I. & Webber, R. (2006). *High crime: High disorder neighbourhoods* [Electonic version] Spatial Analysis and Geodemographics, report submitted to The Audit Commission. (London: UCL). Retrieved from www.spatial-literacy.org

Boba, R. (2003). *Problem analysis in policing*. (Washington, DC: United States Police Foundation)

Chainey, S. (2004). Using geographic information to support the police response to community cohesion. *Proceedings of the* 2004 Association for Geographic Information Conference, London, 12–14 October

Chainey, S. & Ratcliffe, J. (2005). *GIS and crime mapping*. (Chichester: Wiley)

Clarke, R.V. & Eck, J. (2003). *Become a problem solving crime analyst* [Electronic version]. (London: The Jill Dando Institute of Crime Science, London). Retrieved from www.jdi.ucl.ac.uk

Cope, N. (2004). Intelligence led policing or policing led intelligence?: Integrating volume crime analysis into policing. *British Journal of Criminology*, 44(2): 188–203

Dallison, M. (2005, April). Assessing the level of community cohesion within the Pennine Division of Lancashire Constabulary. (Paper presented at the 3rd UK Crime Mapping Conference, London)

Eck, J., Chainey, S.P., Cameron, J. & Wilson, R. (2006). *Mapping crime: Understanding hotspots*. (Washington, DC: United States National Institute of Justice)

GMAC (2006a). *Vulnerable communities* [Electronic version]. Greater Manchester Against Crime Europe Project Report. Retrieved from www.gmac.org.uk

GMAC (2006b). *The greater Manchester's strategic assessment 2006* [Electronic version] Greater Manchester Against Crime. Retrieved from www.gmac.org.uk

Gill, P. (2000). *Rounding up the usual suspects? Developments in contemporary law enforcement intelligence*. (Aldershot: Ashgate)

Goldstein, H. (1979). Improving policing: A problem-oriented approach. *Crime and Delinquency*, 25, 236–258

Gottlieb, S.L., Arenberg, S. & Singh, R. (1994). *Crime analysis: From first report to final arrest*. (Montclair, CA: Alpha Publishing)

Harris, R., Sleight, P. & Webber, R. (2005). *Geodemographics, GIS and neighbourhood targeting*. (Chichester: Wiley)

Home Office (2004). *Building communities, beating crime: A better police service for the 21st century*. Home Office White Paper; CM6360. (London: HMSO)

Home Office (2005). *Crime mapping: improving performance*. (London: Home Office Police Standards Unit)

Longley P.A. (2005) A renaissance of geodemographics for public service delivery. *Progress in Human Geography*, 29, 57–63

Longley, P.A., Goodchild, M.F., Maguire, D.J. & Rhind, D.W. (2005). *Geographic information systems and science* (Second Edition). (Chichester: Wiley)

Innes, M. & Fielding, N. (2002). From community to communicative policing: Signal crimes and the problem of public reassurance. *Sociological Research Online*, 7(2)

National Social Marketing Centre for Excellence (NSMC) (2005). *Social marketing pocket guide*. National Consumer Council

ODPM (2003). *The English indices of deprivation*. (London: Stationery Office)

Openshaw, S. (1984). The modifiable areal unit problem. *Concepts and techniques in modern geography* 38. (Norwich: GeoBooks)

Rose, S. (2005, April). Using a composite spatial index of multi-agency information to assess partnership priorities. (Paper presented at the 3rd UK Crime Mapping Conference, London)

Rose, S. (2006, May). COSMOS: Mapping reassurance using survey data. (Paper presented at the 4th UK Crime Mapping Conference, London)

Sampson, R.J. & Raudenbush, S.W. (1999). Systematic social observation of public spaces: A new look at disorder in urban neighborhoods. *American Journal of Sociology*, 105(3): 603–651

Squires, P. (1998). Cops and customers: Consumerism and the demand for police services. Is the customer always right? *Policing and Society*, 8: 169–188

Statistics Commission (2006). *Crime statistics: User perspectives. Report No. 30* [Electronic version]. (London: Statistics Commission). Retrieved from www.statscom.org.uk

United States Office of Homeland Security (2002). *The national strategy for homeland security*. [Electronic version]. (Washington DC: United States Office of Homeland Security). Retrieved from www.whitehouse.gov/homeland/book/nat_strat_hls.pdf

Vickers D.W. & Rees, P.H. (2007). Creating the national statistics 2001 output area classification. *Journal of the Royal Statistical Society, Series A*, 170(2): 379–403

Chapter 5
Measuring and Mapping Conflict-Related Deaths and Segregation: Lessons from the Belfast 'Troubles'

Victor Mesev[1], Joni Downs[2], Aaron Binns[1], Richard S. Courtney[3], and Peter Shirlow[4]

Abstract Commonly known as the 'Troubles,' the disputes between Irish republicans (mostly Catholics) and British unionists (mostly Protestants) in Northern Ireland have lasted for decades and since the late 1960s have claimed around 3,600 lives. Military intervention by the British Army eventually undermined the activities of the main paramilitary groups (Irish Republican Army that sought the unification of Ireland and the Ulster Volunteer Force and Ulster Freedom Fighters who wished to maintain Northern Ireland's constitutional position within the United Kingdom). Northern Ireland is now slowly transforming out of conflict, but as it does so, more debates become concerned with interpreting the past and the nature and meaning of victimhood.

This chapter maps the spatial distributions of conflict-related deaths in Belfast (Northern Ireland's principal city) in an attempt to unravel the complex social, political, and ethno-religious underpinnings of the Troubles. Religious segregation is claimed by many analysts to be a major contributory variable to explaining the pattern of conflict-related deaths, and as such we explore a modification of the spatial segregation index to examine the distribution of Catholic and Protestant neighborhoods in Belfast. After analyzing the extensive database of deaths and their spatial occurrence, the chapter ends with a series of lessons. Most notably, politically motivated attacks can be unpredictable but also seem to cluster within highly segregated and low social class neighborhoods located within close proximity to interfaces between Catholic and Protestant communities. In addition, paramilitary attacks are difficult to profile demographically, and the vast majority involve civilian casualties.

Keywords Conflicted-related deaths, religion, sectarianism, segregation

[1] Florida State University

[2] University of South Florida

[3] Kutztown University of Pennsylvania

[4] Queens University Belfast

D.Z. Sui (ed.) *Geospatial Technologies and Homeland Security*,

5.1 Introduction

Northern Ireland remains synonymous with political unrest and sectarian-motivated violence. The legacy of the brutal early 1970s, the unpredictable 1980s, and the frustrating 1990s has survived despite the relative calm of recent years, culminating in the peace process and the de-commissioning of some weaponry (Fig. 5.1). Since 1970 approximately 3,600 deaths have been attributed to the Troubles, of which around 1,650 have occurred within the city limits of the capital city, Belfast (Fig. 5.2). In Belfast, the Troubles further divided communities along ethno-religious boundaries, resulting in higher levels of residential segregation than were evident before the onset of conflict. Parallels have been drawn with other segregated cities in the United States and religious divisions in Lebanon, the former Yugoslavia, and the Middle East (Fallah 1996; Adair et al. 2000). The Northern Irish political, social, cultural, and ethno-religious struggles are interpreted differently. Irish republicans view their engagement as being centred upon an anti-colonial struggle, whereas those who are pro-state view conflict as an internal civil war. The conflict in Belfast reflects of a deep-seated division between Catholic and Protestant ethno-religious communities, as well as between the Catholic community and the British State (Boal and Douglas 1982).

It is important to remember that ethno-religious segregation in Belfast is not new. Divisions have endured for almost 200 years, but it was only during the re-emergence of civil unrest in the late 1960s that inter-community violence intensified segregation so much that by September 1969 the euphemistic ‘peace-lines’ were erected to keep communities apart (Boal 1982; McPeake 1998). These peace-lines are not only physical barriers designed to restrict mobility and social interaction, but also present social barriers fueling sectarian mistrust even during the present cease-fire. The Northern Ireland Housing Executive (NIHE) recognizes at least 15 such physical interfaces,

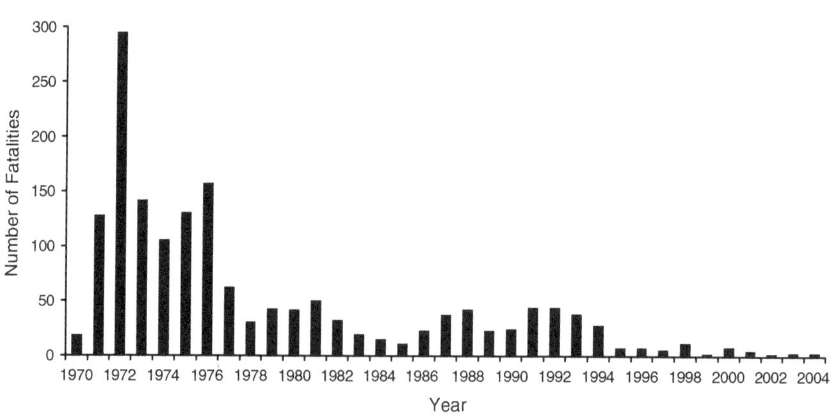

Fig. 5.1 Conflict-related deaths in Belfast (1970–2004)

Fig. 5.2 Location of Belfast, Northern Ireland

many exceeding 6 m in height and some as long as 1 km. Arguments for the deployment of peace-lines and defensive walls focus on the maintenance of social stability, but in practice they have also undoubtedly contributed to increased ethno-religious segregation, even 'ghetto-izing' some of the more isolated and deprived communities (e.g. Ardoyne in north Belfast and Suffolk in west Belfast). This divisive image is echoed by the 1991 Population Census, which indicated that over two-thirds of enumeration districts (census tracts) in Belfast were classed as having 90 percent of their population as either Catholic or Protestant.

There have been numerous commentaries on political violence in Northern Ireland, which have concentrated on trying to establish the facts or data about the levels, distribution, and sources of violence. Such work includes McKittrick et al.'s (1999) book, *Lost Lives*, Amnesty International's (1994) report on *Political Killings in Northern Ireland*, and Sutton's (1994) *An Index of Deaths from the Conflict in Ireland 1969–1993*. In contrast, there have been relatively few attempts by geographers to explore the spatial aspects of political violence in Northern Ireland. Notable exceptions include Schellenberg's (1977) article, which examined shooting incidents, explosions, injuries, and deaths; Murray's (1982) study of deaths and explosions in Northern Ireland between 1969–1977; and Poole's (1983) work on Troubles-related deaths. Although Fay et al.'s more recent (1998) *Mapping Troubles-Related Deaths in Northern Ireland 1969–1998* examines the geographical distribution of deaths, it does so by broad postal delivery zones and as such the arenas which they study are less capable of showing and explaining the locality of politically motivated deaths and the subtleties of place. In all of these the goal has been to try to understand and explain conflict-related deaths *without* necessarily addressing the spatial dimension of their occurrence and instead relying on traditional discourse of their social, political, and ethno-religious underpinnings.

This chapter introduces a comprehensive database of deaths in Belfast attributed to the Troubles (known as conflict-related deaths) up until 2004. It was compiled by extensive negotiation with paramilitary and victim groups in Belfast. Recently published lists of deaths that have emanated from paramilitary and community groups were also used, as was information supplied by Glenravel (a small community group aiming to account for all deaths in Northern Ireland). The data collected were also cross-referenced with newspaper and other public announcements. The resulting database provides objective data on the spatial distribution of conflict-related deaths and violence in Belfast and as such would facilitate a better understanding of the impact of violence upon sensitive community interfaces. It provides a forum with which to build on cross-community dialogue on the shared nature of harm and victimization, with the objective of helping to promote a stable and functional society (Imrie et al. 1996). Moreover, a better understanding of the impact of the violence upon places will allow government departments, statutory agencies, and service providers to identify concentrated areas of deprivation for targeting more resources and to contribute to the development of reconciliation, mutual understanding, and respect between and within communities and traditions.

Our intention is not only to represent the spatial distribution of all conflict-related deaths in Belfast by producing a series of maps, but also to statistically measure whether their geographic patterns can be explained by variables pertaining to economic deprivation and ethno-religious segregation. We also focus on the measurement of segregation—probably the most cited cause for inciting violence and fueling deep-seated mistrust between communities (see Johnston 1984). In response to the established index of similarity, we outline a modified spatial segregation index that measures the degree to which predominantly Catholic and predominantly Protestant groups are geographically distributed at the smallest scale of aggregation: the enumeration district for the 1991 UK Census of Population and the output area for the 2001 UK Census of Population.

The political and ethno-religious geography of Belfast and Northern Ireland is complex and highly emotive, and the reasons behind conflict-related violence and segregation are equally multifaceted and sensitive. Although fully acknowledging these sentiments, this chapter only reports on the use of GIS for understanding the *spatial* distribution of violence and segregation—for political and ethno-religious discussions refer to Shirlow and McGovern (1997), Douglas and Shirlow (1998), and Shirlow and Murtagh (2006).

The chapter aims to address the legacy and perpetuation of conflict in several ways. These include:

- Locating and mapping conflict-related deaths and violence in Belfast since the beginning of the Troubles
- Outlining a spatial segregation index to measure the evenness of the distribution of Catholics and Protestants across spatial units at different scales
- Analyzing the mapped spatial distribution of conflict-related deaths and violence with relation to other social and political indicators, such as deprivation, segregation, and proximity to peacelines
- Providing some lessons for anti-terrorist homeland security in the US

5.2 Data on Conflict-Related Deaths

Our research is different from what has been undertaken before in that it concentrates on the Belfast urban area and addresses all conflict-related deaths and measures of ethno-religious segregation. We produce a number of GIS maps and relate the spatial distribution of conflict-related deaths to other social and political indicators including deprivation, segregation, and proximity to peacelines. Previous compilations of death data have been known to suffer from inaccuracies and inconsistencies that are symptomatic of manual data entries and virtually non-existent data quality procedures. They are also limited in detail and are unlinked to descriptive statistics, such as social indicators of well-being and geographical barriers to mobility (segregation and peacelines). Furthermore, work such as Fay et al. (1998) is limited by the scale of analysis. Our GIS-based approach has in-built intrinsic procedures that ensure quality and consistency through cross-tabulations, searches, and data queries. But, of course, accuracy is dependent on source data, and to this end we invested in time to ensure quality information was collected from newspaper reports, published material, and prior surveys.

We compiled the database primarily from documentary factual research based on a number of sources, including paramilitary groups, victims groups, media reported incidents of conflict-related deaths and violence in Belfast, police data, and information compiled by non-government organizations (Fig. 5.3). Key variables were identified including prevalence, location, concentration, areas, perpetrators, type of victim (age, gender) and related socio-political developments in Northern Ireland.

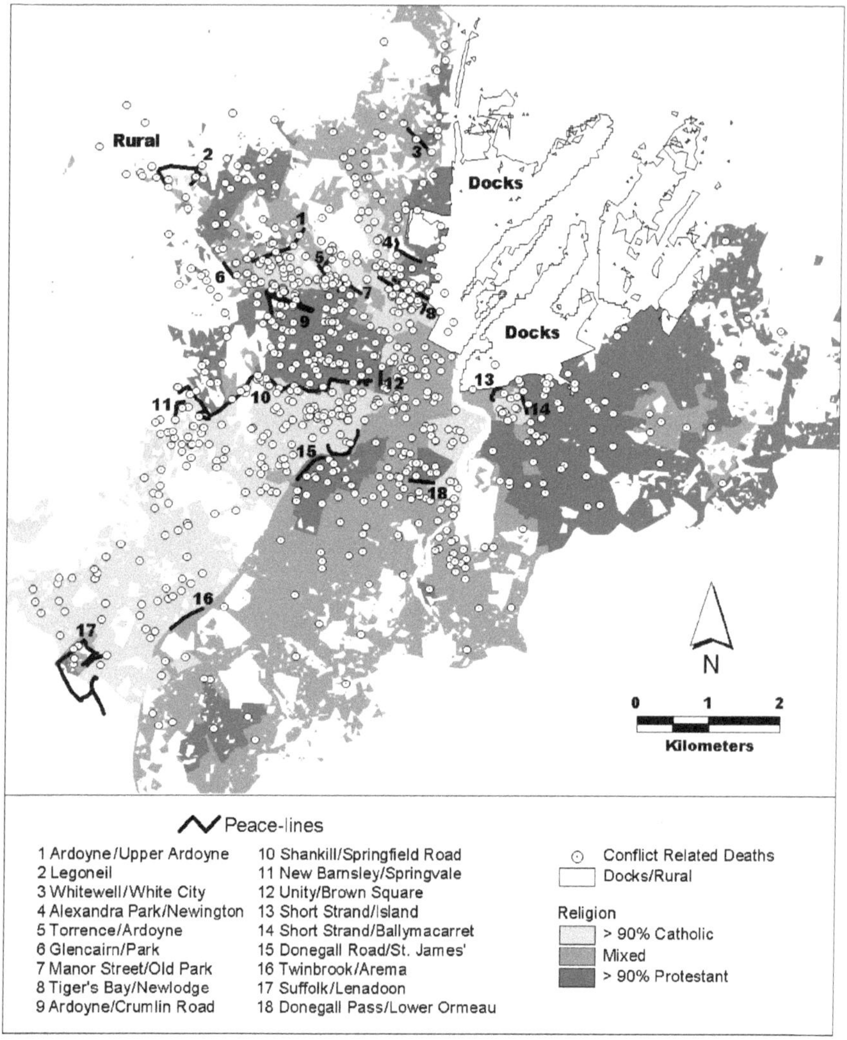

Fig. 5.3 Distribution of postcodes with conflict-related deaths in Belfast (1970–2004). Peace-lines are labeled

The following is a list of the main groups of attributes held in the database:

- Date {Day, month, year}
- Victim[1] {first/last name, gender, age, religion, occupation}
- Home address {postcode centroid}

[1] The above attributes are at the individual level, complete with name and home location of the victim. Although this information is publicly available, we believe there is an ethical duty to hide the identity of victims in our publications.

- Incident address {postcode centroid}
- Death {method, motive, perpetrator, accidental, part of riot}

Previous work (including Fay et al. 1998) has been based on the postal sector or postal district scales of measurement. We believe these scales are too broad and hide internal variations. Remember, the postal sector scale in the UK measures aggregations of around 100 households and postal districts into tens of postal sectors. Our intention is to measure conflict-related deaths at the postal unit scale of measurement, which equates roughly to 14 households. At this scale internal variations (differences in neighborhood structure and behavior, such as segregation) are minimized to the extent of being almost inconsequential. It is only when geographical differences in groups of human behavior are minimized can any relevant analysis be produced and social policy decisions made. Thus, data generated by the project will aid the activities of community groups and voluntary organizations in their attempts to secure funding and additional resources to tackle the legacy of conflict-related violence in their areas. The data will also be of use to service providers in better targeting finite monies/resources to those areas most disadvantaged by the conflict and socially excluded.

5.3 Measures of Segregation

Before we examine the extent of conflict related deaths we outline a technique that measures the scale of ethno-religious segregation in Belfast between Catholics and Protestants at two time intervals: 1991 and 2001. Although segregation is well-established in Northern Ireland, and even more so in Belfast, there have been many disputes as to the severity of social and political isolation across the geographic patchwork of the Belfast city. We first examine the conventional index of dissimilarity before offering a modified spatial segregation index.

5.3.1 The Index of Dissimilarity

The typical index of dissimilarity (D) is a measure of 'evenness' (Taeuber and Taeuber 1976; Massey and Denton 1987; Massey and Eggers 1990; Kaplan et al. 2004). *D* is calculated as:

$$D = \left(0.5 \times \sum_{i=1}^{n} \left| \frac{x_i}{X} - \frac{y_i}{Y} \right| \right) \times 100 \qquad (1)$$

where x_i and y_i are subgroup populations in areal subunit *i*, and *X* and *Y* are subgroup totals across all areal subunits studied. For our study, x_i and y_i represent the number of people expressing a Catholic and Protestant religion, respectively, in areal subunit *i*, and *X* and *Y* are their totals for the larger areal unit for which *D* is being determined.

D is based on the premise that proportional representation of population subgroups across all populated observational units within an area constitutes evenness. '*Evenness is the degree to which the percentage of minority members within residential areas equals the citywide minority percentage; as areas depart from the ideal of evenness, segregation increases*' (Massey and Denton 1989: 373). For example, if subgroups *X* and *Y* constitute 75 percent and 25 percent of an area's population respectively, then perfect evenness would result if these subgroups were found in these proportions in each of the populated areal subunits that make up the larger area. This would result in a *D* of zero. If subgroups *X* and *Y* constituted 100 percent of the population of the subunits within which each were found (that is, the subgroups do not exist with one another in any of the subunits), then perfect unevenness would exist and *D* would go to 100 percent (Fig. 5.4). *D* may also be '*interpreted as the percentage of the minority group that would have to relocate to different neighborhoods [areal subunits] in order to achieve an even distribution*' (Kaplan et al. 2004: 238). This would imply that for each calculation of *D*, the minority and majority subgroups would have to be specifically identified as there is an implied 'directionality' to *D*. This is not the case. '*It [D] is a symmetrical index, so the degree of segregation between X and Y equals that between Y and X*' (Massey and Eggers 1990: 1160). To simplify the interpretation of values of *D*, Kaplan et al. (2004: 239) state that; '*Traditionally, values between 0 and 30 are thought of as low, values between 30 and 70 are thought of as moderate, and values over 70 are thought of as high.*' This ordinal scale interpretation will be used to summarize the results of the analysis of segregation in Belfast using *D*.

Table 5.1 below show the results of the application of the index of dissimilarity to the 1991 and 2001 Belfast data. Columns A and B indicate, respectively, the census year and the areal unit for which *D* was calculated. Column C indicates the areal subunits used to determine *D* for the associated larger areal unit. Values of *D* are found in column *D*. Rows 1 through 4 show the results of determining *D* for the city of Belfast as a whole. Ward data were used to calculate *D* for Belfast in 1991 and 2001 (see rows 1 and 2). The number of wards did not change significantly between 1991 and 2001, and nor did the levels of segregation as measured by the index of dissimilarity. The analyses suggest levels of segregation are in the uppermost portion of the 'moderate' range. Enumeration districts (EDs) and Output Areas (OAs) were also used to determine *D* for Belfast in 1991 and 2001 respec-

$X = 100$ $Y = 0$	$X = 0$ $Y = 0$
$X = 100$ $Y = 0$	$X = 0$ $Y = 0$
$X = 100$ $Y = 0$	$X = 1$ $Y = 100$

Fig. 5.4 Index of dissimilarity, spatial segregation, and exposure

Table 5.1 Changes in segregation as measured by the index of dissimilarity

A	*B*		*C*		*D*
Year	Unit *D* solved for	*n*	Subunits used	*n*	*D* (in percent)
1991	Belfast	1	Ward	52	69.7
2001	Belfast	1	Ward	51	69.2
1991	Belfast	1	ED	582	80.5
2001	Belfast	1	OA	912	80.5
1991	Ward	52	ED	582	51.5[a]
2001	Ward	51	OA	912	54.7[a]

[a]Average *D* for all wards based on ED and OA data
Observed change in average *D* is statistically insignificant

tively (see rows 3 and 4). Although OAs outnumbered EDs, the values of *D* were identical, at least to one decimal place, and suggest 'high' levels of segregation. The dramatic jump in *D* between the analyses using wards to those using EDs and OAs is to be expected since *D* is sensitive to the sizes of subunits used. '*Smaller areas... will generate higher values of the index even if the underlying residential distribution is identical*' (Kaplan et al. 2004: 239). Given this tendency of *D*, we should only concern ourselves with changes in D from 1991 to 2001 and not between analyses. The final analyses depicted in Table 5.1 (rows 5 and 6), involve the determination of average values of *D* for the set of wards in 1991 and 2001 using EDs and OAs. Whereas the previous two analyses indicate little or no change in segregation, the final analyses suggest an increase in average levels of segregation from 1991 to 2001, although both remain in the 'moderate' range. However, a difference in means testing (using independent samples tests) indicated observed differences were statistically insignificant, to a point where segregation levels in Belfast (as measured using the index of dissimilarity) have not changed appreciably over the 1991 to 2001 period and have remained at moderate levels.

Although *D* is considered a measure of evenness, the index is impacted by exposure which is, '*the degree of potential contact between minority and majority members; [and] it reflects the extent to which groups are exposed to one another by virtue of sharing neighborhoods in common*' (Massey and Denton 1989: 373). Figure 5.4 shows why this is true. Here, four of the six areal subunits (i.e. EDs or OAs in a ward) are populated by subgroups *X* and *Y* (where $X \neq 0$ and $Y \neq 0$). Note that subgroup *Y* is segregated into a single areal subunit. *D* for this example is 99.7 percent. The index is not 100 percent despite the fact that subgroup *Y* is totally spatially segregated. This is because a single member of subgroup *X* is present in the only subunit occupied by subgroup *Y*. This small degree of exposure prevents *D* from going to 100 percent. Importantly, the presence of areal subunits not populated by either of the subgroups being compared does not affect *D* in the slightest (*D* would remain unchanged if the two unpopulated subunits were removed from the calculation). What is needed, then, to get at the issue of spatial segregation is an index that is unaffected by exposure and is influenced by unpopulated observational units. Such a measure is developed by Courtney (1998) and is discussed below.

5.3.2 The Spatial Segregation Index

The spatial segregation index (SSI), although not inherently spatial, may be considered a measure of spatial evenness. Spatial evenness occurs when a phenomenon is distributed equally across all spatial units that make up some larger area (e.g., EDs or OAs that make up wards). This state also represents minimal distributional variability. Spatial unevenness is maximized when the totality of a phenomenon is segregated into a single spatial unit. This state equates to maximal distributional variability. The SSI measures spatial evenness in terms of distributional variability. The SSI is a standardized coefficient of variation (*CV*). The *CV* is a measure of relative distributional variability and is calculated as the ratio of the standard deviation to the mean of a distribution. As a measure of relative variability, the *CV* may be used to compare distributions of any size. When a phenomenon is distributed evenly across all observational units, *CV* is zero. When the totality of a phenomenon is found in a single observational unit, *CV* will be maximized. Equation (2) shows *CV* in the maximum distributional variability state, or the complete unevenness case.

$$CV_{Max} = \frac{\sqrt{\dfrac{1\left[x-\left(\dfrac{x}{n}\right)\right]^2+(n-1)\left(0-\dfrac{x}{n}\right)^2}{n}}}{\left(\dfrac{x}{n}\right)} \tag{2}$$

Here, x is the totality of the phenomenon and n is the number of observational units over which x is distributed. The overall numerator represents the maximum variability case for the population form of standard deviation (σ). There will be one occurrence of the square of the totality of x minus the mean, and $n-1$ instances of the square of zero minus the mean. The overall denominator is the mean of x. The maximum *CV* simplifies to equation (3).

$$CV_{Max} = \sqrt{n-1} \tag{3}$$

If the sample form of standard deviation (s) is used, then the maximum *CV* equates to $\sqrt{n}$. This simplification shows that the maximum possible *CV* depends solely upon the number of observational units. This also means that the *CV* calculated for a distribution occurring over 50 observational units would not be directly comparable to the *CV* for a distribution occurring over 10 observational units. Direct comparisons of these distributions could be made, however, if the *CV*s were expressed as a percentage of their maximum possible values given the numbers of observational units in each case. This is what the SSI accomplishes.

$$(a)\ SSI_{\sigma} = \frac{CV_{observed}}{\sqrt{n-1}} \times 100 \quad \text{or}$$

$$(b)\ SSI_{s} = \frac{CV_{observed}}{\sqrt{n}} \times 100 \qquad (4)$$

SSIσ is the SSI for use with σ and (4b) is for use with *s*. SSIs calculated with either formulation will be identical. The SSI may be used to compare distributions of any size occurring over any number of observational units, although one must be careful with the interpretation of the results. An SSI of 65 percent means that the observed variability in the distribution of a phenomenon is 65 percent of what it would be if all of the phenomena were segregated into a single observational unit. However, this number does not indicate the percentage of the phenomenon that would have to change observational unit in order to achieve perfect spatial evenness.

Overall SSI values were calculated for Belfast using 1991 and 2001 ward and enumeration district data. The results are found in Table 5.2. Columns A and B show SSIs for Belfast based on ward data and show that Catholics were more spatially segregated than Protestants in 1991 and in 2001. It is further shown that the overall level of spatial segregation for Catholics dropped from 13.6 percent to 12.5 percent, while a 0.1 percent increase in spatial segregation occurred for Protestants over the same period. Columns C and D have SSI values for Belfast based on ED and OA data. The EDs and OAs are much smaller and, therefore, more numerous than wards. Since the spatial segregation index is expressed as a percentage of the maximum possible coefficient of variation (*CV*), and the maximum possible *CV* is positively related to the number of observational units used, the SSI values for Belfast calculated using the entire set of EDs and OAs (shown in columns C and D respectively) are smaller than those derived using ward data. The results based on wards and EDs may not be directly compared, but the results in columns C and D may be compared despite the differing numbers of EDs and OAs used since they are similar units. The SSIs derived for Belfast using EDs and OAs show, again, that Catholics were more spatially segregated than Protestants in 1991 and 2001 and that Catholics have experienced a drop in spatial segregation (from 4.8 percent to 3.5 percent). The ED derived SSIs do, however, suggest that the level of spatial segregation for Protestants decreased from 1991 to 2001.

Table 5.2 Spatial segregation indices (percent) for Belfast

	A 1991 Ward $n = 52$	*B* 2001 Ward $n = 51$	*C* 1991 ED $n = 582$	*D* 2001 OA $n = 912$
Catholic	13.6	12.6	4.8	3.5
Protestant	9.7	9.8	3.1	2.6

As a final analysis of spatial segregation, 1991 and 2001 Catholic and Protestant population data at the ED and OA levels were used to calculate SSIs for the wards of Belfast. Differences in average levels of spatial segregation for the wards of Belfast were then tested for significance. Table 5.3 below shows the results. Reading across the rows in the table shows time series results, or within-group changes in average levels of segregation. Since the level of segregation of a given group in one time period is not independent from the group's level of segregation in a previous time period, paired samples tests were used to determine the significance of within-group changes in average segregation. Catholics show a significant decrease in their average level of spatial segregation dropping from 29.2 percent to 21.9 percent over the 1991 to 2001 period. The small drop in average level of spatial segregation for Protestants is insignificant. Between-group differences are shown in Table 5.4. Between group differences may be seen as independent, so independent samples tests were performed. Table 5.4 shows that Catholics were significantly more segregated on average than were Protestants (28.8 percent and 18.3 percent respectively) in 1991. By 2001 the observed differences in average levels of segregation were insignificant. Table 5.5 presents minimum and maximum SSIs for Catholics and Protestants for 1991 and 2001. Minimum levels of spatial

Table 5.3 Paired sample significance testing of changes in average spatial segregation index values for wards

	Avg. SSI (%) 1991	Avg. SS1 (%) 2001	*n* wards for paired samples	Significance of Within Group Differences (2-tailed *t* tests)
Catholics	29.2	21.9	51	0.002
Protestants	17.9	17.5	51	0.815

Table 5.4 Independent samples significance testing of between group differences in average spatial segregation index values for wards

	Avg. SSI (%) 1991	Avg. SSI (%) 2001
Catholics	28.8	21.9
Protestants	18.3	17.5
n wards	52	51
Significance of Between Group Differences (2-tailed *t* test)	0.013	0.207

Table 5.5 Maximum and minimum ward spatial segregation index values[a] for Catholics and Protestants 1991 and 2002

	1991 Minimum	1991 Maximum	2001 Minimum	2001 Maximum
Catholics	3.2	98.3	3.1	82.5
Protestants	3.0	78.2	3.2	53.1

[a]SSI values in percent

segregation remained in the low 3 percent range while maximum SSIs showed marked decreases. Catholics exhibited the most dramatic drop in maximum SSI from the high nineties to the low eighties. Protestants showed a decrease in their maximum level of spatial segregation, dropping from the high seventies to the low fifties.

The results of the analyses using the index of dissimilarity suggest that very little change in residential segregation among Catholics and Protestants occurred in Belfast over the 1991 to 2001 period. This means that evenness as measured as proportional representation across observational units has not changed much. The results using the spatial segregation index, however, suggest a more positive trend. Catholics have experienced a significant drop in average level of spatial segregation, as measured by the SSI, and have achieved an average level of segregation that is not significantly different from that of Protestants. This suggests the beginning of the diffusion of Catholics among the enumeration districts and wards that make up Belfast.

5.4 Regression Analyses

Given the severity of ethno-religious segregation, we now investigate the extent to which religious segregation, population density, deprivation, and other factors explain the pattern of conflict-related deaths in Belfast.

5.4.1 Methodology

For this analysis, we examined the pattern of conflict-related deaths in Belfast for 1980 to 1997 ($n \approx 475$). We selected this time period such that the time of death roughly corresponded with socio-economic characteristics from the 1991 UK Census of Population. We used a generalized linear model (GLM) to explain the number of deaths per ED based on socio-economic and geographic factors. The analysis was conducted using the R statistical package.

The socio-economic factors we selected included religion, deprivation, and population density. We used the percent of persons that were Catholic as a measure of the religious makeup of each ED. Since the sum of percent Catholic and Protestant equals approximately 1.0 for each ED, we could have used either variable for the analysis but chose percent Catholic arbitrarily. We used the Robson Index as a measure of deprivation (Robson et al. 1994). At the ED level, the Robson Index uses nine indicators of health, shelter, physical environment, education, family, income, and jobs to generate a measure of deprivation, where larger Index values indicate increased levels of deprivation.

We also used two geographic variables in the analysis: distance to nearest interface (peace-line) and area of the ED. Distance to the nearest interface was measured from the ED centroids, with the expectation that deaths were more likely to occur near interfaces between Catholic and Protestant neighborhoods. ED area was included in the model, because the EDs vary widely in size, and it is possible that the count of deaths per ED may be affected by how the districts are delineated.

Each of the independent variables was first regressed upon one another to detect any presence of collinearity. Because percent Catholic and population density were collinear ($R^2 = 0.46$), population density was not included in the final model. Collinearity was not found between any other variables. The final linear regression model included: percent Catholic, the Robson Index, distance to nearest interface, and ED area.

5.4.2 Results

Each of the four socio-economic and geographic variables displayed a significant relationship with the count of conflict-related deaths per ED in the final GLM (Table 5.6). The overall model explained approximately 15 percent of the observed deaths as measured by the R^2 value. Larger counts of deaths were associated with higher percentages of Catholic residents, higher levels of deprivation, shorter distances to Catholic-Protestant interfaces, and increased sizes of EDs (Table 5.1).

The Robson Index alone explained the largest proportion of variation in counts of conflict-related deaths (6.7 percent), followed by distance to nearest interface (6.1 percent) and percent Catholic (4.7 percent). These results indicate that a larger proportion of conflict-related deaths occurred in deprived, Catholic neighborhoods that were located near interfaces with Protestant neighborhoods than occurred elsewhere in the city. In fact, approximately 45 percent of the deaths occurred in EDs that were composed of at least 90 percent Catholic residents. These findings are consistent with the observed pattern of deaths that is illustrated in Fig. 5.3.

Perhaps less intuitive is the strong association of conflict-related deaths with ED size. The GLM results indicate that larger counts of deaths were associated with larger EDs. Because ED area is partially a function of population density, this suggests that more deaths occurred in rural areas with low populations. However, this is not consistent with the pattern of conflict-related deaths in Belfast, as all of the deaths occurred within at least 50 m of urban areas. The relationship between the count of deaths and ED size is better explained as a function of underlying ED geography. A large number of very small EDs occur within the central urban area of Belfast. While these EDs were located in very close proximity to conflict-related

Table 5.6 Results of the linear regression model used to explain the count of conflict-related deaths in Belfast per enumeration district as a function of religion, the Robson Index of deprivation, distance to nearest interface, and ED area

Variable	Estimate	*t*-value	*p*-value
Percent Catholic	0.0475	3.17	0.001
Robson Index	0.0521	3.85	<0.001
Distance	−0.2401	−3.06	0.002
Area	0.2582	5.32	<0.001

deaths, due to their small size, deaths were not recorded within many of them, and this explains why smaller EDs tended to have fewer deaths.

5.5 Discussion

Based on the spatial segregation index the majority of conflict-related deaths tends to cluster on the western half of Belfast—in the administrative areas, Shankill (predominantly Protestant), West Belfast (predominantly Catholic), and in North Belfast (a patchwork of smaller areas of predominantly Catholic and predominantly Protestant). There is a close link between conflict-related deaths and severe segregation (over 90 percent of either Catholic or Protestant). Figure 5.5a illustrates the distribution by number of conflict-related deaths per postcode on a graduated circle scale (remember, postcodes typically contain around 14 delivery addresses). Some postcodes are notorious black-spots and contain as many as 15 conflict-related deaths over the 34-year period. Conversely, some postcodes have never been linked to any deaths, and these are displayed as open circles. The vast majority of these zero postcodes are located on the east side of the Belfast urban area, as well as parts of the south and the periphery. Figure 5.5b–d illustrate the effect of paramilitary affiliation, i.e., whether the victim was a member of loyalist (Protestant), or repub-

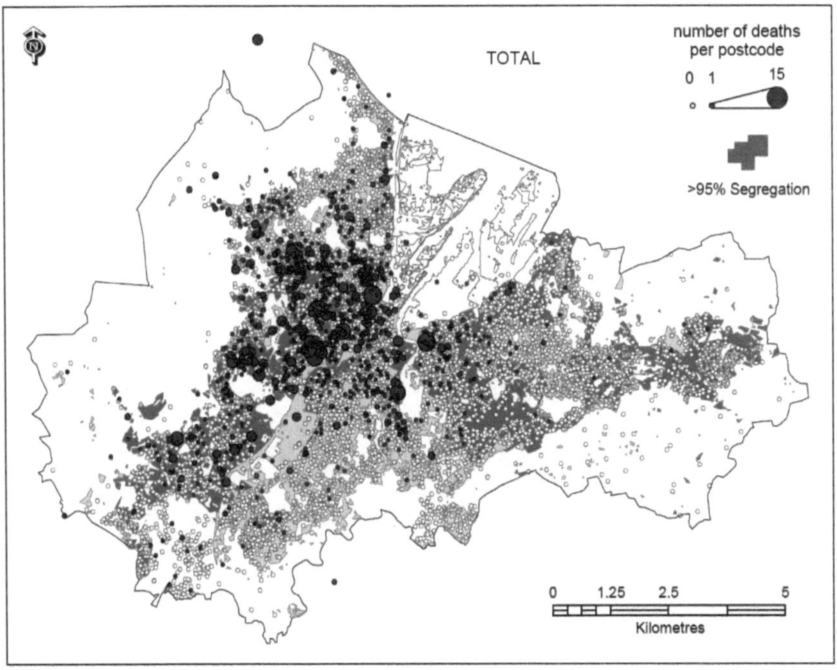

Fig. 5.5a Conflict-related deaths per postcode; (a) total deaths

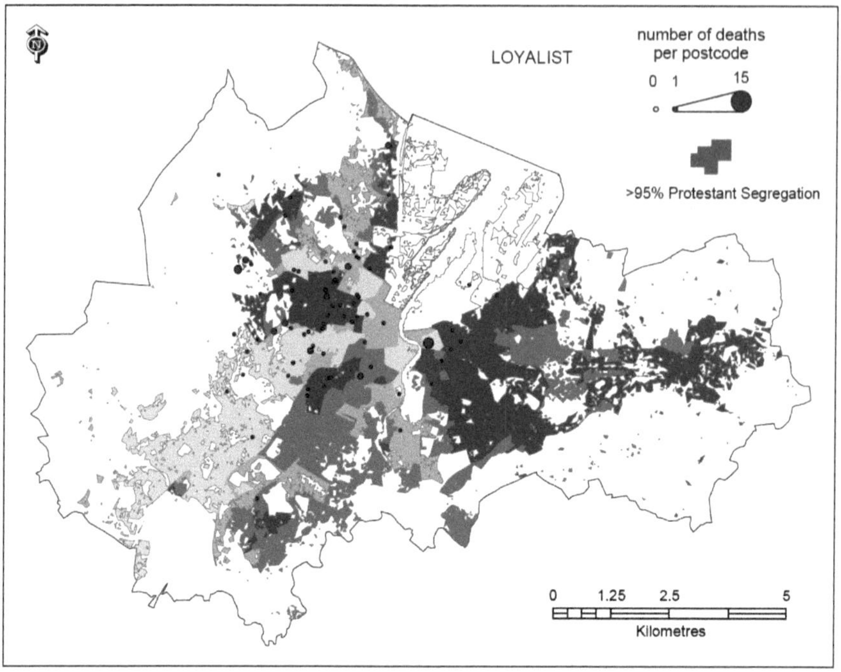

Fig. 5.5b Conflict-related deaths per postcode; (b) loyalist deaths

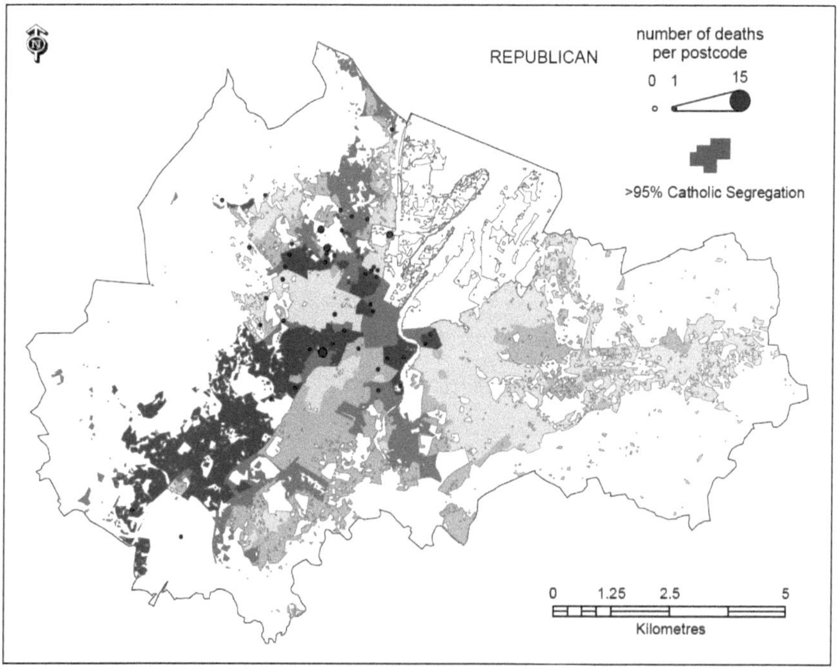

Fig. 5.5c Conflict-related deaths per postcode; (c) republican deaths

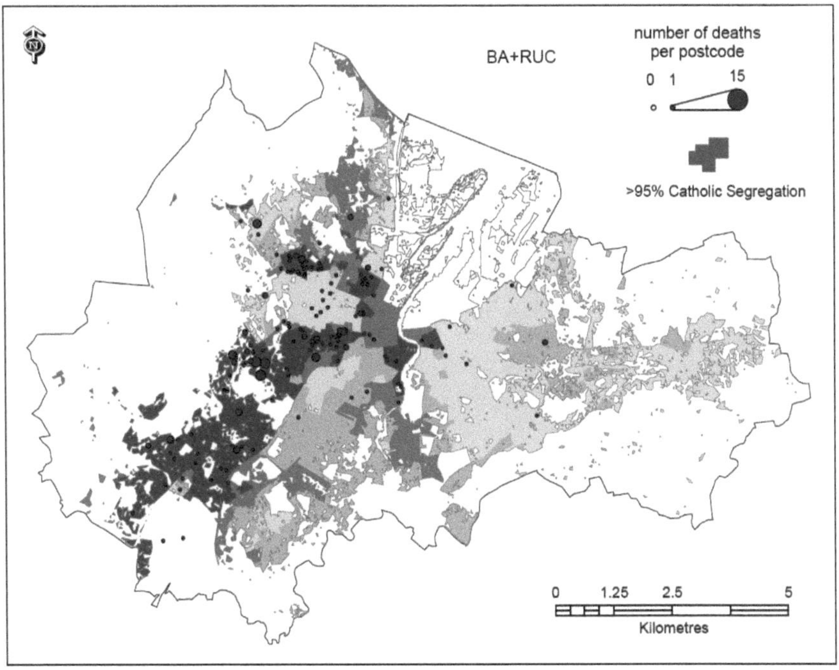

Fig. 5.5d Conflict-related deaths per postcode; (d) BA/RUC deaths

lican (Catholic) paramilitary group, or a member of the RUC (Royal Ulster Constabulary), or the British Army. Victims with loyalist or republican affiliations tended to be located in highly segregated Protestant and Catholic areas respectively. In terms of the British Army/RUC victims, the location tended to follow highly segregated Catholic areas because they were more likely to be assigned to keep the peace in those places and were targets for Republican paramilitary offenses.

From a simple visual comparison of Fig. 5.5a–d, it is plainly apparent that paramilitary deaths account for a small proportion of all conflict-related fatalities. Indeed, according to our database, of the 1,577 conflict-related deaths that occurred in Belfast between 1970 and 2004, 72 percent were civilians, and only 9 percent loyalist paramilitaries, 10 percent republican paramilitaries, and 9 percent British Army/RUC. Moreover, 68 percent of all civilian victims were Catholics.

5.6 Conclusions

Although the present political situation in Northern Ireland is the most stable in decades, the legacy of the 'Troubles' includes deep resentment by some Catholic and Protestant residents. Over the 1970–2004 period, many lives have been lost to sectarianism violence, especially in the most deprived and segregated areas of Belfast. So what lessons can be learned from years of hostility and fear? First of all our database demonstrates that although attacks are highly unpredictable in terms of when they occur, they tend to cluster in the poorest and most segregated areas, and within close proximity to interfaces between communities. Secondly, attacks are difficult to profile demographically, and the vast majority involve civilian casualties. In terms of lessons for national security and specifically US homeland protection, it seems the duration and severity of the Northern Ireland conflict is testament to the intractable nature of unstructured and sporadic bouts of violence. Terrorist attacks on the US may follow similar levels of unpredictability unless the underlying causes are understood and mitigated within a coherent framework that places geotechnology at the core of analyzing the flow of people across the US border and the health of inter-ethnic community interaction.

References

Adair, A.S., Berry, J.N., McGreal, W.S., Murtagh, B. & Paris, C. (2000). The local housing system in Craigavon N. Ireland: Ethno-religious residential segregation, socio-tenurial polarisation and sub-markets. *Urban Studies*, 37, 1079–1092

Amnesty International (1994). *Political killings in Northern Ireland*. (London: Amnesty International)

Boal, F. (1982). Segregation and mixing: Space and residence in Belfast. (In Boal, F. & Douglas, N. (Eds.) *Integration and division: Geographical perspectives on the Northern Ireland problem* (pp. 249–280). London: Academic)

Boal, F. & Douglas, N. (1982). The Northern Ireland problem. (In Boal, F. & Douglas, N. (Eds.) *Integration and division: Geographical perspectives on the Northern Ireland problem* (pp. 1–18). London: Academic)

Courtney, R.S. (1998). Trends in racial segregation in the Commonwealth of Pennsylvania: 1980–1990. *The Pennsylvania Geographer*, 36(1), 24–38

Douglas, N. & Shirlow, P. (1998). People in conflict in place: The case of Northern Ireland. *Political Geography*, 17, 125–128

Fallah, G. (1996). Living together apart: Residential segregation in mixed Arab-Jewish cities in Israel. *Urban Studies*, 33, 823–857

Fay, M.T., Morrissey, M. & Smyth, M. (1998). *Mapping troubles-related deaths in Northern Ireland 1969–1998*. (Londonderry: INCORE)

Imrie, R., Pinch, S. & Boyle, M. (1996). Identity, citizenship and power in cities. *Urban Studies*, 33, 1255–1261

Johnston, R.J. (1984). *Residential segregation, the state and constitutional conflict in American urban areas*. (London: Academic)

Kaplan, D., Wheeler, J. & Holloway, S. (2004). *Urban geography*. (New York: Wiley)

Massey, D.S. & Denton, N.A. (1987). Trends in the residential segregation of Blacks, Hispanics, and Asians: 1970–1980. *American Sociological Review*, 52, 802–825

Massey, D.S. & Denton, N.A. (1989). Hyper-segregation in U.S. metropolitan areas: Black and Hispanic segregation along five dimensions. *Demography*, 26(3), 373–391

Massey, D.S. & Eggers, M.L. (1990). The ecology of inequality: Minorities and the concentration of poverty, 1970–1980. *American Journal of Sociology*, 95(5), 1153–1188

McKittrick, D., Kelters, S., Feeney, B. & Thornton, C. (1999). *Lost lives*. (Edinburgh: Mainstream Publishing)

McPeake, J. (1998). Religion and residential search behaviour in the Belfast urban area. *Urban Studies*, 13, 527–548

Murray, R. (1982). Political violence in Northern Ireland 1969–1977. (In Boal, F.W. & Douglas, J.N.H. (Eds.) *Integration and division: Geographical perspectives in the Northern Ireland problem*. London: Academic)

Poole, M. (1983). The demography of violence. (In Darby, J. (Ed.) *Northern Ireland: The background to the conflict*. Belfast: Appletree Press)

Robson, B., Bradford, M. & Deas, I. (1994, September). *Relative deprivation in Northern Ireland*. Centre for Urban Policy Studies, Manchester University. (Policy Planning and Research Unit Occasional Paper No. 28)

Schellenberg, J.A. (1977). Area variations of violence in Northern Ireland. *Sociological Focus*, 10, 69–78

Shirlow, P. & McGovern, M. (Eds.) (1997). *Who are the people? Protestantism, Unionism and Loyalism in Northern Ireland* (pp. 72–94). (London: Pluto Press)

Shirlow, P. & Murtagh, B. (2006). *Belfast: Segregation, violence and the city*. (London: Pluto Press)

Sutton, M. (1994). *An index of deaths from the conflict in Ireland 1969–1993*. (Belfast: Beyond the Pale)

Taeuber, K. & Taeuber, A. (1976). A practitioner's perspective on the index of dissimilarity. *American Sociological Review*, 41, 884–889

Chapter 6
Internal Security for Communities: A Spatial Analysis of the Effectiveness of Sex Offender Laws

Douglas F. Wunneburger[1], **Miriam Olivares**[1], **and Praveen Maghelal**[2]

Abstract Among the many internal national security issues facing society today, significant focus has been placed on means to protect our nation's children from sexual predators. A number of laws have been passed, typically named after the tragic child victims of sexual attacks. Individually, these laws mandate creation and maintenance of registries for convicted offenders, establish safety zones from which known offenders are restricted, and require notification to neighbors of the presence of registered offenders in neighborhoods.

Almost all of these laws have spatial impacts. Registries identify sex offenders' home addresses. Safety zones are formed by buffers of varying distances around places where children typically gather. Notification laws require that neighbors within specific distances of an offender's home must be notified of their presence. All of these laws specifically apply to offenders who are known, convicted, and registered.

In each case, these laws impose economic and societal costs for implementation and enforcement. Considering their inherent spatial nature, little is understood regarding spatial justification for implementing these laws or impact from their enforcement. This chapter describes the use of spatial analysis to examine the effectiveness of sex offender laws in Brazos County, Texas.

Keywords Child safety zone (CSZ), geographic information system (GIS), geospatial technology, registered sex offender (RSO)

6.1 Introduction

Without question, many homeland security issues are inherently spatial in nature. Toward preparation for untimely crises, hazards mitigation planning and risk assessment rely strongly on geospatial database query and analysis. Inevitably,

[1] Texas A&M University

[2] Florida Atlantic University

D.Z. Sui (ed.) *Geospatial Technologies and Homeland Security*,

response to crisis events requires rapid accumulation, evaluation, and dissemination of geospatial knowledge. While the most feared unnatural threats derive from groups outside our country's borders, we cannot ignore the potential for internal hazards, either human-made or natural. Many of these risks can be mitigated or avoided by careful planning. Further, some risks are self-inflicted, introduced by popular laws and policy implementations which results in unexpected homeland threats that may be slow in impact, and difficult to detect, track or study.

Like homeland security issues, many laws introduce effects that are best regarded spatially. When the notion of enacting laws to identify protected areas is applied, it follows that these laws impose measurable spatial impact to achieve equally measurable spatial beneficial effects. At some point with laws of any nature, the question must be asked, 'Are these laws effective, and at what cost?' Regardless of the likelihood of beneficial impacts, studies of their anticipated spatial effects are seldom undertaken. When they occur at all, such studies of public policy typically are pursued after enactment. In either case, geographic information systems (GIS) analysis and scientific visualization techniques provide a valuable means of clearly communicating the spatial impacts of these actions to law and policy-making groups.

Recently, much concern has been directed toward sex offenders as an internal security risk. Current laws allow convicted sex offenders, after being released from prison, to return and reintegrate into communities. Efforts to manage the risk they pose to society are reflected in a collection of federal, state, and local laws. These require states to create sex offender registries, to require lifetime registration of recidivists and certain aggravated offences, and to establish community notification systems. Additionally, known offenders are required not to live within specified distances from premises such as schools, parks, daycare centers, and other places where children congregate. With great variability in application, authority lies with individual jurisdictions to define this distance either as assigned by a judge according to a determined level of risk of individual Registered Sex Offenders (RSOs), or as a uniform distance regardless of the infraction committed. Nevertheless, each of these laws is described in a manner that identifies specific addressed locations and other areas of varying size.

Imposition of safety buffers and assignment of categorical risk to locations of RSOs' homes provide ready components for spatial evaluation of effectiveness of these laws. Assuming, in the case of sex offenders, that laws are passed to avoid the opportunity for these same offenders to commit recidivist crimes, each individual law edicts a specific spatial extent within which we should expect some beneficial impact. Further, as multiple laws impose separate regulations, each uniquely spatial in nature, urban areas become covered by a legal mosaic of restrictions. As these designated spaces overlap, assuming each law provides unique spatial impact, the singular benefits of individual laws should show cumulative effects. Otherwise, the potential exists that the collection of laws wastes communities' resources because of redundant, competing, or conflicting statutes.

To conduct an evaluation of current sex offender laws, the historical record of related laws must be understood. From this record, specific spatial parameters are extracted to create a legal critical risk zone model to identify the graduated levels of risk suggested by these laws. Using this model, a simple analysis of the addresses of sex crime locations and perpetrators' dwellings serves to compare crime statistics with the risk levels. Since these laws are oriented to limit sex offenders' dwelling opportunities within the general population, the distribution of crimes and perpetrators should differ from the distribution analysis of the general population in the same area. Further, by relating crime locations to perpetrators' addresses, specific travel distances are simulated. It follows that if perpetrators typically travel long distances to commit crimes, then the size of the restricted residency zone may be irrelevant.

This paper includes a study of spatial impacts and effectiveness of legislation passed in reaction to well known acts of violent sex crimes on society. Applying spatial analysis techniques using publicly available data, the study focuses on Bryan, Texas, the seat of Brazos County. Through this study, we seek to illustrate the importance of education in and application of geospatial technology toward policy making, analysis, and refinement.

6.2 The Legal Landscape

In the last decade, several studies analyzed sexual abuses to males and females under the age of 18 in the United States (Tjaden and Thoennes 1998; Greenfield 1997). Finkelhor (1994) calculated that one in five females and one in seven males become victims of sexual abuse by the age of 18. Personal and altruistic fear of sexual abuse is a strong motivator in the nation's psyche. This fear is periodically rekindled by the unfortunate events of violent sex-related crimes, especially by those crimes committed against children.

In an attempt to mitigate the impacts from such occurrences, lawmakers have enacted various restrictive statutes to manage the risk posed by sex offenders. While the past four decades have seen a number of new statutes, the Jacob Wetterling Crimes Against Children and Sexually Violent Offenders Registration Act (42 USC 14071 et seq.) of 1994 reshaped the way law enforcers managed RSOs in the US. This law required the convicted sex offenders to register descriptive information (including name, age, gender, height, weight, race, and details of offense) and report their movements to law administrators. These details must be provided to the state's Department of Public Safety.

6.2.1 Immediate and Lateral Risk Areas

Subsequent to the death in New Jersey of Megan Kanka at the hand of a convicted sex offender who lived across the street from her home, President Clinton signed Megan's Law as an amendment to the Wetterling Act. This law requires all states to make the information of pedophiles and rapists available to the general public (Beck and Travis 2004; Engeler 2005). Upon its enactment in May of 1996, all citizens gained access to the information and whereabouts regarding convicted sex offenders in their communities. The act also requires notification by mail to households within the zone of influence, defined in two levels depending on the location of the urban or rural setting of an RSO's dwelling: Immediate Risk—within three city blocks (or approximately 0.33 miles) in urban or subdivided settings, and Lateral Risk—within one mile in rural settings (areas outside subdivisions). This notification system supports the mandate that offenders must inform respective state authorities regarding their movements anywhere in the United States.

6.2.2 Child Safety Zones

The Wetterling Act asserts minimum federal standards for creating safety buffers around areas children frequent; however, states reserve the freedom to impose more stringent restrictions. In Texas, the Code of Criminal Procedure, SB1054, Article 42.12, Section 13B (Texas Legislature Online, Seventy-Eighth Legislature) mandates the Child Safety Zone (CSZ) for the State of Texas to be 'within 1,000 feet of premises such as a school, day-care facility, playground, public or private youth center, public swimming pool, or video arcade facility, places where children generally gather.' As per the regulation, sex offenders were restricted from going on or within this CSZ until the recent amendment in May of 2003 (Texas Legislature Online, Seventy-Eighth Legislature; SRC-LBB S.B. 1054 78(R) Bill Analysis). According to this amendment, several allowable exceptions were identified. Notably, restrictions do not apply while the defendant is in or traveling between: a community supervision and corrections department office; premises at which the defendant is participating in a program or activity required as a condition of community supervision; a residential facility in which the defendant is required to reside as a condition of community supervision, if the facility was in operation as a residence for defendants on community supervision on June 1, 2003; or a private residence required of the defendant as a condition of community supervision. Otherwise, offenders are restricted from residing within CSZs.

6.2.3 RSO Personal Risk

Presently, the State of Texas allows judges to supersede the usual 1,000 foot CSZ by stipulating separation distances as low as 200 feet from places where children gather, following the drug-free zone requirements applied in the state. Judges assign required distances on a case-by-case basis according, in part, to an assessment of risk particular to each offender. In addition, the state classifies RSOs according to assessed categories of high, medium, and low risk. Enacted for the purpose of limiting chance contacts between sex offenders and unrelated children, these laws follow common crime management procedures by identifying and assessing offenders and moving them away from proximity to potential victims. These restrictions have received wide attention recently. Robertson suggested that:

> ...An understanding of geographic trends of registered sex offenders, especially as they relate to schools and daycare facilities, may help police narrow their suspect lists in open cases, to those individuals contained within their registration database, living within a close proximity to the victim procurement site, who pose a high risk of recidivism. (2000, p. 109)

Unfortunately, personal risk levels were not assigned until enactment of Megan's Law in 1996. Hence RSOs convicted of their crimes prior to its inception are essentially grandfathered into a low risk assessment. For this reason, some local law enforcement agencies develop and rely on their own risk level assessments.

6.3 Spatial Characteristics of Child Safety Zones and Residency Restrictions

As laws implementing child safety zones and residency restrictions evolved during the past decade, they mirrored previous laws creating drug-free zones around places children congregate, e.g. schools and day care centers. In 1995, Florida was the state leader to establish the first residence restriction for sex offenders under probation, followed by fourteen other states in a period of nine years (Levenson 2007a). Soon, some jurisdictions began expanding limitations through larger safety buffers and additional designated safe places, including parks, churches, and school bus stops. In some cases, laws essentially banished RSOs, because residency restrictions meant that none could live legally within a city's limits. Recently, developers have used deed restrictions in an attempt to prevent sex offenders from purchasing homes within new subdivisions. In many large urban areas such policies, being politically difficult to oppose, became a competitive calling card.

It is clear that the thrust of laws passed with the intent to limit the societal impact of sex offenses is toward reducing recidivism by avoiding chances for contact with children by registered (i.e. known) offenders. Further, in promoting and passing these laws, positive effects are assumed without regard to any notion of analysis to verify their effectiveness.

However, negative effects have been well documented. Again, the state of Florida was the first passing 'Jessica's Law' in 2005, requiring electronic monitoring and increasing sentences (Levenson et al. 2007b). Replicating Jessica's Law, California passed a law in 2006 which, among other requirements, prohibits sex offenders from living within 2,000 feet of any park or school. Wartell (2007) conducted a residential parcel analysis in San Diego, CA to analyze the impacts of this law on residence availability. It was found that only 27.4 percent of San Diego's residential parcels were under compliance for sex offenders to live. Considering that only a fraction of those parcels would be vacant, the options are very scarce.

In a spatial analysis conducted in Hamilton County, Ohio to determine geographic exclusion of sex offenders due to Megan's Law, the authors reported that 'there are concerns that the overlapping restriction zones are so large that offenders cannot find suitable housing outside school restriction zones' (Grubesic, Mack, and Murray 2007). Zandbergen and Hart (2006) conducted a study of housing options for sex offenders in Orange County, CA. Their findings report only five percent of suitable parcels for sex offenders within the urban area.

Thomas (2003) suggested that laws such as the UK's Sex Offender Act of 1997 had unforeseen consequences. Most noteworthy of these consequences are lowered awareness of unknown offenders, driving sex offenders underground, and acts of vigilantism. This law dictated development of a mechanism of registration to help police decide the risk due to individual dangerous sex offenders. This idea of registration of sex offenders presumed three arguments. These were: to help the police identify the suspects after a crime, to prevent crime, and to act as deterrent. However, in the case of Doe v. Miller and White (2004), the US District Court of Appeals judge in Iowa cited the lack of research indicating the relationship between proximity and recidivism (Levenson and Cotter 2005). Further, 'There is no research to support the idea that residence restrictions prevent repeat sex crimes.' (ATSA 2005a, Public Policy/Position Paper Section, ¶ 21) Hence, studies of the effectiveness of these laws should address the following questions: Is there a relationship between the locations of sex crimes and RSOs' home addresses? At what distances from CSZs are sex offenses likely to occur? Do safety zones identify areas of higher (or lower) risk? Are the effects of the current mosaic of laws positive and cumulative?

6.4 Spatial Analysis of Effects

In 2005, Olivares and Maghelal sought to evaluate the locations of RSOs' dwellings with respect to CSZs. By comparing their geo-coded addresses to a uniform buffer of 1,000 feet around schools and daycare centers, they found that in Brazos County, Texas, fifty-three percent of RSOs lived within CSZs, presumably in violation of state law.

Their study included all sex offenders falling within the zip codes of Brazos County. All sex offenders convicted under parole or detentions were geo-coded. The Texas Department of Public Safety (TXDPS) listed 164 registered sex offenders living in the zip codes of Brazos County during Spring 2005. Included on this list are the names of each offender (including alias names), date of birth, gender, race, current residential address, information pertinent to the offense, and a latest photo with other information. During the geocoding process, the unmatched addresses were matched interactively. Twelve out of the 164 addresses of the registered sex offenders were either located out of Brazos County or could not be located in the Brazos Parcels file. The layers with information on parks, day cares, and public schools in Brazos County were buffered for a distance of 1,000 feet. These layers were appended and merged together to form the new dissolved layer of all the buffers that formed the Child Safety Zone (Maghelal and Olivares 2005).

Among the registered sex offenders in Brazos County included in this study, 73 percent of the offenders were White. Over ten percent of offenders had committed a crime at least two or more times, about 5 percent of the offenders were females, and over ten percent of the offenders were high risk offenders. 44.2 percent of offenders were aged 15–30 years old when the crime was conducted, and 35 percent of them were currently 30–45 years old. Over 50 percent of victims are aged 15–25 years old, and over 80 percent (131) of the victims were females.

The outcome of the study resulted in the creation of a database containing the locations of RSOs, parks, schools, and daycare centers in Brazos County. Using the database compiled for this study, the City of Bryan, Texas created an on-line mapping service through its existing public GIS web site.[1] The service includes display of the residence addresses of RSOs in the county, each holding a hyperlink to their personal data stored on the Texas Department of Public Safety web site. Through this public access point, community members gained the ability to easily identify residences of RSOs that may fall near their children's walking routes or play areas. Due to publicity created by this innovative use of web-GIS, the city received recognition through several national awards and news reports.

6.4.1 Graduated Critical Risk Model

Seeking to evaluate the singular and additive effects of the legal mosaic effecting RSOs, Olivares and Maghelal (2005) developed a spatial model of graduated risk, the Critical Risk Zones (CRZ) Classification. Spatial themes are built representing the areas under influence by each law. The most basic components of the model include (1) the CSZ buffers (2) the proximity areas of risk—immediate and lateral, and (3) the RSO personal risk levels—low/medium/high (Fig. 6.1). By overlaying these themes according to the local legal dictates, the model integrates three critical variables creating a comprehensive risk assessment (Maghelal, Olivares, Wunneburger and Roman). Twelve CRZ classes (Fig. 6.2) result by identifying the order of expected risk graduated according to the spatial coincidence of these components (Olivares and Maghelal 2005).

The first implementation of this model was conducted in Bryan, Texas. Bryan is a city of approximately 67,000 and is the seat of Brazos County. It is adjacent to the City of College Station, which is approximately the same population size. Brazos County is the home of Texas A&M University and is centrally located in the State of Texas (Fig. 6.3).

[1] http://www.bryantx.gov/departments/?name = sex_offenders

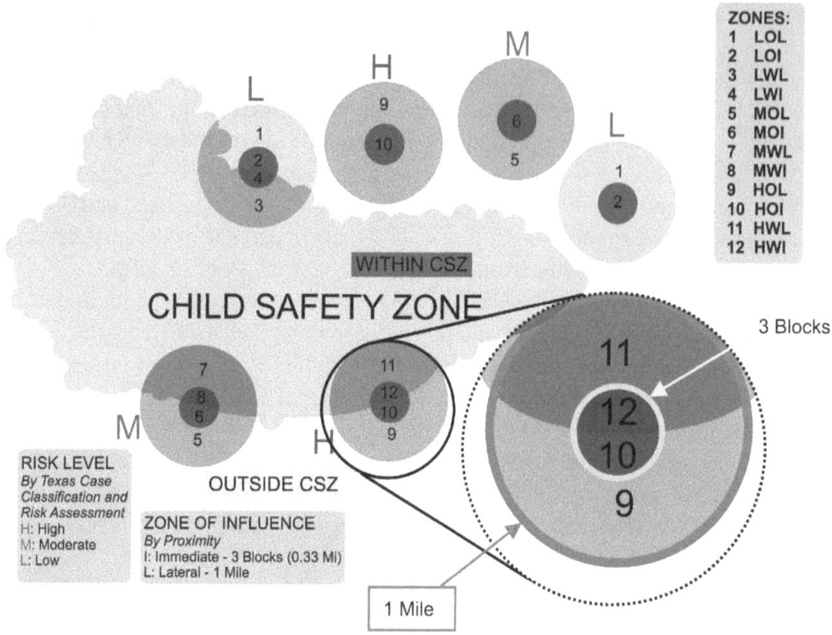

Fig. 6.1 Schematic of spatial characteristics of Critical Risk Zones according to Olivares and Maghelal (2005)

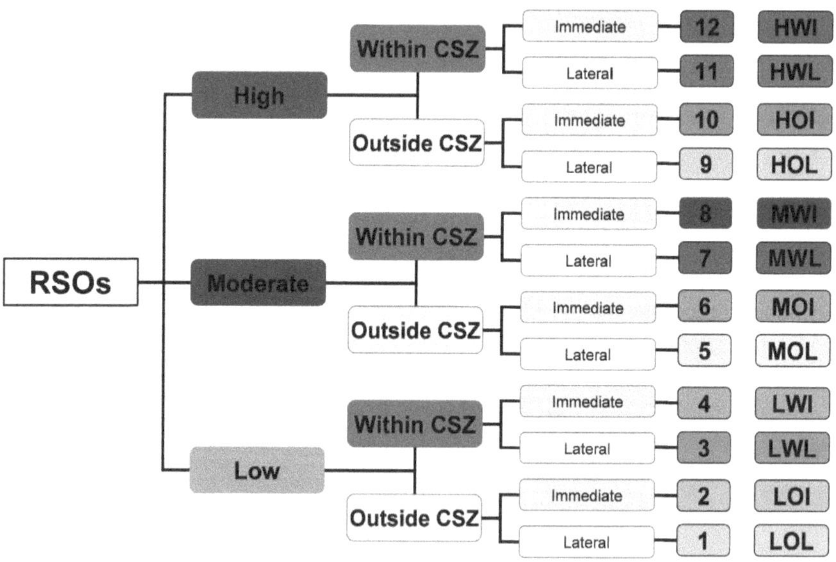

Fig. 6.2 Critical Risk Zone hierarchy (Olivares and Maghelal 2005)

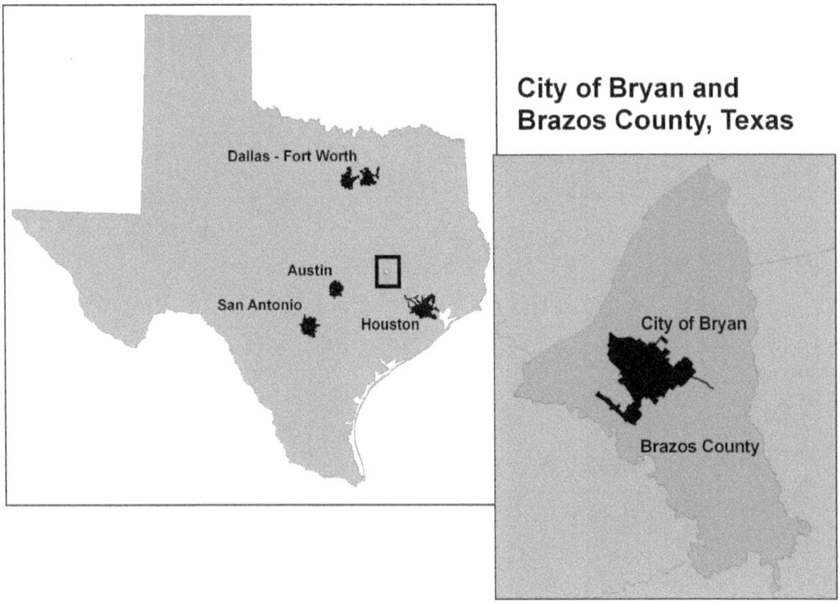

Fig. 6.3 Location of City of Bryan and Brazos County, Texas

The CSZs consist of 1,000-foot buffers surrounding the parcel boundaries of geo-coded day care centers, public parks and schools (Fig. 6.4).[2]

To derive the immediate and lateral risk areas, RSO addresses were geo-coded and buffered based on the Megan's Law Parameters for immediate risk (0.33 miles) and lateral risk (one mile). For each RSO, an assessment of personal risk level is determined—low, moderate, high. This designation provides a risk measure assigned to the immediate and lateral risk areas pertaining to each RSO (Fig. 6.5).

In compiling the model (Fig. 6.6), risk is assessed only in areas proximate to the locations of RSOs' dwellings. The total area covered by all the classes is equivalent to the area under the dissolved boundaries of lateral risk for all registered offenders in the county. Of the total county area (601 sq. mi.), zones of some level of risk cover 20 percent (Table 6.1). CSZs cover 33 sq. mi. within the City of Bryan. Approximately twenty-two percent of the areas within CSZs fall outside areas of any proximate risk determined by CRZ level.

[2] The list of licensed day care centers was provided by the Brazos County Health Department. It is important to note that the BCHD has no responsibility to monitor RSO movement within the county. Further, there is no requirement to locate their addresses prior to awarding a license for opening a new day care center

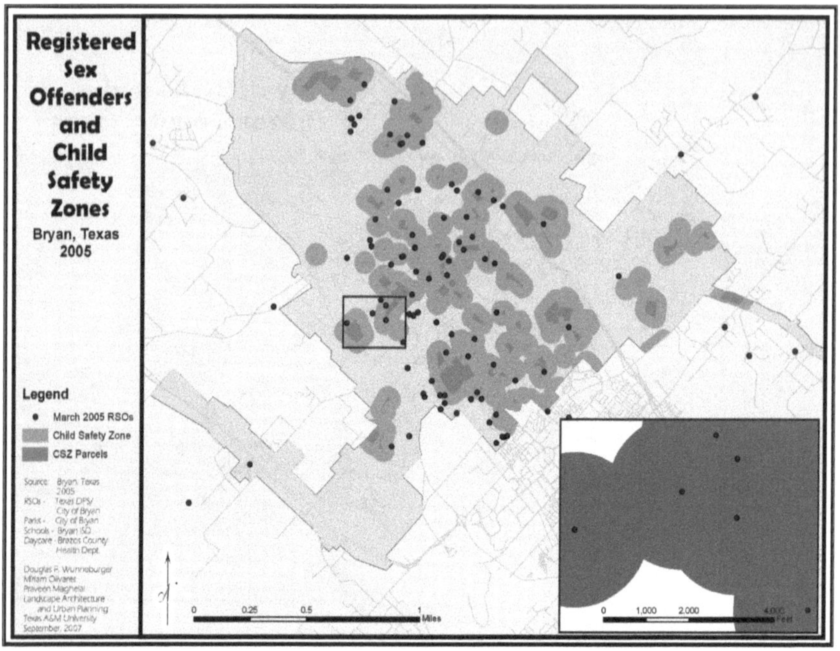

Fig. 6.4 Registered sex offenders and child safety zones

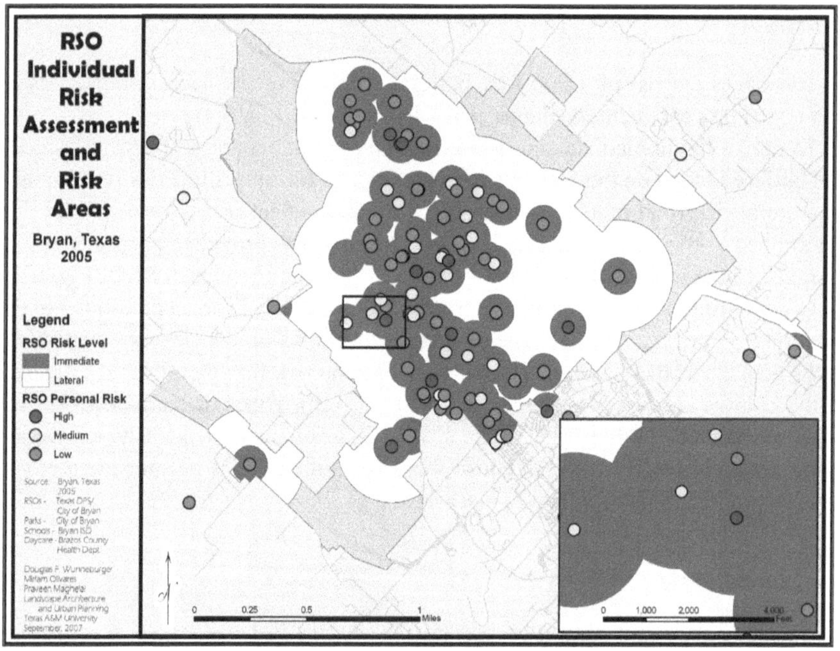

Fig. 6.5 Registered sex offenders individual risk assessment with immediate and lateral risk areas

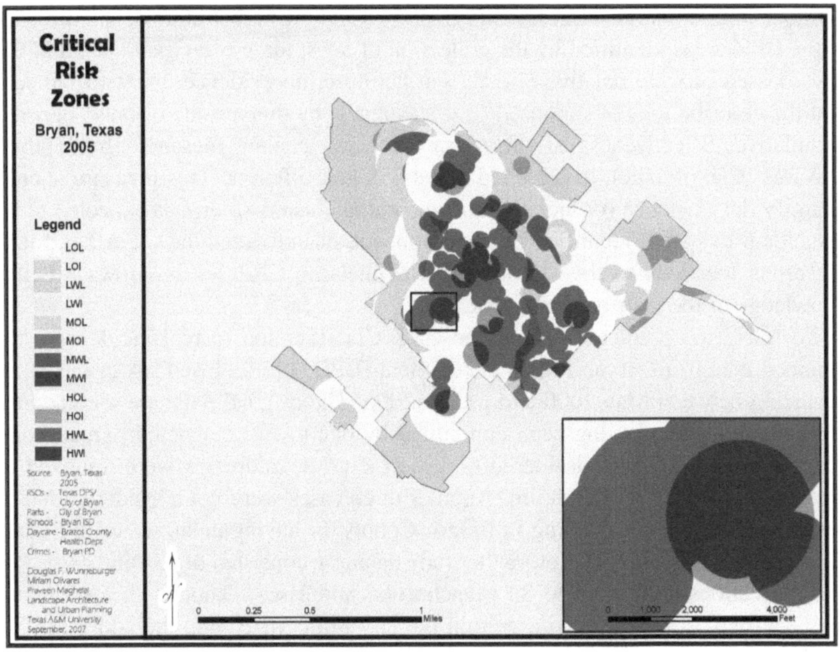

Fig. 6.6 Critical Risk Zones

Table 6.1 Area and percent coverage by CRZ level

Critical Risk Zones	Area in Sq. Mi.	Percent Zone Coverage	Coverage by RSO Risk	
1	LOL	40.0	32.8%	–
2	LOI	4.7	3.8%	–
3	LWL	7.4	6.0%	–
4	LWI	0.6	0.5%	43.1%
5	MOL	20.8	17.1%	–
6	MOI	2.6	2.1%	–
7	MWL	6.5	5.3%	–
8	MWI	1.8	1.5%	25.9%
9	HOL	22.1	18.2%	–
10	HOI	2.2	1.8%	–
11	HWL	10.5	8.6%	–
12	HWI	3.0	2.4%	31.0%
Total Risk Area	122.0			

6.4.2 Risk Evaluation Using the Critical Risk Model

The result of CRZ Classification provides information regarding coverage and proportion of different risk levels for Brazos County. However, the construction methodology makes the graduated classification scheme ordinal only in a logical sense.

Without comparing the addresses of sex crimes and the perpetrators of sex offenses to the CRZ areas identified by the collection of laws, there is no assurance that the laws passed provide positive effects. Furthermore, no evidence exists to indicate that the contributions of supposed risk, as suggest by this mosaic of laws, become cumulative. Effectiveness of the mosaic of laws can be measured by relating Olivares' Classification to recorded arrests for sex offenses. This measure should quantify the existence (or lack thereof) of spatial clusters of crimes expected to be coincident to spaces identified by the composite of individual laws. Further, relating arrest locations to the various areas delineating CRZ levels allows a better knowledge of the distribution of risk.

To relate the predicted risk of the CRZ Classification to real incidences, we obtained data from Bryan Police Department (BPD) of all closed sex crime cases occurring between May 2000 and March 2006. The original database included 69 crimes, each crime having been committed by one to five convicted perpetrators. Three crimes were not considered because the crime addresses were removed to protect the identity of the victim. Another three cases were not considered due to perpetrators either not residing in Brazos County, or having unknown or unlocatable recorded addresses. Therefore the study database consisted of 63 valid cases (63 crime locations connected to 83 perpetrators' addresses). Though the BPD data included only crimes occurring within the city limits of Bryan, the addresses for perpetrators of those crimes tended to be distributed throughout Brazos County.[3]

Twenty-nine crimes (46 percent) were committed during the daytime hours from 7:00 a.m.–6:59 p.m., and 34 crimes (54 percent) were committed between 7:00 p.m.–6:59 a.m. Data provided by BPD regarding closed cases for sex offenses occurring within the city were geo-coded for crime scenes and dwelling addresses of convicted perpetrators (Fig. 6.7). These geo-coded addresses were mapped overlaying the CRZ mosaic to identify the number of crimes and perpetrator dwellings falling within each CRZ classification level. Relating the CRZ classification to geo-coded crime data for Brazos County, Texas, we seek to validate restrictions applied by law and illustrate their effectiveness.

In the Bryan study, there was no knowledge that known registered offenders committed any of the recorded crimes. Despite the lack of a specific association between the addresses of RSOs' dwellings and the addresses of either crimes or suspects, a strong proximal relationship appeared in the case of high risk RSOs (Fig. 6.8). However, the relationship between crime data and CSZs showed little spatial dependency. The analysis indicated that most perpetrators' dwellings and most crime scenes fell away from areas near places where children gather (Fig. 6.9). A slightly higher proportion of perpetrators' addresses fell within the CSZ

[3] All crimes reported by BPD fall within the city's jurisdiction. In some cases, perpetrators' addresses fall outside the city. Those addresses within the entirety of Brazos County were included in the analysis; however, records identifying perpetrators that lived outside Brazos County were excluded from consideration. Further, crimes which occurred in victims' homes did not contain records of the crime location; therefore these records were eliminated as well.

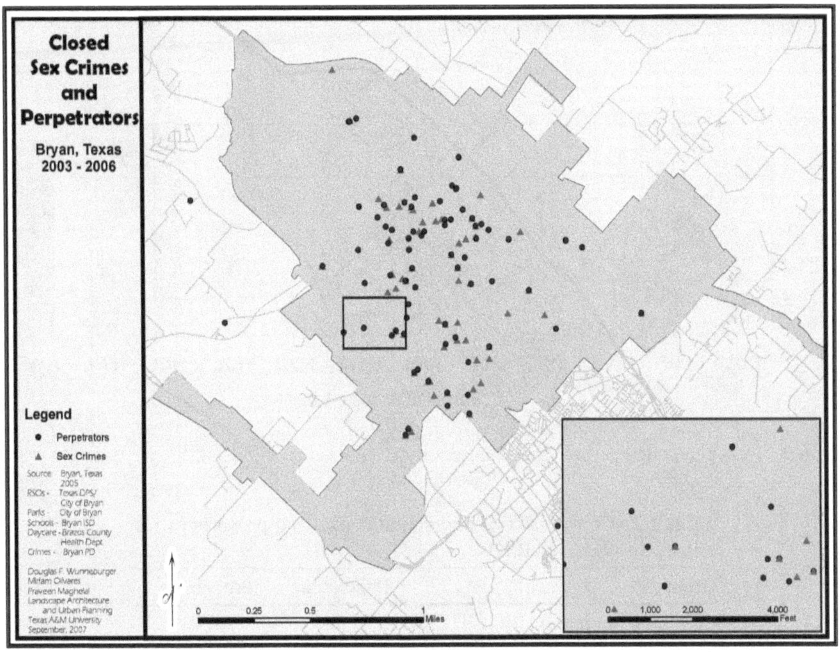

Fig. 6.7 Geo-coded addresses of reported sex crimes and suspected offenders

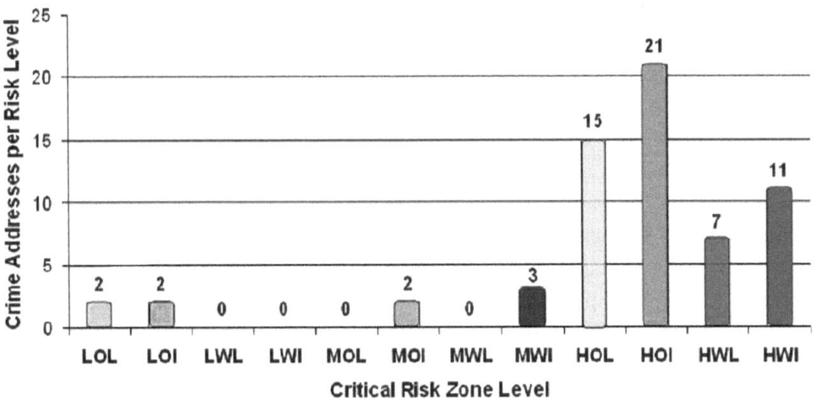

Fig. 6.8 Total geo-coded crime addresses by CRZ level

limits than did proportion of crimes committed. According to the CRZ Classification, most crimes occurred and most perpetrators lived within the immediate risk zones around high-risk offenders. The CSZ areas show no correlation with crime risk; in fact, most crimes occur outside the CSZs. A slightly higher, but still low, proportion of perpetrator dwellings lie in the CSZs.

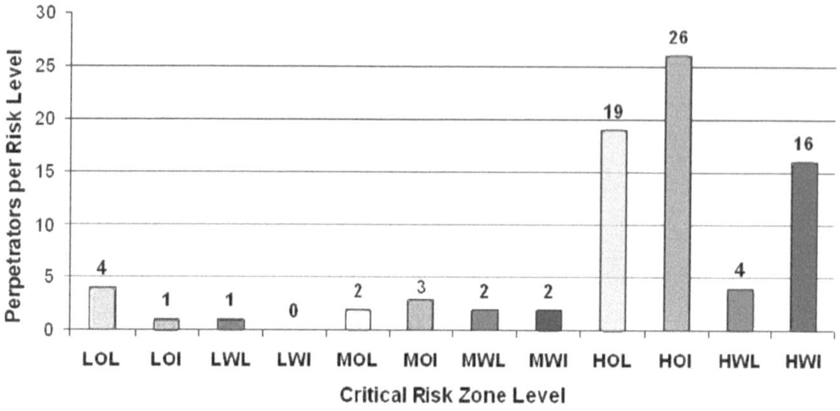

Fig. 6.9 Total geo-coded perpetrator addresses by CRZ level

Table 6.2 Comparison of population counts and proportions for crimes, suspects and RSOs

Group	Population	Proportion
City of Bryan (COB)	65,450	–
Suspects in COB	80	–
In COB CSZs	37,829	58%
Suspects in CSZ	48	60%
RSOs in COB	92	–
RSOs in CSZ	53	57%
Crimes in COB	63	–
Crimes in CSZ	39	62%
Crimes in Immediate Risk Area	51	81%
Crimes in Lateral Risk Area	63	100%

Notably, most crimes occur near the residences of high risk offenders, even though the RSOs typically do not represent the perpetrators of new crimes. Proximity to RSOs appears most significant with regard to the immediate risk areas. Of the three components of the CRZ classification, CSZs appear to have little relationship with either perpetrators' addresses or crime scenes.

Comparing populations of the various geographies, the proportions within the CSZs of total population, RSOs, and crimes are close to equal (Table 6.2). This suggests that the locations of sex crimes are influenced more by population distribution than safety zones around schools, parks, and daycare centers. However, 78 percent of crimes occur within the immediate risk areas of RSOs (almost all crimes occur within the lateral risk area). Because of the strong similarity between the distributions of crimes and the population, the relationship between crimes and immediate risk areas becomes questionable. It is possible the size of the 0.33-mile buffer casts too wide a brush over the urban landscape to provide a meaningful predictor and once again merely represents the distribution of the total population.

In 63 cases, addresses for both the crime scene and perpetrator's address were successfully geo-coded. Of those cases, 17 occurred in the perpetrator's home. Presumably, these crimes were typically victimless offenses, or offenses committed to a victim who was a family member, friend or other acquaintance. In each of these cases, CSZ areas offer little consequence toward preventing a crime.

Observation of the measured distance from perpetrators' addresses to matched crime locations indicated only twelve cases where the distance was within the limits of current CSZ buffers. In five cases, crime scenes were within 500 feet of a perpetrator's home (usually adjacent properties); however, none of these fell within CSZs in Bryan. One case involved a perpetrator traveling between 500 and 1,000 feet, and in six cases, perpetrators traveled between 1,000 and 2,500 feet. These values (500, 1,000 and 2,500 feet) are important because they represent popular CSZ buffer distances. However, it is critical to emphasize that of the several crimes that occurred within CSZs, none of these were close to perpetrators' addresses. Instead, the crimes were committed by offenders that lived outside the specific CSZ surrounding the scene of the crime and typically traveled some distance to perpetrate their crime.

6.5 Proximity of Perpetrators to Crimes

To study further the travel patterns of perpetrators, their geo-coded addresses and crimes were linked (Fig. 6.10). Euclidean distances were measured from the locations for crimes to nearest RSO residence, crimes to nearest CSZ property limit,[4] perpetrators to RSO residences, perpetrators to CSZ properties, and for perpetrators to crimes committed. Evaluation of the measure representing the distance perpetrators traveled from their dwellings to the corresponding crimes committed indicates no valid relationship with CSZ buffer size. The median distances from crime locations to the property limits of schools and for perpetrators dwellings to the same limits were similar (775 feet and 731 feet). Distances from crime scenes to nearest RSOs' residences (1,097 feet) and from perpetrators' residences to nearest RSOs' residences (1,059 feet) were greater. In comparison, the median distance traveled by perpetrators from their dwelling to their crime scene is greater than two miles. Regardless of the size of areas included as CSZs or as immediate and lateral risk zones, perpetrators traveled much farther distances

than the designated buffers creating child safe zones (Fig. 6.11), indicating the CSZs offer little impact as a preventative measure.

[4]CSZ areas are determined by placing a 1000 foot buffer outside each property boundary for schools, parks and daycares. Distance measures to CSZ, relate to the edge of the property at the heart of the CSZ.

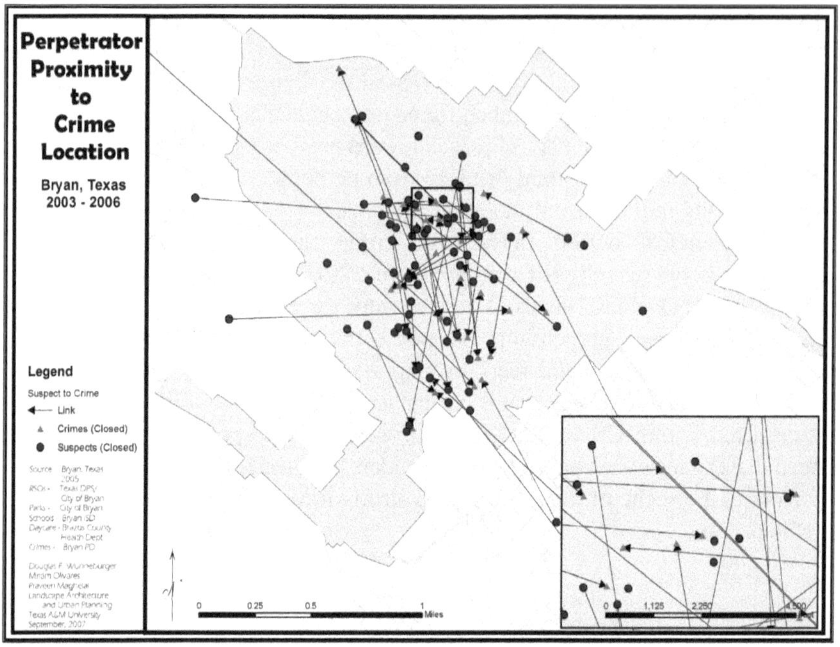

Fig. 6.10 Proximity of perpetrators' dwelling to location of crime they perpetrated than the designated buffers creating child safe zones (Fig. 6.11), indicating the CSZs offer little impact as a preventative measure.

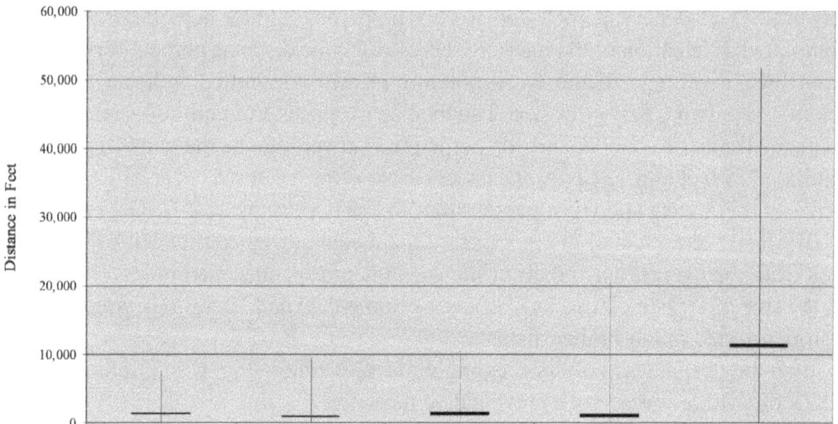

	Crime to RSO Residence	Crime to CSZ	Suspect to RSO Residence	Suspect to CSZ	Suspect to Crime
Mean	1,388	973	1,573	1,351	11,555
Max	7,730	9,474	10,581	20,441	51,674
Min	40	4	75	7	40
Median	1,097	775	1,059	731	11,047

Fig. 6.11 Distances between crimes and perpetrators to RSOs in comparison to distances traveled to commit a crime

6.6 Conclusions

The spatial characteristics of sex offenders' travel patterns are apparently unknown to law and policy makers. It is very likely that well-intentioned legislation may be exacerbating the problem. Results of this study indicate that perpetrators typically travel distances greater than those for child safety zone buffers. Some studies indicate that strong buffer laws adversely affect crime rates. In Iowa, the associations of law enforcers and district attorneys have advocated repeal of some of these laws, yet the popular response advocating the need to prevent sex crimes calls for more restrictions, despite the lack of previous research employing spatial analysis to study the question.

The public perception persists over the impacts of child safety zones on sex crime prevention. It is a typical community response to the fear of sex crimes that spurs mandates to register sex offenders and create safety zone buffers around children from which RSOs are kept safely away. In some cases, communities compete for greater impact through larger buffers. Usually, buffers greater than 1,000 feet serve the purpose to justify effectively banning RSOs from entire cities. Little evidence exists supporting the perception that these buffers reduce the impact of crimes against children by sex offenders.

If any relationship between locations of crimes and RSO dwellings exists, it is so only because the distribution of RSO dwellings and crimes mimic the distribution of the total population. While median distance from crime scenes to CSZs is less than 1,000 feet, crime scenes and perpetrators were typically separated by much greater distances. CSZs have no apparent effect on the distribution of sex crimes; instead crimes tend to follow the distribution of the general population. The mosaic of current laws offers no apparent positive or cumulative effects. Though a large proportion of perpetrators apparently reside within the CSZs, except for those who commit crimes against related victims, perpetrators do not commit their crimes either in the CSZs or near their homes. Nevertheless, new legislation tends to move toward even more restricted area in the forms of larger buffers and more identified safe zones.

In the 2005 Texas legislative session, it was proposed to include school bus stops among the locations to be protected by CSZs. Advocates of the nationwide passage of Jessica's Law propose a mandatory 2,000 foot buffer. Creating maps to indicate the spatial extent of such laws is a simple task (Fig. 6.12). The map clearly shows that viable dwelling places for RSO's become difficult to find and would tend to concentrate them in a very few rural locations. However, when confronted with a map of proposed restricted zones in the San Francisco area prior to the State's passage of Proposition 83 (Jessica's Law) in 2006, a co-author of the amendment called the map an exaggeration[5] (Jaffe 2006).

[5] Passage of Proposition 83 prohibits registered sex offenders from living within 2,000 feet of any school or park. Opponents of Proposition 83 claimed the provision would force sex offenders into rural areas. When confronted with a map showing the spatial impact of the proposition, State Sen. George Runner, a co-sponsor of Proposition 83, said that 'such maps are exaggerated.' (Jaffe 2006)

Fig. 6.12 Map to show the 2,000 feet proposed buffer around premises where children congregate including school bus stops

A close look at the analysis for Bryan, TX under this scenario indicates that the areas outside of the safety zone are mostly rural or industrial, but upscale residential subdivisions tend to fall outside the zone restricted to RSOs, as well. In a singular twist of fate, one residential area lacks bus stops because schools for all grade levels and their adjacent parks are within walking distance, yet a major portion of the neighborhood falls more than 2,000 feet away from each school. Despite the likelihood that most of the children in the area walk to school (school bus service is not provided within two miles), a small portion of the neighborhood satisfies the current and proposed residency restrictions for sex offenders. By achieving the goal of neighborhood schools, a small haven for RSOs may have been created.

Child Safety Zones tend to promote a false sense of safety. Communities must remember these restricted areas attempt to limit only the risk posed by known sex offenders and their likelihood to commit another sex crime. Considering that most sex offenses are committed by criminals related to the victim, citizens should be educated that, at best, restriction zones offer only one tool among several others needed to keep communities safe. Regardless of any perception of the best tactics to address sexual offenders in society, neither legally mandated safety buffers, registration requirements, nor personal risk assessments serve a deterrent effect to those who have not yet committed their offense, have not been arrested for an offense, or have victimized members of their own families. With regard to internal threats to homeland security, the most dangerous sex offender is the one we do not know. Yet typical legislation is targeted specifically toward the known, convicted offender.

Significant law enforcement resources are spent protecting, tracking, monitoring a very small proportion of the total population of sex offenders. Because of the nature of laws imposing community notification and residency restrictions, RSOs become impacted by being kept away from rehabilitative environments and resources. It is likely that these impacts result in interfering with the stated goals of the laws, that of enhancing public safety, by 'exacerbating the stressors (e.g. isolation, disempowerment, shame, depression, anxiety, lack of social supports) that may trigger some sex offenders to relapse.' (ATSA 2005a) In many cases, known offenders find it easier to get lost than to abide by the law. Hence marginalized and unsupervised, offenders face reduced incentives to avoid new offenses. And communities, in assuming that they know who the offenders are and that CSZ areas are safe, drop their guard for the unknown offender. It should be of little surprise that in areas of the most restrictive limits to RSOs, the rate of sex crimes seems to be increasing (Stromberg 2007).

One must consider the individual offender to whom our current and proposed laws are directed (Fig. 6.13). Rather than addressing the risk posed by the total population of sex offenders, in reality most laws focus on the potential recidivism of only a very small percentage of easily identified, known offenders. These trends are well documented:

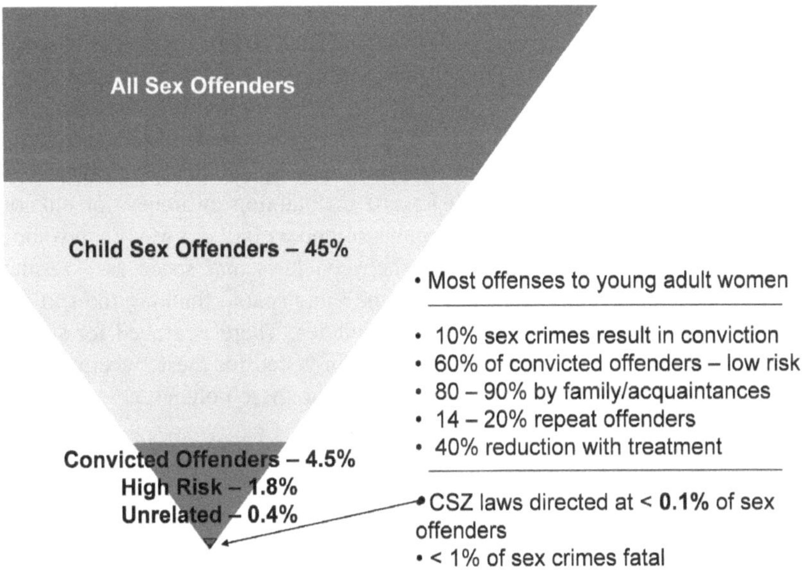

Fig. 6.13 Illustration of population logically addressed by current sex offender registration and restriction laws

1. Most sex offenses do not involve juveniles, but are committed against young adult women (ATSA 2005b).
2. Very few offenses are reported and only 10 percent of reported crimes result in a conviction (ATSA 2005a).
3. Sixty percent of convicted offenders commit crimes that do not directly affect victims (e.g. downloading internet pornography) or are otherwise considered to be low risk (e.g. statutory offenses committed as young adults) (Lane COG 2003).
4. In 80–90 percent of child sex offenses, the perpetrator is an acquaintance of the victim (Lane COG 2003).
5. Child safety buffers do not deal with any of these cases. Further:
6. Recidivism rates for sex offenders are lower than for any other crime except capital murder. Research indicates that 14–20 percent of one-time offenders are likely to commit a new sex offense (ATSA 2005a).
7. With contemporary cognitive-behavioral treatment, this rate has been shown to decrease by as much as 40 percent (Hanson et al. 2002; ATSA 2005a).

The offender must be registered for laws to apply; therefore laws imposing residency requirements on sex offenders are logically applied to prevent recidivist offenders. As shown in Fig. 6.13, child safety zone laws are directed toward this group, less than 0.1 percent[6] of all sexual offenders. The expense of enforcing such laws returns only very small benefits.

Regardless of evidence indicating laws directed at sexual offenders may be counterproductive, policy-makers bowing to political pressure continue to tighten restrictions rather than seek more realistic solutions. To make the spatial impacts of good and bad laws apparent, GIS analysis and visualization of results offer great resources for communicating and understanding such issues. In another chapter of this book, the case is made for the need to satisfy increasing demands for analysts by developing primary education in geospatial technologies. However, an equally strong case can be made to learn to use geospatial technologies to educate.

In the case of managing sex offenders in society, spatial analysis indicates that we need to commit resources more toward maintaining awareness of unknown offenders rather than toward pushing known offenders farther away. In developing an understanding of such relationships between laws and space as a result of geospatial analysis, it is incumbent to use the same spatial thinking tools to communicate the impacts of inherently spatial policies. There is a need for standard methods in developing policies which have been vetted for these, where there are spatial impacts. Spatial thinking is a key to refining these policies.

[6] Proportion of sex offenders who are affected by notification and residency restriction laws calculated from: 45% Child Sex Offenders * 10% Conviction Rate * 40% High Risk * 20% Unrelated * 20% Recidivism = 0.07%

References

ATSA (2005a). *Facts about adult sex offenders*. The Association for the Treatment of Sexual Abusers. [Electronic version]. Retrieved July 6, 2007, from http://www.atsa.com/ppOffender-Facts.html

ATSA (2005b). *The registration and community notification of adult sexual offenders*. The Association for the Treatment of Sexual Abusers. [Electronic version]. Retrieved July 6, 2007, from http://www.atsa.com/ppnotify.html

Beck, V. S. & Travis III, L. F. (2004). Sex offender notification and fear of victimization. *Journal of Criminal Justice*, 32, 455–463

Doe v. Miller & White (2004). U.S. District Court, Southern District of Iowa

Engeler, A. (2005). Is your child a target? The sex offender next door. *Good Housekeeping*, 240(5), 192–197

Finkelhor, D. (1994). Current information on the scope and nature of child sexual abuse. *Child Abuse and Neglect*, 4, 31–53

Greenfield, L. (1997). *Sex offenses and offenders: An analysis of data on rape and sexual assault*. (Washington, DC: U.S. Department of Justice, Bureau of Justice Statistics)

Grubesic, T.H., Mack, E. & Murray, A.T. (2007). Geographic exclusion: Spatial analysis for evaluating the implications of Megan's Law. *Social Science Computer Review*, 25(2) 143–162

Hanson, R.K., Gordon, A., Harris, A.J.R., Marques, J.K., Murphy, W., Quensey, V.L. & Seto, M.C. (2002). First report of the collaborative outcome data project on the effectiveness of treatment for sex offenders. *Sexual Abuse: A Journal of Research and Treatment*, 14(2), 169–194

Jaffe, Ina. (2006). Calif. follows trend with sex-offender crackdown. *Morning Edition*. Radio Program November 2, 2006. National Public Radio. [Online transcript]. Retrieved on January 7, 2007, from http://www.npr.org/templates/story/story.php?storyId = 6418295

Lane Council of Governments (Lane COG) (2003). *Managing sex offenders in the community: A national overview*. (Eugene, OR: Report funded by U.S. Department of Justice)

Levenson, J.S. & Cotter, L.P. (2005). The impact of sex offender residence restrictions: 1000 feet from danger or one step from absurd? *International Journal of Offender Therapy and comparative Criminology*, 49(2), 168–178

Levenson, J.S. (2007a). Residence restrictions and their impact on sex offender reintegration, rehabilitation, and recidivism. *ATSA Forum*, XVIII(2)

Levenson, J.S., Brannon, Y.N., Fortney, T. & Baker, J. (2007b). Public perceptions about sex offenders and community protection policies. *Analyses of Social Issues and Public Policy*, 7(1), 1–25

Maghelal, P. & Olivares, M. (2005). *Critical risk zones: Violators of Megan's Law*. 25th ESRI International User Conference, San Francisco, CA

Maghelal, P., Olivares, M., Wunneburger, D. & Roman, G. (Final Review). Where are they? A spatial inquiry of sex offenders in Brazos County. *Urban and Regional Science Information System*. [Electronic preprint version]. Retrieved on September 25, 2006 from http://207.145.30.84/pm_anonymous

Olivares, M. & Maghelal, P. (2005). Sex *offenders and critical risk zones*. National Institute of Justice's MAPS 8th Annual Crime Mapping Research Conference, Savannah, GA

Robertson, L.S. (2000). *Assessment of the proximity of registered sex offenders to schools and day cares: St. Louis City, Missouri*. Thesis, Department of Administration of Justice, Center for the Study of Crime, Delinquency, and Corrections. Southern Illinois University at Carbondale, Missouri

Stromberg, M. (2007). Locked up, then locked out: Experts say residency restrictions for sex offenders may create more problems than they solve. *Planning*, 73(1), 20–25

Thomas, T. (2003). Sex offender community notification: Experiences from America. *The Howard Journal of Criminal Justice*, 42(3), 217–228

Tjaden, P. & Thoennes, N. (1998). *Prevalence, incidence, and consequences of violence against women: Findings from the national violence against women survey*. (Washington, DC: U.S. Department of Justice, National Institute of Justice)

Zandbergen, P.A. & Hart, T.C. (2006). Reducing housing options for convicted sex offenders: Investigating the impact of residency restriction laws using GIS. *Justice Research and Policy*, 8(2), 1–24

Chapter 7
Remote Sensing-Based Damage Assessment for Homeland Security

Anthony M. Filippi

Abstract For natural or anthropogenic disasters, rapid assessment is critical for an appropriate and effective emergency response. Remote sensing has served—and will continue to serve—a vital function in disaster damage-assessment activities. This includes disaster-mapping of natural and agricultural ecosystems and human settlements, which may involve assessments of structural damage, contamination, and affected populations. Single- and multi-date (change detection) analyses can be employed, and a need to exploit both spectral and spatial information in order to delineate damage regions from remote sensor imagery is identified. This chapter provides a brief overview of some of the remote-sensing damage-assessment applications that are of utility in the realm of homeland security. Specific attention is given to remote sensing-based detection of vegetation damage and soil contamination, including a discussion of the remote-sensing implications of artificial radionuclide contamination, as well as damage to urbanized areas and other human settlements.

Keywords Contamination, damage assessment, disaster, human settlements, remote sensing

7.1 Introduction

Many natural and technological hazards—accidental or intentional—entail the potential to significantly adversely affect natural and agricultural ecosystems, the human-built environment, and human populations. Remote sensing can be beneficial to the study of such hazards (Jensen and Hodgson 2006). In an analysis of remote-sensing articles (published between 1972 and 1998) involving hazard and disaster research, Showalter (2001) concluded that remote sensing has most commonly been employed as a method for the detection, identification, mapping, and

---

Texas A&M University

D.Z. Sui (ed.) *Geospatial Technologies and Homeland Security*,

monitoring of hazards and/or the effects of disasters. A secondary research focus has dealt with enhancing emergency management via time-sensitive damage-assessments. Basic hazard/disaster-related remote-sensing research is well-recognized; however, to maximize the utility of remote sensing, it should be combined with modeling for prediction (Showalter 2001). Model results and other data, in turn, can be used as inputs to decision support systems (DSSs) (Levy et al. 2005), for instance. For disaster risk management, such inputs need to satisfy accuracy and spatio-temporal resolution and coverage requirements (Tralli et al. 2005).

7.1.1 Data Collection and Access

Remote sensing is of great utility in synoptically assessing damage over wide geographic areas in a time- and cost-effective manner. For disaster-monitoring, satellites can potentially yield low-quality data due to atmospheric pollutants or weather conditions (e.g., thick cloud cover, atmospheric haze), particularly for optical sensors. Satellite vehicles may also not be properly positioned for ideal timing of image-acquisition of the event. However, some satellites have off-nadir (i.e. pointable) viewing capabilities (e.g. SPOT); thus, in the event of a disaster, for example, areas not directly below the satellite vehicle can be imaged even if the satellite track is not at an optimum (Jensen 2007). Additionally, even high spatial-resolution multispectral remote-sensor satellite imagery (e.g. QuickBird, IKONOS, etc.) constitutes a data source that has a high-probability of both pre- and post-disaster data availability, relative to other potential data sources of utility (e.g. aircraft-based data). Satellite remote sensing is also possible over dangerous or hostile areas, where planes may not have ideal access. Airborne remote sensors do possess the advantage of flexibility in the timing of data-acquisition, which can be advantageous relative to satellite sensors, assuming that aircraft resources are within an effective distance range such that a timely data-acquisition mission is possible. Rapid, yet accurate, processing of remote sensor data is very useful for effective emergency response. Additionally, remote sensing can be used to minimize the risk of disaster, though the emphasis of this chapter will be upon its use in damage assessment of environmental and human-built systems.

The utility of remote sensing in assessing a varied array of disasters is recognized internationally and has been extensively employed by spacefaring nations. In order to facilitate data access for a broader community, the 2000 Disaster Charter (i.e. Charter on Cooperation to achieve the Coordinated Use of Space Facilities in the Event of Natural or Technological Disasters) was initiated by the European Space Agency (ESA) and the Centre National d'Etudes Spatiales (CNES), and it represents the first international means of cooperatively sharing remote-sensing resources on a global basis for disaster management (Ito 2005). Remote sensing and the Charter are further discussed in Bally et al. (2005) in the context of humanitarian aid in response to natural disasters and conflicts. Utilizing software such as Google Earth can facilitate the delivery of remotely sensed data to disaster-relief workers, as well as the general public (Nourbakhsh et al. 2006). Some satellite systems have been designed for disaster applications, such as the Disaster Monitoring Constellation (DMC), an international program involving five low Earth-orbit microsatellites, and the participant list includes the United Kingdom, Algeria, Nigeria, Turkey, and China (Ebinuma et al. 2005).

7.1.2 *Change Detection*

A generally very useful and commonly employed technique in remote-sensing damage assessments is multitemporal analysis, or change detection (Hall et al. 2006). Pre- and post-disaster images can be registered such that manual and/or digital change detection can be performed (Jensen 2005; Boucher et al. 2006). In addition, damage information can then be correlated with a GIS database of property values, providing a financial impact assessment (Jensen 2007). Jensen (2005) noted an example of visual on-screen change detection and digitization, where pre- and post-disaster digitized photographs or images are displayed concurrently, and they are ideally topologically linked via objected-oriented programming such that a polygon drawn on one image to delineate an object will also appear around the same feature in the other image. This method was used to assess Hurricane Hugo-induced damage; buildings with no damage, partial damage, and complete damage were delineated, as were those buildings that had been moved. Many other change-detection algorithms exist, some of which are more automated in nature (e.g., Jensen 2005; Warner 2005). It is important to underscore that baseline data are important to change detection and effective disaster management tasks, and such data can be derived from a variety of sources, including remote sensing, GIS maps/data layers, architectural and structural drawings, photographs, textual-based building information, among many others. And as Laefer et al. (2006) note, such baseline data should be accurate, as comprehensive as possible, timely, and accessible.

7.1.3 *Scope of the Chapter*

Both natural and anthropogenic hazards and disasters can be considered as threats to homeland security and can be addressed with remote sensing. One key component of homeland security lies with securing and maintaining a healthy, sustainable environment, and regardless of the cause of a given disaster scenario, the motivation to safeguard air, water, vegetation and soil resources is similar (Houser 2002). Dependent upon these environments are human populations and settlements, which are susceptible to a suite of natural and technological threats. In this chapter, the discussion of remote-sensing damage assessment in a homeland security context is implicitly concerned with a population being able to sustain itself in the long-term, which includes food and environmental security, as these issues have implications for the maintenance of social order. Therefore, in addition to localized disasters, large-scale, high-impact events that have the potential to persist over long periods of time are also considered in this chapter.

While a broad definition of damage is assumed here, this chapter should be considered as an introduction, rather than a comprehensive overview of this application domain. In terms of natural hazard-oriented remote sensing, a variety of review articles (e.g. Jayaraman et al. 1997; Rycroft 2000) and research articles exist. For instance, Tralli et al. (2005) provides an overview of satellite remote sensing applications in earthquake, volcanic, flood, landslide, and coastal inundation hazards. Kerle et al. (2003) considered the specific case of detecting volcanic-based mudflow (lahar) disaster areas via optical and radar remote sensing, and Kääb et al. (2005) addressed the remote sensing of glacier- and permafrost-related hazards. San Miguel-Ayanz et al. (2000) reviewed remote-sensing in assessing forest fires, droughts, and floods. In Islam and Sado (2002) remote sensing and GIS-based modeling were employed for flood hazard mapping and priority map-generation for flood countermeasures. Other remote-sensing studies have considered flood detection (GläBer and Reinartz 2005), hydrogeological disasters (e.g. landslides) (Rizzo and Iodice 2006), and subsidence due to excessive groundwater-pumping (Tomás et al. 2005). Regarding technological- or terrorism-based disasters, a wide variety of chemical, biological, radiological, nuclear, and explosive (CBRNE) threats are relevant to homeland security (e.g. Kwan et al. 2006). In the present treatment, environmental remote sensing-based detection of vegetation damage and soil contamination is specifically considered, including a discussion of the remote-sensing implications of artificial radionuclide contamination, as is damage to urbanized areas and other human settlements (e.g. from radiological, nuclear, and explosive threats). Nuclear or radiological terrorism is the primary concern of the counterterrorism community, and it remains a global threat (Bunn and Wier 2006; Fox 2007). For example, cesium 137 is present in more than one thousand irradiation devices in US hospitals and research facilities. If one of these sources were used to create a radiological bomb, it could render 25 square kilometers of land uninhabitable for more than 40 years (Global Security Newswire 2007). However, an attempt to cover all aspects of potential threats in the present venue would be untenable; therefore, chemical and biological detection and assessment are covered only briefly here. And whereas remote sensing can be used for epidemiological purposes (Herbreteau et al. 2007), such disease-monitoring applications (natural or man-made) are outside the scope of this chapter. Also, atmospheric and water contamination are not treated in detail here, though such remote assessments could be of importance in a homeland security context (Houser 2002). The objective of this chapter is to consider several types of damage or contamination that can result from various types of natural or anthropogenicevents, as well as applicable remote-sensing approaches. Environmental/biophysical damage assessment using remote sensing is considered first.

7.2 Environmental/Biophysical Remote-Sensing Damage Assessment

The maintenance of vegetative health and soil contamination assessment in the aftermath of a natural or anthropogenic disaster are important in terms of the sustainability of natural/unmanaged ecosystems, as well as for continued agricultural production and food security. Remote sensing is not only of utility in agricultural monitoring and modeling (e.g. Vincente-Serrano 2007), it is also of critical importance in climate change science (e.g. Suzuki et al. 2007), and climate change may translate into significant effects in terms of food (agricultural production) and water security. For instance, the US Department of Defense (DoD) recently released a report examining the link between climate change and United States national security (Schwartz and Randall 2003). The report noted the potential for climate change to lead to a destabilization of the international geopolitical environment, possibly resulting in future conflicts and wars based upon related food and water resource shortages (Schwartz and Randall 2003; Shearer 2005). Other studies have also examined this relationship (e.g., Barnett 2003; [Anon] 2004; Associated Press 2007). Regardless of the feasibility of such a scenario (Shearer 2005), remote-sensing assessment of vegetation stress or damage, stemming from variety of sources, over various time and space scales is of importance.

One agricultural threat to homeland security and international stability may take the form of Ug99, a virulent strain of black stem rust fungus (*Puccinia graminis*). Discovered in Uganda in 1999, it has since spread via airborne spores, though it could also be transported by travelers. The *Puccinia graminis* fungus affects wheat, which is significant since wheat feeds more people than any other single food source, and Nobel laureate Norman Borlaug has noted its 'immense potential for social and human destruction' (Mackenzie 2007). Research into rust fungus-resistant varieties is being conducted (e.g. Beteselassie et al. 2007), and remote sensing can be used to detect and monitor wheat rust. For instance, Moshou et al. (2005) fused *in situ* multispectral and hyperspectral fluorescence image data to detect the winter wheat disease yellow rust (*Puccinia striiformis*). Disease presence could be detected by comparing fluorescence at 550 and 690 nm, and the overall classification error of their approach was 1 percent.

Another very important agricultural threat is drought, which is the costliest natural hazard in the United States (Wilhite 2000). Unlike some other hazards, drought develops slowly; however, it can exert severe environmental, economic, and social consequences, the effects of which can persist over a long time period (Ross and Lott 2003).

Given an event that damages crops or other vegetation, the notion of what constitutes vegetation damage must be considered. Sometimes 'vegetation damage' entails a very specific meaning, such as defoliation due to a pest infestation (Harris 1974). Often in the literature, though, it is a collective term referring to dead and dying vegetation that expresses a deviation, in terms of morphology and/or physiology, from the normal vegetation pattern; alternatively, the term can indicate vegetation in waning health. Murtha (1976) defined vegetation damage as 'any type and intensity of an effect on one or more plants, or parts thereof, produced by an external agent, that temporarily or permanently reduces the financial or aesthetic value, or impairs or removes the biological capacity for growth and reproduction, or both.' Since damage types can be specified for particular applications, Murtha (1976) additionally presented four damage types based on appearance in remote sensor imagery: (I) completely defoliated vegetation; (II) vegetation exhibiting some defoliation; (III) vegetation with foliage color inconsistent with normal foliage color of the species; and (IV) vegetation with no visible damage, but exhibits a deviation from its normal spectral reflectance in non-visible wavelengths. A dichotomous key was suggested with 29 damage types within the four general damage categories, including morphological and physiological damage. This scheme is based on the realization that a particular agent can cause a variety of damage symptoms, and the converse is true as well in that a given syndrome/symptom can be caused by a variety of agents (Murtha 1976), which is a complicating issue. Note though that chemical responses of plants may serve as reliable indicators of various environmental stimuli (e.g. soil, water, and light conditions, temperature, mechanical or insect damage, exposure to pathogens (which may affect animals and humans), and exposure to airborne and other chemicals), and this property may have significant implications for remote sensing in terms of detection (De Moraes et al. 2004).

In short, remote evaluation of vegetation damage in a hazards context is of direct importance in the case of agricultural production, and it can be used as a proxy indicator of environmental contamination, which may result from natural or anthropogenic disasters. Widespread drought or fungal or insect infestation could threaten the capability of the increasing national and global human population to feed itself. As for contamination, some contaminants may persist for many years, potentially rendering large tracts of land unusable for human activity. These phenomena are clear threats to homeland security.

7.2.1 Natural Disaster-Induced Vegetation Damage

Many types of natural disasters can adversely affect vegetation communities, including—but not limited to—floods, hurricanes and severe windstorms (e.g., McMaster 2005), drought (San Miguel-Ayanz et al. 2000; Volcani et al. 2005), insect infestation (Moskal and Franklin 2004), and fire (San Miguel-Ayanz et al. 2000), resulting in vegetation structural damage (Lévesque and King 1999), stress (Volcani et al. 2005), or mortality (Murtha 1976). Remote assessment of forest damage has been considered in Ghitter et al. (1995), Ranson et al. (2003), and Moskal and Franklin (2004). For agricultural crops, hyperspectral spectrometry has been used to discriminate between uninfested and infested winter wheat in Mirik et al. (2006). Silleos et al. (2002) utilized a joint satellite remote sensing/GIS approach to assess crop damage.

Many remote-sensing studies have addressed vegetation abundance or health/stress mapping. Vegetation indices have been widely employed to estimate relative abundance and photosynthetic activity of green vegetation (e.g. leaf-area-index (LAI), chlorophyll content, green biomass, and absorbed photosynthetically active radiation (APAR)) (Jensen 2007). The normalized difference vegetation index (NDVI) (Tucker 1979) is a very commonly used index, but vegetation indices specifically designed for narrow-band analyses have been developed, which may be more useful for certain applications. Several derivative-based vegetation indices were developed in Elvidge and Chen (1995), Adams et al. (1999) discuss the yellowness index (YI), which describes leaf chlorosis exhibited by stressed vegetation, and the indices given in Kogan (1990) and (Gao 1996) have been used for drought assessment.

Other types of algorithms can be utilized to assess potential damage, whether it is vegetative damage or another damage type. For instance, remote sensor imagery can be classified, possibly at a high level of specificity (Filippi and Jensen 2006, 2007), which, while difficult, is more likely with the use of hyperspectral (high-dimensional) data. Hyperspectral classification may identify vegetation stress categories, for example (Jensen and Filippi 2005). Hyperspectral remote sensing involves processing potentially hundreds of contiguous bands, and endmember determination can aid information-extraction. An endmember is an idealized, pure signature for a given class (Schowengerdt 1997; Chang 2003). Once endmembers have been identified, per-pixel fractional abundances of various materials can be estimated (Rogge et al. 2006). Hyperspectral data-processing in a time-sensitive environment (e.g. disaster damage-assessment) may necessitate automated endmember extraction for rapid-response since manual endmember selection methods are typically quite temporally-inefficient, and the output varies by the human analyst. Examples of autonomous endmember algorithms include N-FINDR (Winter 1999) and the sequential maximum angle convex cone (SMACC) (Gruninger et al. 2004), among others.

Once laboratory-, *in situ*-, or image-derived endmembers have been identified, such spectra can be employed for material-mapping and fractional abundance estimation (i.e. at the subpixel level). For instance, linear spectral mixture analysis (LSMA) assumes that a given spectral mixture can be modeled via a set of linearly independent endmember spectra (Adams and Smith 1986; Allard and Jouan 2005). Whereas LSMA assumes that the spectral mixing is linear, it is often nonlinear in nature (Borel and Gerstl 1994). LSMA also depends upon locating all endmembers in a given image, which may not be possible.

The Spectral Angle Mapper (SAM) algorithm is an alternative endmember-based mapping method. SAM compares angle α between reference spectrum r and the unknown pixel measurement vector t in n-space and assigns it to the reference class with the smallest angle. The arccosine of the dot product of the spectra is calculated; the algorithm computes α between t and r (Kruse et al. 1993):

$$\alpha = \cos^{-1}\left(\frac{\sum_{i=1}^{n} t_i r_i}{\left(\sum_{i=1}^{n} t_i^2\right)^{\frac{1}{2}}\left(\sum_{i=1}^{n} r_i^2\right)^{\frac{1}{2}}}\right) \tag{7.1}$$

where n in the number of channels.

As a demonstration of automatic endmember identification used in conjunction with the SAM algorithm, post-Hurricane Katrina remote vegetation damage is considered here. This method is also potentially applicable to other types of

Fig. 7.1 Isometric view of a hyperspectral AVIRIS image cube, New Orleans area, LA, USA, with bands 42 (739.35 nm), 30 (645.58 nm), and 18 (529.81 nm) as R, G, B for the top-plane image, converted to grayscale. The scene was acquired on 8 September 2005 with a pixel size of ~9.7 m. The image cube displays, in the z-dimension, 224 spectral channels on the 370–2500-nm wavelength interval

disaster-mapping. Hurricane Katrina made landfall on 29 August 2005, and Airborne Visible/Infrared Imaging Spectrometer (AVIRIS) (Green et al. 1998) image data were collected on 8 September 2005 over a portion of New Orleans, LA, USA. The hyperspectral AVIRIS entails 224 spectral channels (370–2500 nm), and the spatial resolution of the image analyzed was 9.7 m. Figure 7.1 shows an AVIRIS image cube visualization constructed from this data set. For this experiment, the overall scene was subset to the area given in Fig. 7.2. This area was partially

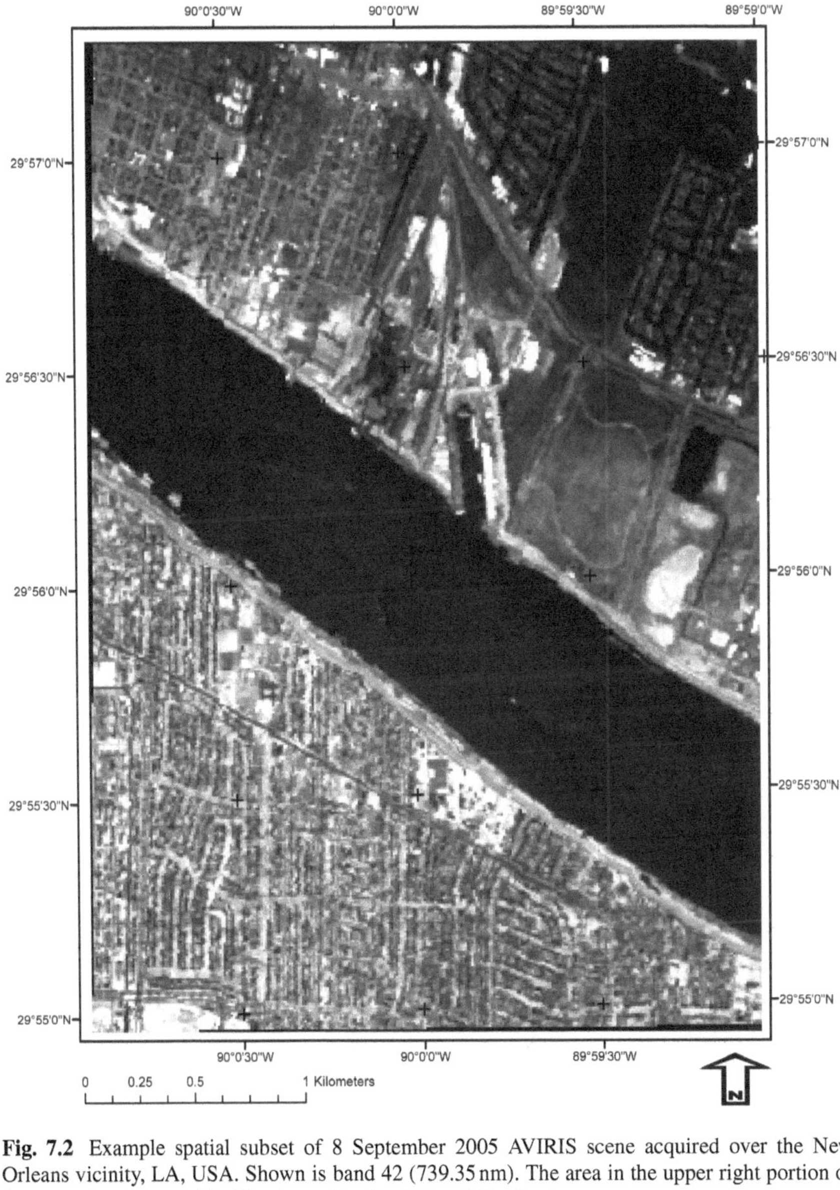

Fig. 7.2 Example spatial subset of 8 September 2005 AVIRIS scene acquired over the New Orleans vicinity, LA, USA. Shown is band 42 (739.35 nm). The area in the upper right portion of the scene is submerged by Hurricane Katrina-induced floodwaters

affected by Hurricane Katrina-induced floodwaters resulting from the breaching of the levees and floodwalls in the region (Esworthy et al. 2005); thus, this particular subscene includes both inundated and non-inundated areas. The image was atmospherically corrected to apparent surface reflectance using ACORN (vers. 4.15) (ImSpec LLC 2002). The SMACC algorithm (Gruninger et al. 2004) was then used to autonomously extract image endmembers, assuming the following parameters: number of endmembers = 15; RMS error tolerance = 0.01 (recommended starting value for reflectance data); unmixing constraint = sum to unity or less; and SAM coalesce value = 0.10 (to coalesce redundant endmembers). Example SMACC-derived endmembers for green turfgrass and potentially stressed/photosynthetically impaired/sparse turfgrass classes are given in Fig. 7.3. Note that there may be some soil background reflectance contribution to the potentially stressed/photosynthetically impaired/sparse turfgrass endmember spectral signature. Note further that these were not the only endmembers extracted by SMACC. These endmembers were then input to SAM; using a SAM threshold of 0.10 radians, a hard SAM classification was generated, shown in Fig. 7.4. Note that for vegetation damage-mapping, multitemporal remote-sensing change detection can be utilitarian (e.g., Rogan et al. 2002), but natural changes in vegetation spectral and structural properties (e.g. senescence associated with natural vegetation phenological cycles)

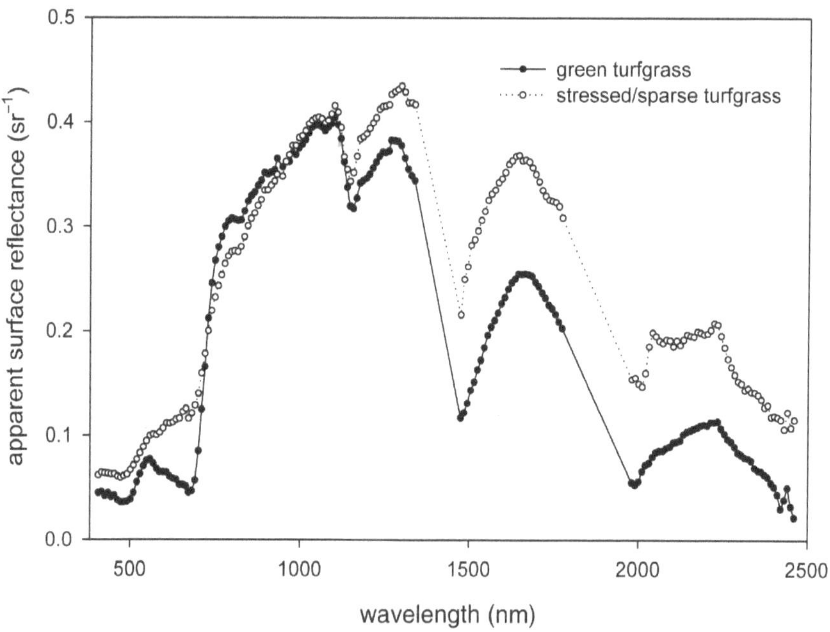

Fig. 7.3 SMACC-derived spectral endmember plot for (visually) green and potentially stressed/sparse turfgrass classes

Fig. 7.4 SMACC image endmember-based SAM classification including green turfgrass (shown in white) and potentially stressed/photosynthetically impaired/sparse turfgrass (shown in gray) classes. Areas in black are unclassified. Endmembers used as input to SAM are given in Fig. 7.3

should be accounted for when distinguishing between disaster-related and growth cycle-related vegetation changes (Jensen 2005).

While the identification of vegetation endmembers was demonstrated here, automated endmember extraction methods may also be invoked for urban materials, which may be more critical in a time-sensitive disaster scenario. However, urban areas do present some complications in terms of high within-class spectral diversity and spectral similarities among various urban material types (Herold et al. 2003), discussed below. Note that whereas various natural disasters can induce vegetation stress/damage, a nuclear/radiological disaster can also result in a vegetative spectral response detectable via remote surveillance, and can be monitored with gamma-ray or multi-/hyper-spectral spectrometers.

7.2.2 Nuclear/Radiological Event-Induced Vegetation Damage and Soil Contamination

Many of the concepts discussed above in the context of natural disaster-induced vegetation damage assessment also apply to man-made scenarios. For example, the effect of pollutants on vegetation spectral characteristics has previously been considered (Williams and Ashenden 1992). However, in the case of remotely assessing nuclear/radiological event-induced vegetation damage and soil contamination, some special considerations should be noted. Also note that remote assessment of nuclear/radiological contamination should be complemented with laboratory analyses of field samples.

7.2.2.1 Use of Gamma-Ray Spectrometric Sensors

In the event of a release of artificial radionuclides from a nuclear facility—either accidental or the result of a terrorist strike—gamma spectrometric sensors are of great utility in providing rapid, broad-scale mapping of nuclide-specific contamination. *In situ* and airborne gamma spectrometric sensors can be employed in surveying point releases from nuclear facilities, monitoring radioactive fallout over large regions, as well as for detecting lost or stolen radioactive sources (Aarnio et al. 1998; Winkelmann et al. 2004). However, as will be discussed, optical remote sensor data also represent an important information resource for associated assessments (Ridd et al. 2006).

In situ gamma-ray sensors can provide field data and can be mobile or fixed. Mobile *in situ* gamma spectrometers can be deployed for sampling at specific locations as required. Aarnio et al. (1998) described a mobile gamma spectrometer that can be operated on-foot or by car. Compared with controlled laboratory analyses, less-than- optimal gamma spectrometric field measurements may result from low-quality power, temperature fluctuations between measurements, and vibrations. Some systems (e.g., Aarnio et al. 1998) are designed to address such noisy field conditions. Fixed monitoring networks can continuously measure the gamma dose rate (Winkelmann et al. 2004). One example of a sensor network is the SensorNet Program, which is being designed and developed by Oak Ridge National Laboratory (ORNL) and partners. SensorNet involves more than sensor network, however. It is actually a 'vendor-neutral interoperability framework for Web-based discovery, access, control, integration, analysis, and visualization of online sensors, sensor-derived data repositories, and sensor-related processing capabilities' (Gorman et al. 2005). The core of SensorNet began with the concept of mounting wireless sensors on cell-phone towers throughout the United States, where sensors can notify command centers for emergency response. The current focus of the program is upon the development of flexible systems that operate in accordance with specific standards and that are tunable to accommodate local circumstances (Munger 2006). SensorNet includes radiological/nuclear sensors, but the program is more comprehensive in scope. The goal is to support a national system of systems for the detection, identification, and assessment of chemical, biological, radiological, nuclear, and explosive (CBRNE) threats in real-time (Gorman et al. 2005).

If a CBRNE event is detected, the modeling system will generate a real-time plume model, predict the impact on populations, and recommend actions (Kulesz 2004). SensorNet thus provides for real-time data collection and modeling, and it minimizes single-application or inflexible sensor network problems, as well as those associated with irreconcilable sensor standards (Gorman et al. 2005). Although chemical and biological detection and identification sensors/methods (e.g. Kwan et al. 2006) are only addressed very briefly in this chapter (see below), an overview of some approaches is given in Javidi (2006).

Another distributed sensor network oriented toward detecting CBRNE threats is the Department of Homeland Security 'Cell-All' program, which is currently in the research phase. The idea is to incorporate chemical, biological, and radiological sensors into cellphones; upon detection of a possible threat, spatio-temporal information would be transmitted to emergency responders. While there may be privacy concerns, such an approach would enable near-ubiquitous detection capability, which is useful since it would extend the coverage afforded by fixed sensor networks (Hall 2007a).

In addition to *in situ* operations, gamma-ray spectrometric sensor systems have also been operated remotely from airborne platforms, providing for time-efficient artificial radionuclide detection over broad areas (Winkelmann et al. 2004). For instance, NaI(Tl)-detector arrays have been installed in concert with germanium-semiconductor (HPGe) detectors on helicopters. A NaI(Tl)-spectrometer can yield high-resolution measurements, and a HPGe-detector may generate radionuclide-specific indentifications. If a nuclear event broadly disperses artificial radionuclides, NaI(Tl)-detectors are of some utility in gamma dose rate (GDR) determination. However, complex radioclide mixtures and excessively high counts may render NaI(Tl)-detectors unsuitable for assessing soil contamination type and degree. Higher flight altitudes can ameliorate such effects, and in this respect, HPGe-detectors are particularly advantageous (Winkelmann et al. 2004). Note that GDR measurements made via various methods (e.g. different dosimeters and flying heights) may need to be normalized (Brajnik et al. 1992). The lower detection limit for ^{137}Cs for a HPGe-detector at 100 m above-ground level (AGL) and a measurement time of 60 s is ~2 kBq/m^2. In a Black Forest experiment, GDR and ^{137}Cs-induced soil contamination were determined. GDR entailed a measurement error of ~40 percent. Only ~10 percent of the GDR in the study area was attributed to ^{137}Cs deposited following 1950s and 1960s atmospheric nuclear weapons testing, as well as after the 1986 Chernobyl accident. Uranium and thorium decay series radionuclides in the soil, as well as potassium, accounted for the rest of the contribution. A much higher ^{137}Cs contribution would exist in a disaster release scenario (Winkelmann et al. 2004).

Other airborne gamma-ray spectrometer systems have been discussed. The gamma spectrometer system detailed in Aarnio et al. (1998) can also be operated from an aerial platform. In order to accurately map radioactive fallout, a calibration process was developed to compensate for attenuation associated with a moving aircraft (or ground vehicle). Bresnahan (1998) demonstrated the use of an airborne gamma-ray spectrometer to detect possible hazardous waste units at the Savannah River Site (SRS); remotely sensed gamma spectra comprise a complex mixture of all isotopes present, natural or artificial. The presence of specific isotopes—in sufficient quantities—can be identified in energy spectra as diagnostic peaks; however, for a quantitative information-extraction, statistical and matrix methods are utilized to isolate each isotope (Bresnahan 1998). Schwarz et al. (1997), Billings and Hovgaard (1999), Hyvönen et al. (2003), and Sanderson et al. (2004) describe other airborne gamma-ray spectrometer systems and applications.

7.2.2.2 Use of Multispectral and Hyperspectral Radiometers

For environmental damage assessments associated with radiological disasters, useful information can also be extracted from multispectral and hyperspectral remotely sensed imagery. Importantly, such airborne- and spaceborne-derived data types are commonly available to the civil/scientific sector. If sufficient and appropriate data readily exist, this increases the likelihood that a rapid, broad-area assessment can be performed, assuming of course that such data are of utility for the application of interest. As will be explained, radionuclide contamination-induced vegetation stress can be discerned in reflectance spectra, which may be useful in remote radiological monitoring. Vegetation is a notable instrument of radioactive isotope cycling and thus represents a utilitarian gauge of radionuclide contamination (Gaso et al. 1998; Boyd et al. 2006). In the event of a radiological disaster, such assessments are of critical importance in terms of studying human exposure and ecosystem health, and for global food security, which was considered in the wake of the Chernobyl nuclear accident (United States, Congress, Senate, Committee on Agriculture, Nutrition, and Forestry 1986; United States, Congress, Joint Economic Committee, Subcommittee on Agriculture and Transportation 1987). Assessment and monitoring of such contamination is also important in terms of analyzing potential secondary contamination from, for example, fire events (e.g. forest or grassland fires), which can transport contaminants via ash and smoke (Abduragimov and Odnolko 1993).

On 26 April 1986, the Chernobyl Nuclear Power Plant (CNPP) (Ukraine) sustained a reactor failure (reactor number 4), with no containment structure surrounding it. A reactor core explosion decimated part of the reactor building and ignited graphite in the core. The graphite fire was exposed to open air, as there was no containment structure, resulting in the release of a large amount of radioactive material into the troposphere, as well as deposition near the CNPP and across the northern hemisphere (Sadowski and Covington 1987; Davids and Tyler 2003). In an early study of the CNPP nuclear accident, Sadowski and Covington (1987) analyzed Landsat-5 TM and SPOT HRV data, which included pre- and post-disaster dates. They found that TM data from 29 April 1986 contained moderate and high brightness values (BVs) for the damaged CNPP in TM bands 5 and 7, respectively, associated with emitted/thermal radiation from the burning graphite. This is consistent with the utility of TM bands 5 and 7 for high-temperature target detection in volcanic and wildfire monitoring. TM band 6 was found to exhibit no high-temperature response to the damaged CNPP, likely due to the limited radiometric range and spatial resolution of this band. This is consistent with Givri (1995), who also noted a problem with the single TM thermal band for thermal emission monitoring of the CNPP accident. However, band 6 could detect water-temperature gradients in the adjacent cooling pond in the pre- and post-disaster images. An interpretation of the TM imagery also enabled detection of the blast effect and possible debris (Sadowski and Covington 1987).

Using data acquired in 2001 in the Chernobyl exclusion zone, Davids and Tyler (2003) found that radionuclide contamination could be detected using measured spectral reflectance of vegetation. The deposition within the exclusion zone was dominated by radiologically significant quantities of ^{90}Sr, ^{137}Cs, $^{238-241}Pu$, and ^{241}Am, with especially high levels of ^{137}Cs (e.g., exceeding 30 MBq m^{-2} within 10 km of the CNPP) (Skuterud et al. 1994; Davids and Tyler 2003). Radiogenic output of ^{137}Cs was >85 PBq (half-life = 30.2 years), and it also dominated radionuclide deposition in Belarus, upon which 70% of the fallout impinged (Soukhova et al. 2003; Boyd et al. 2006). Davids and Tyler (2003) specifically used *in situ* and laboratory spectroscopy to detect radionuclide contamination-induced vegetation stress in two forest species—silver birch (*Betula pendula*) and Scots pine (*Pinus sylvestris*). Scots pine exhibited a high mortality rate following the Chernobyl event; in the time interval since the disaster, pine in some areas has been replaced by silver birch, which is less radiosensitive (Skuterud et al. 1994). Vegetation stress-detection methods include monitoring photosynthetic rate—i.e., respiration and transpiration—and measuring relative and absolute pigment concentrations in plant leaves, as plant stress and mortality result from some condition or substance adversely affecting plant metabolism, development, or growth. Change in relative and absolute pigment concentrations can translate into shape alterations in reflectance spectra (Lichtenthaler 1996).

Within the CNPP exclusion zone, radionuclide (^{90}Sr and ^{137}Cs) concentrations in pine needles and birch leaves were closely related to soil contamination and γ-dose rates. For instance, for birch, significant negative correlations were found between the chlorophyll red edge and the natural logarithmic transformations of ^{90}Sr and ^{137}Cs concentrations in the leaves, ^{137}Cs concentration in the soil, and γ-dose rates, where $r = -0.88$, -0.76, -0.92, and -0.90, respectively. The Three Channel Vegetation Index (TCHVI) similarly proved useful. While radionuclide contamination resulted in discernable effects in the reflectance spectra of silver birch and Scots pine, the spectral responses were different. In birch leaves, radiological contamination was strongly related to chlorophyll *a* concentration; however, for Scots pine, the effect was less clear, with differences between laboratory- and *in situ*-based results. The relative concentrations of various leaf pigments affected the laboratory spectra, whereas biomass (which is expected to decline with rising contamination) was an important determinate of spectral reflectance with the *in situ* data (Davids and Tyler 2003). The varied spectral response as a function of vegetation type/species may suggest the importance of an initial vegetation classification in an airborne- or spaceborne-based remote-sensing analysis, and mapping techniques such as those discussed above (e.g. LSMA, SAM, etc.) could be utilized.

Boyd et al. (2006) also analyzed hyperspectral data (acquired with a field spectroradiometer) and considered forests of Belarus, which experienced much lower levels of contamination (principally from fission products with half-life ≈ 30 years) relative to the CNPP exclusion zone. These areas could thus return to use (e.g., timber and fuel wood production) within the next several decades. The highest spectral reflectances, particularly in the visible and near infrared, were in old-growth forest areas with the highest mean ^{137}Cs-specific activities, and this spectral condition suggests plant stress. The authors removed the continuum, discussed below, from laboratory-based hyperspectral data and found that older forests (*c.* 35 years) exhibited significant differences between sites with low- and high-level contamination at wavelengths casually associated with foliar biochemicals. (Six continuum-removed absorption features were distinguished at approximately 470 nm, 670 nm, 920 nm, 1200nm, 1360nm and 1840 nm). For younger forests, there were no significant differences. Thus, contamination level may be able to be inferred from vegetation spectra, partitioned by age (Boyd et al. 2006).

Continuum removal, a technique employed in Boyd et al. (2006) for hyperspectral analysis, is of utility in a variety of application domains (e.g. vegetation (Filippi and Jensen 2007), geology (Van der Meer 2004), etc.). Continuum removal facilitates analysis of individual (diagnostic and other) spectral absorption features (Clark et al. 1990). Spectral absorption features are comprised of the continuum and individual features, where individual absorption features are superimposed onto a ‘background absorption’ (i.e. the continuum). The depth of a given absorption feature is related to the abundance of the absorbing material. For a given spectrum, local maxima are fitted with a continuum of straight-line segments. This continuum is divided into the reflectance spectrum, normalizing absorption bands to a common reflectance (Clark and Roush 1984; Kruse 1990).

For the New Orleans hyperspectral AVIRIS image introduced above, the *continuum-removed*, SMACC-derived vegetation endmembers that were extracted from the AVIRIS subscene are given in Fig. 7.5; the spectra in Fig. 7.3 served as the input to the continuum-removal algorithm. This figure demonstrates continuum-removal over the majority of the AVIRIS wavelength range (sans UV, noise-dominated, and spectral overlap bands). However, it has been reported that continuum-removed hyperspectral data may not provide optimal vegetation classification accuracies, relative to continuum-intact vegetation spectra, if the continuum is removed over a wide spectral wavelength range (Filippi and Jensen 2007). Therefore, it is expected that continuum removal will be most beneficial when applied only to vegetation spectra in diagnostic wavelength regions (e.g. Kokaly et al. 1994; Clark et al. 2003).

For certain purposes, extensions of Davids and Tyler (2003) and Boyd et al. (2006) would be of interest. From a food security standpoint, remote-sensing studies similar to these experiments, but specific to agricultural crops, would be useful.

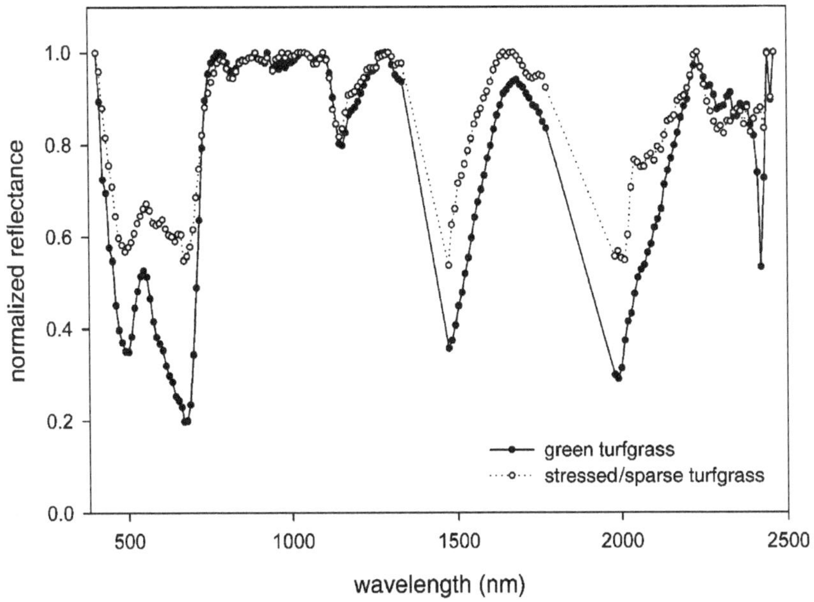

Fig. 7.5 AVIRIS image-derived endmembers (from Fig. 7.3) with the continuum removed

The effects of radiological contamination on crops depend on a variety of variables, including the dosage, type of isotope contamination, stage of crop phenological cycle, crop canopy cover, soil type and quality, rainfall, topography, and drainage/run-off characteristics. Short-lived isotopes (e.g. ^{131}I and ^{140}Ba) dissipate on the order of days; however, long-lasting isotopes, such as ^{90}Sr and ^{137}Cs, are resident after decades (United States, Congress, Senate, Committee on Agriculture, Nutrition, and Forestry 1986).

Davids and Tyler (2003) and Boyd et al. (2006) were field- and/or laboratory-based studies, and further research is needed from spaceborne and airborne platforms. Remote sensing can also be utilized as part of a modeling framework for radiological studies. For instance, Hasager and Thykier-Nielsen (2001) used IRS-1C LISS III satellite data to extract land cover classes, and the aerodynamic roughness for each class was assigned, which was used as input to a real-time radiological deposition model. As for airborne remote sensing, note that Unmanned Aerial Vehicles (UAVs) could be used for flight in extreme weather conditions and can be flown over areas of high radiological contamination without exposing personnel to dangerous radiation levels (Grasty et al. 1997; Skyttner 2002). Skyttner (2002) proposed the use of remote senor-equipped UAVs for high-resolution data-acquisition as part of an integrated system for monitoring and early warning, with example application to the Chernobyl Nuclear Site. The existing sarcophagus encasing the reactor is not well-sealed, is unable to prevent ground water contamination, and is at risk of roof collapse. Frequent/continual monitoring of the sarcophagus and surrounding plant area is necessary until these issues are addressed. The spatial resolution of civilian satellite imagery was deemed unacceptable for this application, which explains the utility of the UAV-acquired imagery (Skyttner 2002).

Given that UAVs are of utility for various types of damage assessment (including bomb/battle damage), in addition to their anti-surface warfare roles, multiple UAVs may be required to serve this function, particularly in a time-sensitive, disaster-response environment. Smith et al. (2001) gives a possible solution to the UAV availability problem (i.e. the performance of a configuration of UAVs). Such considerations are important in terms of optimizing resources for efficient and effective remote damage assessment.

7.2.2.3 Fluorescence-Based Detection

Other remote data-collection methods can be employed for detecting radiological materials. For example, Time-Resolved Laser-Induced Fluorescence (TRLIF) has been used for uranium determination in an environmental context (i.e. in the analysis of vegetation, soil, and water), as well as in other application domains. TRLIF entails laser excitation and subsequent temporal resolution of the uranium fluorescence signal, which excludes short-lifetime fluorescence. TRLIF thus exploits the four characteristic uranium fluorescence peaks at around 494, 516, 540, and 565 nm (Moulin et al. 1993).

This section has presented a discussion of various remote sensing-based environmental/biophysical damage and contamination assessment methods and considerations. It indicates the range of potential environmental disaster scenarios and corresponding remote-monitoring solutions.

7.3 Remote-Sensing Damage Assessment of Human Settlements

7.3.1 Estimation of Disaster-Affected Human Populations

An important dimension of disaster assessment in populated areas involves estimating the population affected by a given natural or technological event. In the event of a disaster, a first priority is typically to assess and limit human injuries and fatalities; a second priority is often to assess structural damage and recovery (Adams et al. 2007). The former priority will be discussed briefly in this subsection, whereas the latter priority will be examined in subsequent subsections. Note that such assessments can potentially be accomplished—at least in part—via remote-sensing, though a more robust and accurate method likely entails the joint utilization of GIS-based modeling.

In order to estimate the population affected by a given disaster, the compilation of a population database is necessary, and remote sensing can be an important tool in this endeavor. Regarding the problem of remote sensing-based population estimation, various approaches have traditionally been followed. For example, human population has been estimated at the local level by counting individual dwelling units (e.g. Forster 1985; Haack et al. 1997). This method assumes: 1) sufficient spatial resolution for detecting individual structures and the ability to discern the residential, commercial, or industrial character of particular structures; 2) available approximations of the average number of persons per structure; 3) an estimation of the number of seasonal/migratory workers and homeless individuals; 4) all dwelling units are occupied, with a specified number of individuals per unit, calibrated with *in situ* data. The individual dwelling unit approach is, however, very costly, time-consuming, and impractical if a regional population base is to be compiled (Jensen 2007). Depending on objectives, other approaches may be more appropriate. Lo (2006) and Liu and Herold (2007) provide an extensive overview of various remote-sensing methods for estimating population and census data. Note that a relationship has been observed between urbanized areas and population magnitude:

$$r = a \times P^{b}, \tag{7.2}$$

where r is the radius of the urbanized/populated area; a is an empirically derived constant of proportionality, which varies regionally; P is the population; and b is an empirically derived exponent (Tobler 1969). However, more input data are necessary to accrue a more realistic assessment of ambient population density, a temporally averaged measure of population density that accounts for human mobility (Dobson et al. 2000; Sutton et al. 2003). Furthermore, an approach that employs land cover as a variable may be more accurate than some conventional methods, as land cover class is one of the best indicators of human population density. The LandScan project, developed at ORNL, is a significant example of a methodology that employs such information (Dobson et al. 2000). Specifically, global population at a grid cell size of 1 km is estimated primarily based on land cover, transportation network proximity, topography (slope), and nighttime lights, derived from Defense Meteorological Satellite Program (DMSP) imagery (Dobson et al. 2000; Bhaduri et al. 2002). LandScan is being refined at ORNL to yield LandScan USA, a very high-resolution database, where cell size = 90 m, that includes both nighttime/residential and daytime distributions, which is important given the temporal dimension of potential disasters. LandScan or LandScan USA population data can be intersected with measured or modeled disaster maps (e.g. for contaminant plumes) to determine the threatened population (Bhaduri et al. 2002).

While remote assessment of urban contamination and damage from a nuclear or radiological source is considered below, note here that a population database (e.g., LandScan USA) could be used to refine previous nuclear weapons-based blast casualty rate models (e.g. Hadjipavlou and Carr-Hill 1989). Some previous casualty rate models were based primarily on data from the Hiroshima and Nagasaki bombings (e.g. Defense Civil Preparedness Agency 1973), which were later employed to estimate casualties in American cities following a nuclear attack (Office of Technology Assessment 1980).

Hadjipavlou and Carr-Hill (1989) presented a revised nuclear blast casualty rate model, which can be adapted to a remote-sensing context. They note several variables that could affect blast-induced casualty rates from a proximal explosion, including the shock strength and duration; building environment type (which is important in terms of protection and debris); and human location and posture, among others. Primary blast effects, derived directly from the shock on the body, are primarily important at overpressures where the majority of persons would already be killed due to other effects. The problem then is to assess debris (secondary) and bodily displacement (tertiary, from the blast wind) hazards. Buildings are amalgamations of various materials, and different constituents fail at different overpressures. Building materials and the construction method employed are thus important in estimating failure type and associated hazards to human populations. Assuming statistically independent factors, the overall casualty rate provided by Hadjipavlou and Carr-Hill (1989) is:

$$\Omega\left(A, \Delta P_0, \ldots W, \ldots\right) = 1 - \prod_i \left[1 - \Omega_i\left(A, \Delta P_0, W, \ldots\right)\right], \tag{7.3}$$

where $\Omega_i\,(\cdot)$ is the casualty rate from the ith process, A is the building type, ΔP_0 is the peak overpressure, and W is yield; other important variables are possible, however. The Hadjipavlou and Carr-Hill method (1989) utilizes both models and casualty data. Given the relationship between the injury rate I and κ, where $\kappa = \ln[m/(1-m)]$ and m is mortality rate, the empirical data on severe or serious injuries could be bounded by

$$I(\kappa) = 0.27 \exp\left[-\left(\frac{\kappa + 2}{3.5}\right)^2\right], \tag{7.4}$$

Note that their causality predictions were applicable to the period immediately following a nuclear blast (i.e., within 1–2 days), and further note that the largest proportion of injured would likely be expected at lower overpressures.

A variety of parameters should actually be considered, including weapon yield (to determine blast wave duration); building geometry; distribution of building/house sizes; the distribution of building orientations relative to an approaching blast wave; window distribution, etc. (Hadjipavlou and Carr-Hill 1989). Brode (1968) provided a model for an idealized blast wave; however, real blast waves are modulated as function of the geometry of the surface over which they pass, as well as atmospheric temperature or density gradients. Urban landscape morphology thus distorts the form of the blast wave and related forces, at least on a localized basis (Hadjipavlou and Carr-Hill 1989). Remote sensing can efficiently provide much of this information (e.g., urban materials classifications, building height/geometry/distribution/orientation, topography, temperature data, etc.) (Hepner et al. 1998; Herold et al. 2003; Benediktsson et al. 2003; Miura and Midorikawa 2006; Jensen 2007). In addition, as noted above, population estimates from sources such as LandScan could also be used to enhance blast casualty estimates. Radioactive fallout from a nuclear weapon or a radiological dispersion device (RDD) should also be considered in order to accrue a more complete prediction/assessment of affected human populations (Dombroski and Fischbeck 2006).

As evidenced by GIS-based modeling approaches that assimilate disparate information sources, such as with LandScan (Dobson et al. 2000), remote sensing-derived information can be utilized as part of GIS-based, geographic information-processing framework. Now that human populations have been considered, the discussion below will be constrained to primarily remote-sensing approaches to assessing structural damage to the built environment and contamination from various sources.

7.3.2 Structural Damage Assessment of Human Settlements

This section briefly discusses remote-sensing damage assessment of buildings/structures, or populated places in general, regardless of whether the disaster is natural or anthropogenic in origin (e.g. earthquake- or bomb-induced). Building damage assessment is often conducted *in situ*, especially when interior structural damage is involved (e.g. the 1993 World Trade Center bombing) (Ramabhushanam and Lynch 1994). Osteraas (2006) provides an *in situ*-based blast damage assessment of the bombing of the Murrah Federal Building in Oklahoma City, USA. However, through-wall radar imaging could be used to image targets through building walls, which may aid in post-disaster interior structural damage assessment and in search-and-rescue (Aryanfar and Sarabandi 2004; Song et al. 2005; Yang and Fathy 2005; Griffiths and Baker 2006).

Assuming the more probable circumstance—at least currently—of the absence of building-penetrating technology for a given disaster event, remote sensing is of particular utility when the building damage is outwardly apparent. For example, high-resolution, pre- and post-Hurricane Katrina IKONOS satellite images collected over New Orleans, LA are given in Fig. 7.6, where hurricane-induced damage to individual structures and flooded areas are visually apparent in the post-hurricane image. Additionally, proximal (close-range) remote sensing—in the form of digital video (DV) or *in situ* spectroradiometer data acquisition—can be used in concert with *in situ* mobile GIS data collection, as well as high-spatial resolution airborne or spaceborne images, for the management of urban disasters (Montoya 2003).

Remotely sensed images for disaster-monitoring can be manually/visually interpreted or analyzed via computer algorithms. Regarding manual photographic interpretation, Haack et al. (1997) noted that large-scale color infrared (CIR)

Fig. 7.6 Pre- and post-Hurricane Katrina IKONOS satellite images of a portion of New Orleans, LA, USA, acquired on 28 August 2002 (left) and 2 September 2005 (right), respectively (true-color, converted to grayscale). Building damage and flooded areas are readily apparent in the post-hurricane image on the right. Such outwardly apparent structural damage can be visually interpreted by a human analyst. (IKONOS satellite image courtesy of GeoEye. Copyright 2007. All rights reserved.)

photography is typically preferred over other photographic data for disaster assessment in urban areas. They further note the recommended CIR image scales for various urban disaster-assessment tasks, including determining the number and type of buildings damaged, damage to utilities, and evaluation of access routes to the damaged area and evacuation routes. Information extracted from high-spatial resolution optical image data can vary as a function of image collection time and viewing angle. Shadows, solar illumination variability and geometric errors can minimize the efficacy of automated damage-assessment algorithms (Stramondo et al. 2006). Due to such problems, manual/visual image interpretation is still one of the most commonly used methods of remote damage assessment (Saito et al. 2004; Yamazaki et al. 2004). In terms of digital image-processing, Turker and San (2004) simulated the effect of shadows for airborne optical images. Automated image analysis/feature recognition algorithms (e.g. Levner et al. 2003) would ideally be utilized in order to rapidly process large-volume digital image data streams (Hausamann et al. 2005). An automated method (nonlinear mapping based on the Coincident Enhancement principle) and visual inspection, applied to high-resolution QuickBird image data, were compared in Sakamoto et al. (2004). For the detection of collapsed buildings, pre- and post-earthquake images were utilized; one image was mapped to the other, where the images were related by deformation, yielding a false alarm (over-detection) rate of 20 percent. Matsuoka et al. (2004) discussed damage detection via edge analysis using high-spatial resolution imagery.

The type and characteristics of data acquired in an urban context should be considered. Jensen (2007) notes recommended temporal, spectral, and spatial requirements for various urban remote-sensing tasks. For emergency scenarios, pre-disaster images should be acquired every 1–5 years, with a spatial resolution of 1–5 m and spectral coverage in the panchromatic, visible, and near-infrared (NIR). For post-disaster imagery, panchromatic or NIR high-resolution (≤ 0.25–2 m) data should be collected within 12 hours-2 days (Schweitzer and McLeod 1997), or possibly sooner depending on the nature of the emergency. In the event of significant cloud cover, radar data could be acquired to provide useful disaster information. For post-disaster quantitative damage assessment involving buildings/housing, transportation, and utilities, panchromatic and NIR imagery at a spatial resolution of 0.25–1 m should generally be obtained with 1–2 days (Jensen 2007). Adams et al. (2007) also notes that high-spatial resolution imagery is very useful for detecting damage to individual structures.

As with vegetation and other natural materials, if imaging spectrometer data are acquired over urban targets, hyperspectral algorithms may yield image classifications at a higher level of specificity relative to multispectral classifications (Shippert 2004; Filippi and Jensen 2006). Hyperspectral mapping algorithms, such as those discussed above or others, have been applied to urban scenes. For instance, Bhaskaran et al. (2004) employed airborne hyperspectral image and GIS data for addressing hailstorm damage and the dynamic allocation of resources to vulnerable areas. A spectral library for urban materials was compiled, and the SAM algorithm was used for automatically generating a classification of roofing materials with differing resistances to hailstones. Gomez (2002) discussed the use of hyperspectral imaging for transportation applications, including evaluating road characteristics, impervious surface classification, and detecting and identifying vehicles. Ben-Dor et al. (2001) demonstrated that various urban materials could be successfully mapped using Compact Airborne Spectral Imager (CASI) hyperspectral airborne data in the visible and near-infrared wavelength region. It should be noted that there is a distinction between mapping urban materials and urban land cover and land use; Herold et al. (2007) discusses this issue, as well as other fundamental issues related to urban applications of imaging spectroscopy. In another hyperspectral urban application, Hepner et al. (1998) integrated interferometric synthetic aperture radar and AVIRIS data to generate 3-D surface feature geometry and topography in an urban area.

Urban areas exhibit high spatial complexity, and spectral mixing typically occurs across all spatial scales. Urbanized environments entail high spectral heterogeneity

(Ben-Dor et al. 2001; Herold et al. 2003; Herold et al. 2007); thus, even high-dimensional, high-spatial resolution remote sensor data sets may not necessarily lead to accurate urban cover (or damage) classifications (Herold et al. 2003). Remote-sensing urban information classes of interest often entail high standard deviations and violate the assumption of a multivariate normal distribution (Myint et al. 2004). In addition, characterizing pre-disaster or post-disaster damaged urbanized areas using solely pixel-based approaches can be problematic, particularly when using high spatial-resolution imagery. For example, per-pixel classifiers can produce structural clutter (i.e. a salt-and-pepper effect) in the classified image (Van de Voorde et al. 2007).

Given the high spectral diversity of materials and the high spatial frequency in urban areas (Gamba et al. 2003; Herold et al. 2003), regardless of whether low- or high-dimensional image data are employed, a region-based approach may complement spectral-oriented processing for urban classification, particularly if the objective is rapid damage region-mapping (Shackelford and Davis 2003; Hall et al. 2006). Joint exploitation of spectral and spatial information has previously been investigated. For example, Lehmann et al. (1998) fused high-spectral and high-spatial resolution data for improved urban monitoring, where 126-band HyMap data (3-m spatial resolution) and HRSC-A data (12-cm spatial resolution) were used.

The strong spatial-spectral variability associated with Very High Resolution (VHR) urban image data is often addressed via either image segmentation (applied prior to any classification step) or post-classification processing (applied to a classified image) (Van de Voorde et al. 2007).

Damage region-delineation may be conducted via image segmentation (e.g. Benz et al. 2004; Benediktsson et al. 2003). As the spatial resolution of remote sensors increases, finer landscape objects can be resolved. In such a case, within-object spectral homogeneity decreases, as does the spectral separability between 'image-objects' (Blaschke et al. 2004). A variety of definitions exist for image-objects, but they can basically be seen as contiguous areas within an image (Benz et al. 2004), or 'groups of pixels with a meaning in the real world' (Schneider and Steinwender 1999). The human visual system recognizes objects, rather than individual image pixels. Spatial relationships are additionally accounted for. Recent research has aimed to automate and improve upon human image-interpretation processes (Hausamann et al. 2005), though image segmentation does facilitate the integration of expert knowledge. Based on a homogeneity criterion, a matrix of image values is partitioned (Blaschke et al. 2004).

Segmentation approaches are often classified as pixel-, edge-, or region-based segmentation algorithms. Pixel-based approaches include feature space-thresholding and segmentation. The objective with edge-based algorithms is to locate edges between regions (i.e. boundaries separating image-objects). For such an approach, edge detection is typically employed, where high-pass filtering may constitute an intermediate step (Blaschke et al. 2004). An example of a high-pass filtering-based change-detection approach is given in Adams et al. (2007), where pre- and post-earthquake pan-sharpened QuickBird satellite imagery was acquired, and a 9×9-pixel Laplacian edge detection filter was applied, followed by a 25×25-pixel dissimilarity texture measure. The output images were differenced, and the mean standard deviation was plotted within a 200×200-pixel window. This method was able to detect urban damage in areas with distinctive characteristics of severe structural damage that was visually apparent in the high-resolution imagery: collapsed apartment blocks were identified by zones of debris and rubble piles. Additionally, shape and position alterations were apparent where buildings had 'pancaked' or collapsed sideways (Adams et al. 2007). Finally, region-based procedures consist of region-growing, merging and splitting approaches, and combined methods (Blaschke et al. 2004). While edge-detection and region-growing image segmentation algorithms are widely recognized, Benediktsson et al. (2003) proposed a method for delineating urban image features based on morphological intrinsic characteristics, rather than their boundaries.

A commonly used image-segmentation method has been implemented in the eCognition® software, produced by Definiens (Benz and Schreier 2001; Benz et al. 2004). In this object-oriented approach, a hierarchy of objects is extracted for multi-scale object representation, and objects are described in terms of their neighborhood relations, color information, and size and shape (Hausamann et al. 2005). Rule bases are employed to assign reliability weights via fuzzy logic, and the defined rules can be applied to the image-objects. Expert knowledge can serve as collateral information for segmentation validation; thus, the system facilitates human intervention/interaction (Benz et al. 2004; Hausamann et al. 2005).

Image segmentation can conceivably be applied to a wide variety of damage-detection scenarios. For example, for monitoring gas pipelines for potential damage, Hausamann et al. (2005) discussed the use of UAV-based optical and synthetic aperture radar (SAR) data collection, in conjunction with eCognition (Benz and Schreier 2001) for automated feature recognition. In this case, assuming the collection of multi-temporal remote sensor data, any detected changes to the pipeline can be forwarded to an alarm system (Hausamann et al. 2005). Image segmentation/object-based classification can also be applied to biophysically-orientated damage assessment, such as tree mortality, crop damage, etc. (Blaschke et al. 2004; Guo et al. 2007). Such a methodology could complement previously discussed approaches to vegetation analysis. Object recognition can also be accomplished via an approach such as that described in Bernstein and Di Gesù (1999); their methodology employs principal component analysis (PCA), mathematical morphology operators, and classification of detected objects. The approach can be applied to the problem of post-storm damage-assessment.

Post-classification processing is an alternative method for dealing with Very High Resolution (VHR) urban per-pixel classifications, which may also aid in damage-assessment tasks. For instance, Van de Voorde et al. (2007) proposed neural network-based shadow-removal, knowledge-based rules for classification enhancement, and a region-based filter for minimizing high-frequency structural clutter.

A variety of remote-sensing studies—often based on change detection—has been conducted to detect earthquake damage, though such methods may be applicable to other types of urban disaster assessment tasks (e.g. involving building/infrastructure structural damage, etc.). Tucker and San (2003) used SPOT multispectral and panchromatic data to detect earthquake-related damage in Turkey in 1999. Pre- and post-event images were geometrically and radiometrically calibrated to each other, and the multispectral and panchromatic data were merged. Change was detected by subtracting Band 3 (near-infrared) of the merged pre-earthquake image from that of the post-event image, resulting in an overall accuracy of 83 percent.

Since the pre- and post-event images were acquired one month apart, changes in vegetation growth conditions were separated from earthquake-induced changes via use of the NDVI. To detect damaged building blocks, vector polygons of building blocks were then overlaid upon reference orthophotographs and the change image. The percentage of change pixels within each polygon was computed and then reclassified in accordance with a damage-classification scheme: no damage, low damage, high damage, and fully damaged. In severe earthquake-damaged areas, a dramatic difference in texture and tone was observed between the pre- and post-earthquake images. Collapsed buildings could thus be detected and classified; however, the change detection method used did not provide information regarding the nature of the change. Additionally, vertically-collapsed apartment buildings are generally not detected via this method unless there is apparent damage to the roofs (Tucker and San 2003).

Also for earthquake damage assessment, Miura and Midorikawa (2006) developed a methodology for updating GIS building inventory data through the use of high-spatial resolution satellite remote sensor image data. The distribution of mid- and high-rise buildings (greater than four stories) is obtained through the extraction of building edges (via a Canny edge-detector), shadows, and vegetated areas, whereas low-rise building distribution is assessed using land cover information. Kumar et al. (2006) used Resourcesat-1/LISS-IV and Cartosat-1 stereoscopic data for earthquake damage assessment. As another example, Kaya et al. (2005) compared government statistics of collapsed buildings with SPOT HRVIR-derived estimates, where the SPOT multispectral data-processing involved ISODATA classification of pre- and post-earthquake images (which included a collapsed building class), and SPOT panchromatic imagery was density-sliced to delineate collapsed-building areas, as they exhibited high reflectance. In an attempt to discriminate between change caused by structural damage and change due to other factors, a multi-temporal deterministic approach, based on principal components analysis, normalization, and threshold-setting, was proposed in Rejaie and Shinozuka (2004). In Andrienko et al. (2003), potential earthquake damage was evaluated using computational tools for grid data and visualization-based interactive data exploration. Herring (2007) described the Visualizing Impacts of Earthquakes with Satellites (VIEWS) system, which has been used for post-earthquake satellite image analysis. VIEWS can be used in concert with a digital camera or digital video recorder for vehicle- or on-foot-based field data collection.

Change detection-based damage assessment using 2.5-D/3-D building models can also be conducted, and such models can be constructed via a variety of methods. For instance, Vosselman and Dijkman (2001) used airborne laser altimetry and ground plans for 3-D building model reconstruction, where a 3-D Hough transform was employed to extract planar surfaces from the point data clouds.

Digital photogrammetry can also be used in the generation of urban digital surface models (DSMs). Conventional stereo image-matching is used to determine corresponding pixels in overlapping images, and the resultant surface model entails elevations based on just the visible surfaces of building tops or vegetation (Lu et al. 2006; Jensen 2007). Retrieving a bare earth DEM may or may not be useful for a given damage-assessment task, though Lu et al. (2006) proposed use of Dempster-Shafer theory to extract building areas in overlapping aircraft- or satellite-based images to reconstruct bare earth terrain elevations. This approach does not require the use of additional data (e.g., GIS-based or terrain laser scanner data). Paul et al. (2006) used a semi-automated feature-extraction method where only one QuickBird satellite image, acquired at a viewing angle of 24.5°, was utilized in generating a digital city model (DCM) with a building height accuracy of 1–2 m. While LiDAR data can also be used for such a purpose, this approach was motivated by the fact that LiDAR data are not readily available on a global basis and that LiDAR-processing requires specialized skills. Interferometric synthetic aperture radar (InSAR) can also be used for building characterization, consisting of identification and height estimation (Guillaso et al. 2005). To produce urban elevation maps, which for damage assessment would be particularly useful if generated both pre- and post-disaster, Perissin and Rocca (2006) employed a permanent scatters (PS) method that makes use of a long series of SAR data, which is typically used for monitoring ground deformations with millimeter accuracy. And as noted above, InSAR and hyperspectral data can be combined to produce urban 3-D surface feature geometry and topography (Hepner et al. 1998). SAR is discussed further below in the context of damage assessment.

Whereas remote sensing-based methods of damage assessment for homeland security applications constitute the focus of this chapter, the concept of battle damage assessment (BDA), which applies to military targets, may also be considered here. As with some of the earthquake remote damage-assessment methods, some BDA methods may, at least in some circumstances, be applied to the assessment of urban/building damage arising from a variety causes. Furthermore, it is conceivable that the traditional context of BDA may be applicable to a homeland security-oriented problem. The term 'battle damage assessment' has been regularly used since 1992, upon the standardization of the term and its definition by an Intelligence Community Battle Damage Assessment Working Group (Graffis 1997). This Defense Intelligence Agency (DIA) definition is as follows:

> BDA is the timely and accurate estimate of damage resulting from the application of military force, either lethal or non-lethal, against a pre-determined objective. BDA can be applied to the employment of all types of weapon systems (air, ground, naval, and special forces) throughout the range of military operations. BDA is primarily an intelligence responsibility with required inputs and coordination from the operators. BDA is composed of physical damage assessment, functional damage assessment, and target system assessment (Defense Intelligence Agency 1996; Graffis 1997).

Before 1992, the most similar Department of Defense-wide term was 'bomb damage assessment'; however, the definition only referred to air-based attacks. Battle damage assessment is basically an evaluation of the level of success of an attack on an enemy target. BDA can guide the commander regarding optimal force usage, and it can provide feedback to the soldier regarding his or her efficacy against the target. Note that (at least as of the recent past) the US Army, Air Force, and the joint community have not been agreement with respect to the definition, characteristics, and the customer of BDA (Graffis 1997).

BDA can be conducted via *in situ* data collection by ground teams, though this is often not a practical option, necessitating the employment of other options. Aerial photography has traditionally been used. However, photography-based BDA can encounter spatial resolution problems associated with film and camera performance, and haze, clouds, and smoke can obscure the image target. Thermal infrared sensors, such as Forward Looking Infra-Red (FLIR), can penetrate many atmospheric aerosols better than visible sensors due to the longer wavelengths relative to the size of the aerosol particulates (Kopp 1995; Jensen 2007). Also, Kopp (1995) discussed the use of optical laser remote sensing/LiDAR for bomb damage assessment and reconnaissance. Roper (2005) summarized various geospatial technologies that were applied to the World Trade Center (WTC) ground zero recovery effort, including LiDAR and thermal imaging for hot-spot detection. Of course, many other data sources and techniques for BDA are possible. Rauch (2004) gives an historical overview of BDA.

For a given BDA task, the allocation of data-collection resources may be of concern. Smith et al. (2001) was mentioned above with respect to the problem of determining the optimal UAV configuration, and MacKenzie (2003) used collaborative tasking to automatically task and re-task heterogeneous UAVs. In addition, the joint allocation of long-range weapons and BDA sensors was discussed in Yost and Washburn (2000).

In a battlefield environment, another means of conducting BDA is through the use of a sensor projectile dispensed from a dropped munition near the target, where the projectile is equipped with a camera that enables real-time image acquisition pre- and post-impact. This approach is advantageous relative to satellite or high-altitude suborbital remote sensing, which can be significantly affected by adverse weather conditions and temporal lag between the event and image collection (Frost and Costello 2003). Although this BDA method is possible given an air warfare scenario, in a natural or man-made disaster situation not involving a dropped munition, other data collection methods are necessary, as discussed above (e.g. involving UAVs, other aircraft, or satellites).

SAR can also be employed in urban structural damage assessment (in addition to the through-wall radar imaging application noted above). For example, Ferretti et al. (2000) analyzed the case of building collapse via the permanent scatterers method. Differential InSAR (DInSAR) can be utilized to extract highly accurate deformation measurements of buildings; time-series analysis of collapsed buildings indicated precursory motions. As another example, Yonezawa and Takeuchi (2001) demonstrated the potential of detecting urban disasters via decorrelation of SAR data. Stramondo et al. (2006) fused SAR and optical data to improve urban earthquake damage detection. Stilla and Soergel (2007) discuss a SAR-based method for building recognition and reconstruction in urban areas, which can potentially be very useful in a damage-assessment scenario. In addition to conventional, structural remote damage assessments of buildings and human settlements, contamination and damage assessments associated with nuclear/radiological and other threats can also be considered, which is addressed below.

7.3.3 Nuclear/Radiological Event-Induced Urban Contamination and Damage

With respect to the built environment, this section focuses on evaluating artificial radionuclide signals/contamination associated with a dirty/radiological bomb, a release from a nuclear power plant, or some other source, as well as on assessing a nuclear weapon blast in a populated place, including estimating for the point of detonation. Monitoring atmospheric radionuclide contamination is also briefly discussed here.

In situ (mobile or fixed) and airborne gamma-ray spectrometers were discussed above in the context of assessing artificial radionuclide-induced environmental/vegetation damage and searching for lost radiological sources, and the general principles underlying such technologies also apply to monitoring urbanized or otherwise developed areas. Thus, a recapitulation of such technologies/methods is not given here. Note, however, that many urbanized areas already host a variety of pre-existing radiological sources (e.g. hospitals, research laboratories, previously contaminated areas, etc.), and knowledge of such sources can be useful for prevention. In the United States, the US Department of Homeland Security (DHS) and the US Department of Energy's (DOE) National Nuclear Security Administration (NNSA) are supporting an initiative to conduct baseline surveys of radioactive material within major US cities to facilitate dirty/radiological bomb detection. Such bombs mix radioactive material with explosives. Spatial baseline inventories could enhance searches for potential radiological or nuclear weapons; radiation detected via new scans could be compared with prior known sources (Hall 2007b). The problem of locating and identifying a lost radiation source via mobile gamma spectrometry and an optimization of on-line statistical methods was described in Hjerpe et al. (2001), where the objective was to obtain a high probability of detection while minimizing the false alarm rate due to natural background radiation variability.

Thus, field and airborne gamma-ray spectrometers are of utility in a variety of nuclear/radiological mapping and damage assessment contexts. However, the information-extraction potential from airborne and spaceborne sensors commonly available to the civil/scientific sector (e.g. optical sensors, etc.) should also be considered in nuclear/radiological-related damage-assessment analyses, as such data could allow for rapid, broad-area assessment, at least for certain dimensions of the problem. This was discussed above in the context of vegetation/environmental monitoring, for example.

Remote sensing has long been employed in detecting and monitoring the testing of nuclear weapons, both above- and underground, and for treaty verification (Gupta 1994; Cloud 2001). Jasani (1995) discussed the potential role of civilian remote-sensing satellites in monitoring nuclear testing. Regarding underground nuclear testing, spaceborne interferometric synthetic aperture radar (InSAR) represents one means that has been used to detect such events (Vincent et al. 2003). However, the case of nuclear damage/contamination assessment, rather than testing, is considered here.

Certain aspects or principles of BDA discussed above may also be cross-applicable to assessing the damage associated with a nuclear blast. In addition, as a means of conducting an indirect bomb damage assessment (IBDA) following a nuclear blast, Orr and Spohn (1967) introduced a multivariate statistical model to accrue the best estimate for the point of detonation and other burst parameters of a long-range nuclear weapon. The Orr-Spohn method merges *a priori* weapon system accuracy with sensor array data input (electromagnetic/optical/infrared/other), which provide information regarding observations of ground zero, burst height, and weapon burst yield. (The discussion above concerning sensor networks has particular relevance to this topic.) Such a model is applicable to urban and non-urban areas, though in order to fully exploit the model, sufficient sensor input data are assumed, which may be more likely in an urban/populated environment. However, if no sensors are deployed, the damage assessment may be based only upon the *a priori* knowledge of weapon system yield, and the lethal region around the target can be estimated.

For this method, note that the aim point is described by a set of desired burst parameters, whereas the burst point is the set of actual burst parameter values. The reported point is characterized by a set of observed burst parameters given by a sensor. Also, a weapon system circular error probable (CEP) is the circle radius centered at the aim point within which 50 percent of multiple independently launched missiles would impinge (Orr and Spohn 1967). To compute the probability of destruction, the probability density function for the burst point distribution around the aim point can be integrated over the lethal region. The aim point (i.e., the mean of the distribution) is taken to be the origin of the coordinates. The covariance matrix and the mean characterize this distribution, and weapon system error is considered. The density function of this weapon system is (Guenther and Terragno 1964; Orr and Spohn 1967):

$$f(\Theta;0,\Sigma_w) = \left[(2\pi)^p \left|\Sigma_w\right|\right]^{-1/2} \exp\left(-\frac{1}{2}\Theta'\Sigma_w^{-1}\Theta\right), \tag{7.5}$$

where p = the number of dimensions describing the burst;

$$\Sigma_w = \begin{Vmatrix} \tau_1^2 & & \\ & \ddots & \\ & & \tau_p^2 \end{Vmatrix} = \text{covariance matrix of weapon system error}$$

(τ_j^2 = variance in the jth coordinate); $\Theta = \lfloor \theta_1,\ldots,\theta_p \rfloor$ = vector of burst point coordinates; and Θ' = transpose of Θ.

Given one sensor, when combining the sensor report with *a priori* weapon system CEP, the true burst point Θ can be approximated. If the burst point were known, the sensor distribution density would be (Orr and Spohn 1967):

$$f(X;\Theta;\Sigma_s)=\left[(2\pi^p)|\Sigma_s|\right]^{-1/2}\exp\left[-\frac{1}{2}(X-\Theta)'\Sigma_s^{-1}(X-\Theta)\right], \tag{7.6}$$

where $X = [X_1,\ldots,X_p]$ = sensor report vector; $\Theta = [\Theta_1,\ldots,\Theta_p]$ = burst point vector; and Σ_s = covariance matrix of the sensor system distribution. However, Θ would not be known, but the weapon system density gives the probability of the Θ's taking on any given value (Orr and Spohn 1967):

$$f(X;\Theta;\Sigma_w)=\left[(2\pi^p)|\Sigma_w|\right]^{-1/2}\exp\left[-\frac{1}{2}\Theta'\Sigma_w\Theta\right], \tag{7.7}$$

The parametric form of the sensor report density distribution is thus

$$g(X,\Theta)=f(\Theta;0;\Sigma_w)f(X;\Theta;\Sigma_s), \tag{7.8}$$

which can be used to accrue Θ. This function is the likelihood function and was employed by Orr and Spohn (1967) to further develop maximum likelihood estimates (MLEs) involving one and *n* sensors. Part of what was obtained by this analysis is illustrated in Fig. 7.7. Total CEP gives the probable error in the MLE based on both weapon and sensor information, ultimately providing a more realistic estimate (Orr and Spohn 1967).

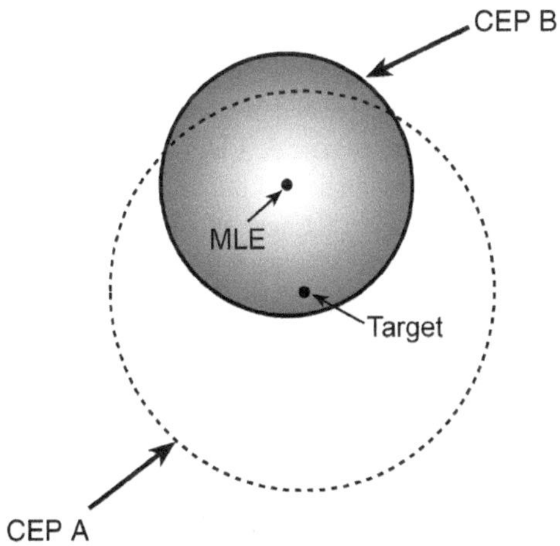

Fig. 7.7 The total system circle of equal probability (CEP), or total CEP, is the radius of the circle of equal probability that includes all sensor reports. 'CEP A' is the CEP for the weapon system alone (target at center), whereas 'CEP B' is the CEP for the weapon system and one sensor (MLE of ground zero at center). As new sensor reports are added, the MLE improves/shifts, and the total CEP contracts (Modified from Orr and Spohn (1967)

IBDA models for nuclear blasts where the delivery method is something other than a long-range nuclear missile can also be considered. In any case, output from models such as these can be intersected with population data layers, which is of particular importance in urbanized areas, and can be jointly inspected with other BDAs and remote-sensing/change detection analyses to yield more comprehensive damage assessments, including estimates of physical structural damage (Tucker and San 2003), contamination (Hall 2007b), and casualty rates from nuclear blasts (Hadjipavlou and Carr-Hill 1989), radiological dispersion devices (RDDs) (Dombroski and Fischbeck 2006), or conventional explosives (Wightman et al. 2001).

For a nuclear/radiological event, atmospheric measurements, as well as atmospheric physical dispersion modeling should also be considered. This would be of particular immediate importance for populated/urban places. Grasty et al. (1997) discussed aircraft-based measurements of ^{131}I and ^{140}La taken in the radioactive plume from the Chernobyl accident; while aircraft contamination presented problems in tracking a radioactive plume, it was determined that a NaI gamma-ray spectrometer could yield useful information concerning a nuclear accident. Papastefanou (2006) considered natural and man-made radionuclide tropospheric aerosol residence times, which is a function of various removal processes (i.e. dry deposition (via impactation, diffusion, sedimentation, and resuspension) and wet deposition by rain drops). Fugitive (airborne) dust generation, transport, and deposition numerical models (e.g. Lancaster and Nickling 1994), can be very useful in a radionuclide context. Note that remote sensing can explicitly be employed by providing classifications of land cover types important to fugitive dust generation, transport, and deposition, and such spatio-temporal data can be used as input to numerical atmospheric models. The expert system-based approach described in Stefanov et al. (2003) is an example of this. In considering dispersion, transport, and deposition following a RDD ('dirty' bomb) event, Dombroski and Fischbeck (2006) used a Gaussian puff model for the initial explosion, and for the aerosolized material from the associated fire, a Gaussian plume model was employed. As noted above in Hasager and Thykier-Nielsen (2001), airborne or spaceborne remote sensing can provide critical inputs to atmospheric dispersion/deposition models (e.g. land cover/aerodynamic surface roughness). The RIsø Mesoscale PUFF (RIMPUFF) model (Mikkelsen et al. 1984) was used for real-time simulation of puff and plume dispersion modeling of a ^{137}Cs-based radiological release event (Hasager and Thykier-Nielsen 2001). Data collected by *in situ* sensor networks can also be used as input for such models (Stull 1991; Gorman et al. 2005; Seaman 2003). For instance, meteorological field masts can provide direct measurements of aerodynamic roughness (Stull 1991).

7.3.4 Remote Detection of Other (Non-radionuclide) Contaminants in Populated Places

The previous section ended with a brief discussion of radionuclide-based contamination; however, remote sensing can be applied to other atmospheric plumes/contaminants from unconventional sources (i.e. chemical and biological agents).

Regarding atmospheric contamination by biochemical weapons of mass destruction, in concert with a network of pre-positioned biochemical detectors, locating the point-of-origin of a chemical or biological weapon release and predicting the downwind plume movement in the atmosphere may be possible with atmospheric nowcasting, which yields precise information regarding the current state of the atmosphere over a region, or a prediction over the next several minutes in time. Data can be assimilated in real-time, resulting in frequent updates to the atmospheric state. High-quality data regarding the current turbulent state becomes critical at very short time scales, and the principal spatial scales of interest range from tens of meters to ~10 km. Very fine urban-scale numerical models and detailed knowledge of the physics and chemistry should enable an effective operational response (Seaman 2003). Aircraft-based data acquisition for detecting and monitoring airborne plumes is also possible; for example, the US Environmental Protection Agency utilizes multi- and hyperspectral passive infrared sensors for emergency response to detect and classify absorption and emission features of chemical vapors (Kroutil et al. 2006). In terms of detection and classification of chemical and biological agents, it is possible for chemical or biological agents to comprise mixture clouds in the event of a terrorist attack, which could complicate detection by existing agent detection systems if such systems are specific to single agents (Kwan et al. 2006). Thus, Kwan et al. (2006) examined the use of spectral unmixing methods commonly employed in hyperspectral imaging for classifying and estimating the abundance fraction, or concentration, of chemical and biological agents in a mixture form. In addition, LiDAR—and particularly DIfferential Absorption LiDAR (DIAL)—when utilized in conjunction with other sensors may provide the ability to detect trace chemical concentrations. A DIAL LiDAR transmits two laser pulses, where one pulse involves the wavelength of interest, and the other pulse is offset somewhat from the first and is not absorbed by the chemical species of interest. This second pulse compensates for transmission path losses; to obtain the absorption attributable to the species of interest, the two backscattered signals are subtracted (Kopp 1995).

Remote sensing—and imaging spectroscopy in particular—can also be utilized to evaluate other disaster-associated contaminants. As an example, following the 11 September 2001 terrorist attack on the World Trade Center (WTC), a laboratory reflectance spectroscopy analysis and airborne hyperspectral AVIRIS-based mapping of WTC-related asbestiform mineralogical contamination was performed by the US Geological Survey (USGS) (Clark et al. 2001). Dust/debris plume maps were also generated. A primary concern was to assess and map possible broadcasting of potentially carcinogenic asbestiform dusts, though other dusts were also of concern with respect to potential respiratory problems. Mineralogical-mapping is facilitated by identifying diagnostic absorption features, and imaging spectroscopy may constitute the only practical method in this application domain.

The WTC area was imaged by the AVIRIS sensor on 16, 18, 22, and 23 September 2001. Dust/debris and beam-insulation samples were analyzed for various chemical and mineralogical parameters; the laboratory analytical techniques utilized included Reflectance Spectroscopy (RS), X-Ray Diffraction (XRD), X-Ray

Fluorescence Spectroscopy (XRF), Scanning Electron Microscopy (SEM) with elemental analysis by Energy Dispersive Spectroscopy (EDS), chemical analysis and leach tests. The image data were analyzed for chrysotile-specific absorption features. Variable amounts of chrysotile asbestos were detected, and AVIRIS-derived material maps were generated using the Tetracorder program. This information assisted with the disaster recovery effort. It should be noted that the Tetracorder maps did not indicate highly concentrated widespread distribution of asbestiform minerals or other toxic materials in the dust/debris (Clark et al. 2001, 2003; Herold et al. 2007). Clark et al. (2001) also note that spectroscopy may not be sensitive to whether a specific mineral has an asbestiform shape—i.e. long needle-like shapes with diameters less than a micrometer. More research is needed to determine whether imaging spectroscopy can successfully differentiate asbestiform from nonasbestiform mineral shapes (Clark et al. 2001).

Remote sensing can also be used for monitoring other contaminants as well. For instance, the Hurricane Katrina-induced floodwaters in the New Orleans area were contaminated with oil and other constituents in some areas (EPA 2005; Esworthy et al. 2005; Pardue et al. 2005; Cobb et al. 2006; Presley et al. 2006; Mielke et al. 2006). Remote sensing has long been useful in the detection and monitoring of oil spills. Brekke and Solberg (2005) give a review of remote sensing-based detection of oil spills on the sea surface, which includes an overview of SAR and optical approaches. Remote sensing of vegetation and other terrestrial materials is also pertinent to this application domain, as it can be used to indicate the type and degree of pollution damage. For instance, Rosso et al. (2005) treated *Salicornia virginica* plants with two metals and two crude oil types and found that the spectral features were significantly correlated with symptoms of cadmium and lightweight petroleum contamination, with distinct spectral signatures for both contaminant types. The use of AVIRIS data for detection oil-induced vegetation stress is examined in Li et al. (2005).

7.4 Summary

This chapter reviewed various areas of remote-sensing damage assessment useful for homeland security operations. Natural hazards/disasters were considered, as were examples anthropogenic in origin, including chemical, biological, radiological, nuclear, and explosive (CBRNE) threats. This review was organized by remote-sensing damage-assessment task, where the general application domains were environmental/biophysical in nature (i.e. the 'natural' environment) and human settlements (i.e. the built environment). A wide range of damage-assessment tasks is possible, many of which are addressable via remote sensing. Likewise, a broad array of remote-sensing technologies can be applied to a given problem, including photogrammetry, multi- and hyperspectral imaging, SAR, and LiDAR, among others. Sensors installed on remote (e.g. airborne or spaceborne) platforms were discussed, as were *in situ* networks of fixed/discrete and dynamic/quasi-ubiquitous sensors. Also addressed here was the integration of remote sensing with GIS-based and mathematical modeling. Many remote-sensing technologies and algorithms useful for damage assessment currently exist and have been proven in real-world scenarios, though a significant margin for improvement still exists. While remote sensing is not a damage-assessment panacea, as sensor and information-processing algorithms continue to evolve and improve, remote sensing will be better situated to respond to—or otherwise address—the range of current and future threats.

Acknowledgements R. O. Green and M. Eastwood, Jet Propulsion Laboratory, supplied the AVIRIS image. IKONOS satellite images were provided by GeoEye.

References

Aarnio, P. A., Ala-Heikkilä, J. J., Hakulinen, T. T. & Nikkinen, M. T. (1998). Gamma spectrometric monitoring of environmental radioactivity using mobile equipment. *Journal of Radioanalytical and Nuclear Chemistry*, 233(1–2), 217–223

Abduragimov, I. M. & Odnolko, A. A. (1993). Fires on radioactively contaminated areas. *Priroda*, 1, 28–30

Adams, M. L., Philpot, W. D. & Norvell, W. A. (1999). Yellowness index: An application of the spectral second derivative to estimate chlorosis of leaves in stressed vegetation. *International Journal of Remote Sensing*, 20(18), 3663–3675

Adams, B., Huyck, C. K., Eguchi, R., Yamazaki, F., Estrada, M. & Herring, C. (2007). Earthquake recovery: QuickBird imagery of Algeria supports damage detection and relief efforts. *Earth Imaging Journal*, 1(1). Retrieved February, 28, 2007 from http://www.eijournal.com/algeria.asp

Adams, J. B. & Smith, M. O. (1986). Spectral mixture modeling: A new analysis of rock and soil types at the Viking Lander 1 site. *Journal of Geophysical Research*, 91, 8098–8112

Allard, Y. & Jouan, A. (2005). Multisensors and contextual information fusion: Application to object detection and automated terrain analysis. (In E. Shahbazian, G. Rogova & P. Valin (Eds.), *Data Fusion for Situation Monitoring, Incident Detection, Alert and Response Management* (pp. 381–393). Amsterdam: IOS Press)

Andrienko, G., Andrienko, N. & Gitis, V. (2003). Interactive maps for visual exploration of grid and vector geodata. *ISPRS Journal of Photogrammetry & Remote Sensing*, 57, 380–389

Anon (2004). Abrupt climate change: Inevitable surprises. *Population and Development Review*, 30(3), 563–574. doi: 10.1111/j.1728-4457.2004.00030.x

Aryanfar, F. & Sarabandi, K. (2004). Through wall imaging at microwave frequencies using space-time focusing. (In *IEEE Intl. antennas and propagation symposium*, Vol. 3 (pp. 3063–3066))

Associated Press (2007). Ex-generals: Global warming threatens U.S. security. *CNN.com*, 15 April 2007, Retrieved April 16, 2007, from http://www.cnn.com/2007/US/04/15/warming.military.ap/index.html

Bally, P., Béquignon, J., Arino, O. & Briggs, S. (2005). Remote sensing and humanitarian aid—A life-saving combination. *ESA Bulletin*, 122, 36–41

Barnett, J. (2003). Security and climate change. *Global Environmental Change*, 13, 7–17

Ben-Dor, E., Levin, N. & Saaroni, H. (2001). A spectral based recognition of the urban environment using the visible and near-infrared spectral region (0.4-1.1 μm). A case study over Tel-Aviv, Israel. *International Journal of Remote Sensing*, 22(11), 2193–2218

Benediktsson, J. A., Pesaresi, M. & Arnason, K. (2003). Classification and feature extraction for remote sensing images from urban areas based on morphological transformations. *IEEE Transactions on Geoscience and Remote Sensing*, 41(9), 1940–1949

Benz, U. C., Hofmann, P., Willhauck, G., Lingenfelder, I. & Heyen, M. (2004). Multi-resolution, object-oriented fuzzy analysis of remote-sensing data for GIS-ready information. *ISPRS Journal of Photogrammetry & Remote Sensing*, 58, 239–258

Benz, U. C. & Schreier, G. (2001). OSCAR – Object oriented segmentation and classification of advanced radar allows automated information extraction. *Proc. IGARSS*, July 2001, Sydney

Bernstein, R. & Di Gesù, V. (1999). A combined analysis to extract objects in remote sensing images. *Pattern Recognition Letters*, 20, 1407–1414

Beteselassie, N., Fininsa, C. & Badebo, A. (2007). Sources of resistance to stem rust (*Puccinia graminis* f. sp. *tritici*) in Ethiopian tetraploid wheat accessions. *Genetic Resources and Crop Evolution*, 54, 337–343

Bhaduri, B., Bright, E., Coleman, P. & Dobson, J. (2002). LandScan: Locating people is what matters. *Geoinformatics*, 5(2), 34–37

Bhaskaran, S., Datt, B., Forster, B., Neal, T. & Brown, M. (2004). Integrating imaging spectroscopy (445-2543 nm) and geographic information systems for post-disaster management: A case of hail-storm damage in Sydney. *International Journal of Remote Sensing*, 25(13), 2625–2639

Billings, S. & Hovgaard, J. (1999). Modeling detector response in airborne gamma-ray spectrometry. *Geophysics*, 64(5), 1378–1392

Blaschke, T., Burnett, C. & Pekkarinen, A. (2004). Image segmentation methods for object-based analysis and classification. (In S. M. de Jong & F. D. van der Meer (Eds.), *Remote sensing image analysis: Including the spatial domain* (pp. 211–236). Dordrecht, The Netherlands: Kluwer)

Borel, C. C. & Gerstl, S. A. W. (1994). Nonlinear spectral mixing models for vegetative and soil surfaces. *Remote Sensing of Environment*, 47, 403–416

Boyd, D. S., Entwistle, J. A., Flowers, A. G., Armitage, R. P. & Goldsmith, P. C. (2006). Remote sensing the radionuclide contaminated Belarusian landscape: A potential for imaging spectrometry? *International Journal of Remote Sensing*, 27(10), 1875–1874

Boucher, A., Seto, K. C. & Journel, A. G. (2006). A novel method for mapping land cover changes: Incorporating time and space with geostatistics. *IEEE Transactions on Geoscience and Remote Sensing*, 44(11), 3427–3435

Brajnik, D., Miklavžič, U. & Tomšič, J. (1992). Map of natural radioactivity in Slovenia and its correlation to the emanation of radon. *Radiation Protection Dosimetry*, 45(1/4), 273–276

Brekke, C. & Solberg, A. H. S. (2005). Oil spill detection by satellite remote sensing. *Remote Sensing of Environment*, 95, 1–13

Bresnahan, P. (1998). Identification of potential hazardous waste units using aerial readiological measurements. *Photogrammetric Engineering & Remote Sensing*, 64(10), 995–1001

Brode, H. L. (1968). Review of nuclear weapons effects. *Annual Review of Nuclear Science,* 18, 153–202

Bunn, M. & Wier, A. (2006). *Securing the bomb 2006*. Project on Managing the Atom, Harvard University, President and Fellows of Harvard College, Cambridge, MA. Retrieved March, 19, 2007 from http://www.nti.org/e_research/stb06webfull.pdf

Chang, C.-I. (2003). *Hyperspectral imaging: Techniques forsSpectral detection and classification.* (New York: Kluwer)

Clark, R. N., Gallagher, A. J. & Swayze, G. A. (1990). Material absorption band depth mapping of imaging spectrometer data using a complete band shape least-squares fit with library reference spectra. (In *Proc. of the Second Airborne Visible/Infrared Imaging Spectrometer (AVIRIS) Workshop*, JPL Publication 90–54 (pp. 176–186))

Clark, R. N., Green, R. O., Swayze, G. A., Meeker, G., Sutley, S., Hoefen, T. M., Livo, K. E., Plumlee, G., Pavri, B., Sarture, C., Wilson, S., Hageman, P., Lamothe, P., Vance, J. S., Boardman, J., Brownfield, I., Gent, C., Morath, L. C., Taggart, J., Theodorakos, P. M. & Adams, M. (2001). *Environmental studies of the World Trade Center area after the September 11, 2001 attack. Version 1.1* [Electronic version] from http://pubs.usgs.gov/of/2001/ofr-01-0429/

Clark, R. N. & Roush, T. L. (1984). Reflectance spectroscopy: Quantitative analysis techniques for remote sensing applications. *Journal of Geophysical Research*, 89, 6329–6340

Clark, R. N., Swayze, G. A., Livo, K. E., Kokaly, R. F., Sutley, S. J., Dalton, J. B., McDougal, R. R. & Gent, C. A. (2003). Imaging spectroscopy: Earth and planetary remote sensing with the USGS Tetracorder and expert systems. *Journal of Geophysical Research*, 108(E12), 5131. doi: 10.1029/2002JE001847

Cloud, J. (2001). Imaging the world in a barrel: CORONA and the clandestine convergence of the earth sciences. *Social Studies of Science*, 31(2), 231–251

Cobb, G. P., Abel, M. T., Rainwater, T. R., Austin, G. P., Cox, S. B., Kendall, R. J., Marsland, E. J., Anderson, T. A., Leftwich, B. D., Zak, J. C. & Presley, S. M. (2006). Metal distributions in New Orleans following hurricanes Katrina and Rita: A continuation study. *Environmental Science & Technology*, 40(15), 4571–4577

Davids, C. & Tyler, A. N. (2003). Detecting contamination-induced tree stress within the Chernobyl exclusion zone. *Remote Sensing of Environment*, 85, 30–38

Defense Civil Preparedness Agency (1973). What the planner needs to know about blast and shock. (In *DCPA attack environmental manual.* Washington, DC: U.S. Defense Civil Preparedness Agency)

Defense Intelligence Agency (DIA) (1996). *Battle damage assessment (BDA) quick guide*. (DIA: Washington, DC) February 1996, E-1

De Moraes, C., Schultz, J. C., Mescher, M. C. & Tumlinson, J. H. (2004). Induced plant signaling and its implications for environmental sensing. *Journal of Toxicology and Environmental Health, Part A*, 67, 819–834

Dobson, J. E., Bright, E. A., Coleman, P. R., Durfee, R. C. & Worley, B. A. (2000). LandScan: A global population database for estimating populations at risk. *Photogrammetric Engineering & Remote Sensing*, 66(7), 849–857

Dombroski, M. J. & Fischbeck, P. S. (2006). An integrated physical dispersion and behavioral response model for risk assessment of radiological dispersion device (RDD) events. *Risk Analysis*, 26(2), 501–514. doi: 10.1111/j.1539–6924.2006.00742.x.

Ebinuma, T., Rooney, E., Gleason, S. & Unwin, M. (2005). GPS receiver operations on the disaster monitoring constellation satellites. *The Journal of Navigation*, 58, 227–240

Elvidge, C. D. & Chen, Z. (1995). Comparison of broad-band and narrow-band red and near-infrared vegetation indices. *Remote Sensing of Environment*, 54, 38–48

EPA (U.S. Environmental Protection Agency) (2005). *Response to 2005 hurricanes: Murphy oil spill*. [Electronic version] from http://www.epa.gov/katrina/testresults/murphy/index.html

Esworthy, R., Schierow, L. J., Copeland, C. & Luther, L. (2005). *Cleanup after hurricane Katrina: Environmental considerations*. CRS Report for Congress, Congressional Research Service, RS33115, 13 October 2005, 30 pp

Ferretti, A., Ferrucci, F., Prati, C. & Rocca, F. (2000). SAR analysis of building collapse by means of the permanent scatterers technique. In *International Geoscience and Remote Sensing Symposium*, Honolulu, Hawaii, 24–28 July 2000, 7, 3219–3221. doi: 10.1109/IGARSS.2000.860388

Filippi, A. M. & Jensen, J. R. (2007). Effect of continuum removal on hyperspectral coastal vegetation classification using a fuzzy learning vector quantizer. *IEEE Transactions on Geoscience and Remote Sensing*, 45(6), 1857–1869. doi: 10.1109/TGRS.2007.894929

Filippi, A. M. & Jensen, J. R. (2006). Fuzzy learning vector quantization for hyperspectral coastal vegetation classification. *Remote Sensing of Environment*, 100, 512–530. doi: 10.1016/j.rse.2005.11.007

Forster, B. C. (1985). An examination of some problems and solutions in monitoring urban areas from satellite platforms. *International Journal of Remote Sensing*, 6(1), 139–151

Fox, J. (2007). Watching nuclear bazaar difficult, spy chief says. *Global Security Newswire*, 11 September 2007, Retrieved from http://www.nti.org/d_newswire/issues/2007_9_11.html#EDEFC957

Frost, G. & Costello, M. (2003). Bomb damage assessment using a dispensed sensor projectile. *IEEE Transactions on Aerospace and Electronic Systems*, 39(3), 813–823

Gamba, P., Benediktsson, J. A. & Wilkinson, G. (2003). Forward to the special issue on urban remote sensing by satellite. *IEEE Transactions on Geoscience and Remote Sensing*, 41(9), 1903–1906

Gao, B.-C. (1996). NDWI: A normalized difference water index for remote sensing of liquid water from space. *Remote Sensing of Environment*, 58, 257–266

Gaso, M. I., Segovia, N., Herrera, T., Perez-Silva, E., Cervantes, M. L., Quintero, E., Palacios, J. & Acosta, E. (1998). Radiocaesium accumulation in edible wild mushrooms from coniferous forests around the nuclear centre of Mexico. *The Science of the Total Environment*, 223, 119–129

Ghitter, G. S., Bowers, W. W. & Franklin, S. E. (1995). Discrimination of adelgid-damage on single balsam fir trees with aerial remote sensing data. *International Journal of Remote Sensing*, 16, 2779–2794

Givri, J. R. (1995). Satellite remote sensing data on industrial hazards. *Advances in Space Research* 15(11), 87–90

GläBer, C. & Reinartz, P. (2005). Multitemporal and multispectral remote sensing approach for flood detection in the Elbe-Mulde region 2002. *Acta hydrochimica et hydrobiologica*, 33(5), 395–403

Global Security Newswire (2007). Pentagon advisers urge nationwide conversion of radioactive-sourced medical, safety gear. *Global Security Newswire*, 10 October 2007

Gomez, R. B. (2002). Hyperspectral imaging: A useful technology for transportation analysis. *Optical Engineering*, 41(9), 2137–2143

Gorman, B. L., Shankar, M. & Smith, C. M. (2005). Advancing sensor web interoperability. *Sensors*, 22(4). Retrieved March 19, 2007 from www.sensorsmag.com

Graffis, J. M. (1997). *Do the Army and Air Force see eye to eye on BDA?* School of Advanced Military Studies, United States Army Command and General Staff College, Fort Leavenworth, KS, 53 p

Grasty, R. L., Hovgaard, J. & Multala, J. (1997). Airborne gamma ray measurements in the Chernobyl plume. *Radiation Protection Dosimetry*, 73(1–4), 225–230

Green, R. O., Eastwood, M. L., Sarture, C. M., Chrien, T. G., Aronsson, M., Chippendale, B. J., Faust, J. A., Pavri, B. E., Chovit, C. J., Solis, M., Olah, M. R. & Williams, O. (1998). Imaging spectroscopy and the Airborne Visible/Infrared Imaging Spectrometer (AVIRIS). *Remote Sensing of Environment*, 65, 227–248

Griffiths, H. D. & Baker, C. J. (2006). Radar imaging for combating terrorism. (In *Imaging for detection and identification* (pp. 1–19), NATO Advanced Study Institute, 23 July–5 August 2006, Il Ciocco, Italy. The Netherlands: Springer). Retrieved March 24, 2007 from http://www.nato-us.org/imaging2006/papers/griffiths.pdf

Gruninger, J., Ratkowski, A. J. & Hoke, M. L. (2004). The Sequential Maximum Angle Convex Cone (SMACC) endmember model. (In Sylvia S. Shen & Paul E. Lewis (Eds.), *Algorithms and technologies for multispectral, hyperspectral, and ultraspectral imagery X.* (pp. 1–14) Proc. of SPIE, Vol. 5425 (Bellingham, WA: SPIE)

Guenther, W. C. & Terragno, P. J. (1964). A review of the literature on a class of coverage problems. *The Annals of Mathematical Statistics*, 35(1), 232–260

Guillaso, S., Ferro-Famil, L., Reigber, A. & Portier, E. (2005). Building characterization using L-band polimetric interferometric SAR data. *IEEE Geoscience and Remote Sensing Letters*, 2(3), 347–351

Guo, Q., Kelly, M., Gong, P. & Liu, D. (2007). An object-based classification approach in mapping tree mortality using high spatial resolution imagery. *GIScience & Remote Sensing*, 44(1), 24–47

Gupta, V. (1994). Remote sensing and photogrammetry in treaty verification: Present challenges and prospects for the future. *Photogrammetric Record*, 14(83), 729–745

Haack, B. K., Guptill, S., Holz, R., Jampoler, S., Jensen, J. R. & Welch, R. (1997). Urban analysis and planning. (In *Manual of photographic interpretation* (pp. 517–553) Bethesda, MD: American Society for Photogrammetry & Remote Sensing)

Hadjipavlou, S. & Carr-Hill, G. (1989). A revised set of blast casualty rates for civil defence use: An overview. *Journal of the Royal Statistical Society. Series A (Statistics in Society)*, 152(2), 139–156

Hall, M. (2007a). Phones studied as attack detector. *USA Today*, 4 May 2007. Retrieved May 5, 2007 from http://www.usatoday.com/tech/news/techpolicy/2007–05–03-cellphone-attack-detector_N.htm

Hall, M. (2007b). Authorities want to survey city radiation. *USA Today*, 16 March 2007. Retrieved March 16, 2007 from http://www.usatoday.com/news/nation/2007–03–15-radiation_N.htm?csp = 34

Hall, S. W., Carolla, M. A. & Deason, J. P. (2006). The use of imagery in environmental disaster preparedness and response. *Federal Facilities Environmental Journal*, Winter 2006, 65–72. doi: 10.1002/ffej.20075

Harris, J. W. E. (1974). Small-scale imagery in forest pest surveys in British Columbia. *The Canadian Surveyor*, 28(2), 155–161

Hasager, C. B. & Thykier-Nielsen, S. (2001). IRS-1C LISS III land cover maps at different spatial resolutions used in real-time accidental air pollution deposition modelling. *Remote Sensing of Environment*, 76, 326–336

Hausamann, D., Zirnig, W., Schreier, G. & Strobl, P. (2005). Monitoring of gas pipelines—a civil UAV application. *Aircraft Engineering and Aerospace Technology: An International Journal*, 77(5), 352–360

Hepner, G. F., Houshmand, B., Kulikov, I. & Brayant, N. (1998). Investigation of the integration of AVRIS and IFSAR for urban analysis. *Photogrammetric Engineering & Remote Sensing*, 64(8), 813–820

Herbreteau, V., Salem, G., Souris, M., Hugot, J.-P. & Gonzalez, J.-P. (2007). Thirty years of use and improvement of remote sensing, applied to epidemiology: From early promises to lasting frustration. *Health Place*, 13(2), 400–403

Herold, M., Gardner, M. E. & Roberts, D. A. (2003). Spectral resolution requirements for mapping urban areas. *IEEE Transactions on Geoscience and Remote Sensing*, 41(9), 1907–1919

Herold, M., Schiefer, S., Hostert, P. & Roberts, D. A. (2007). Applying imaging spectroscopy in urban areas. (In *Urban remote sensing* (pp. 137–161). Boca Raton, FL: CRC)

Herring, C. (2007). When disaster strikes: VIEWS system applies satellite imagery to post-disaster analysis. *Earth Imaging Journal*. Retrieved February 28, 2007 from http://www.eijournal.com/When_Disaster_Strikes.asp

Hjerpe, T., Finck, R. R. & Samuelsson, C. (2001). Statistical data evaluation in mobile gamma spectrometry: An optimization of on-line search strategies in the scenario of lost point sources. *Health Physics*, 80(6), 563–570

Houser, P. R. (2002). Air and water monitoring for homeland security. *Earth Observation Magazine*, 11(8), 33–36

Hyvönen, E., Pänttäjä, M., Sutinen, M.-J. & Sutinen, R. (2003). Assessing site suitability for Scots pine using airborne terrestrial gamma-ray measurements in Finnish Lapland. *Canadian Journal of Forest Research*, 33, 796–806

ImSpec LLC (2002). *ACORN 4.0 user's guide: Stand-alone version*. Analytical Imaging and Geophysics LLC. Boulder, CO: ImSpec LLC, 76 p

Islam, M. M. & Sado, K. (2002). Development priority map for flood countermeasures by remote sensing data with geographic information system. *Journal of Hydrologic Engineering*, 7(5), 346–355

Ito, A. (2005). Issues in the implementation of the International Charter on Space and Major Disasters. *Space Policy*, 21, 141–149

Jasani, B. (1995). Could civil satellites monitor nuclear tests? *Space Policy*, 11(1), 31–40

Javidi, B. (Ed.) (2006). *Optical imaging sensors and systems for homeland security applications*. (New York: Springer)

Jayaraman, V., Chandrasekhar, M. G. & Rao, U. R. (1997). Managing the natural disasters from space technology inputs. *Acta Astronautica*, 40(2–8), 291–325

Jensen, J. R. (2005). *Introductory digital image processing: A remote sensing perspective*. Third Edition. (Upper Saddle River, NJ: Prentice-Hall)

Jensen, J. R. (2007). *Remote sensing of the environment: An earth resource perspective*. Second Edition. (Upper Saddle River, NJ: Prentice-Hall)

Jensen, J. R. & Filippi, A. M.(2005). Thematic information extraction: Hyperspectral image analysis. (In Jensen, J. R., *Introductory digital image processing: A remote sensing perspective*, Third Edition (pp. 431–465). Upper Saddle River, NJ: Prentice-Hall)

Jensen, J. R. & Hodgson, M. E. (2006). Remote sensing of natural and man-made hazards and disasters. (In M. K. Ridd & J. D. Hipple (Eds.), *Remote sensing of human settlements*, Manual of Remote Sensing, Third Edition, Vol. 5 (pp. 401–428). Bethesda, MD: American Society for Photogrammetry and Remote Sensing)

Kääb, A., Huggel, C., Fischer, L., Guex, S., Paul, F., Roer, I., Salzmann, N., Schlaefli, S., Schmutz, K., Schneider, D., Strozzi, T. & Weidmann, Y. (2005). Remote sensing of glacier- and permafrost-related hazards in high mountains: An overview. *Natural Hazards and Earth Systems Sciences*, 5, 527–554

Kaya, S., Curran, P. J. & Llewellyn, G. (2005). Post-earthquake building collapse: A comparison of government statistics and estimates derived from SPOT HRVIR data. *International Journal of Remote Sensing*, 26(13), 2731–2740

Kerle, N., Froger, J.-L., Oppenheimer, C. & Van Wyk De Vries, B. (2003). Remote sensing of the 1998 mudflow at Casita volcano, Nicaragua. *International Journal of Remote Sensing*, 24(23), 4791–4816

Kogan, F. N. (1990). Remote sensing of weather impacts on vegetation in nonhonogeneous areas. *International Journal of Remote Sensing*, 11, 1405–1419

Kokaly, R. F., Clark, R. N. & Swayze, G. A. (1994). Vegetation and cryptobiotic soils mapping in arid regions. (In *Spectral analysis Workshop: The use of vegetation as an indicator of environmental contamination* (Nov. 9–10, 1994). Reno, NV: Desert Research Institute)

Kopp, C. (1995). *Air warfare applications of laser remote sensing*. Air Power Studies Centre Paper No.33, Royal Australian Air Force, June 1995

Kroutil, R., Lewis, P., Thomas, M., Curry, T., Miller, D., Combs, R. & Cummings, A. (2006). Emergency response chemical detection using passive infrared spectroscopy. *SPIE Newsroom*, doi: 10.1117/2.1200604.0157, 3 p

Kruse, F.A. (1990). Thematic mapping with an expert system and imaging spectrometers. (In *Advances in spatial information extraction and analysis for remote sensing: The II/VII ISPRS Commission International Workshop Proceedings* (January 13–17, 1990) (pp. 59–68). University of Maine, Orno, Maine)

Kruse, F. A., Lefkoff, A. B., Boardman, J. W., Heidebrecht, K. B., Shapiro, A. T., Barloon, P. J. & Goetz, A. F. H. (1993). The Spectral Image Processing System (SIPS)—Interactive visualization and analysis of imaging spectrometer data. *Remote Sensing of Environment*, 44, 145–163

Kulesz, J. (2004). *SensorNet fact sheet*. Oak Ridge National Laboratory, 31 August 2004. Retrieved March 19, 2007 from http://computing.ornl.gov/cse_home/datasystems/sensornet.pdf

Kumar, K. V., Martha, T. R. & Roy, P. S. (2006). Mapping damage in the Jammu and Kashmir caused by 8 October 2005 Mw 7.3 earthquake from the Cartosat-1 and Resourcesat-1 imagery. *International Journal of Remote Sensing*, 27(20), 4449–4459

Kwan, C., Ayhan, B., Chen, G., Wang, J., Ji, B. & Chang, C.-I. (2006). A novel approach for spectral unmixing, classification, and concentration estimation of chemical and biological agents. *IEEE Transactions on Geoscience and Remote Sensing*, 44(2), 409–419

Laefer, D. F., Koss, A. & Pradhan, A. (2006). The need for baseline data characteristics for GIS-based disaster management systems. *Journal of Urban Planning and Development*, 132(3), 115–119

Lancaster, N. & Nickling, W. G. (1994). Aeolian sediment transport. (In A. D. Abrahams & A. J. Parsons (Eds.), *Geomorphology of desert environments* (pp. 447–473). London: Chapman & Hall)

Lehmann, F., Bucher, T., Hese, S., Hoffmann, A., Mayer, S., Oschuts, F. & Zhang, Y. (1998). Data fusion of HyMAP hyperspectral with HRSC-A multispectral stereo and DTM data: Remote sensing data validation and application in different disciplines. *First EARSeL Workshop on Imaging Spectroscopy, University of Zurich, Switzerland, 6–8 October 1998* (Paris: EARSeL), pp. 105–117

Lévesque, J. & King, D. J. (1999). Airborne digital camera image semivariance for evaluation of forest structural damage at an acid mine site. *Remote Sensing of Environment*, 68, 112–124

Levner, I., Bulitko, V., Li, L., Lee, G. & Greiner, R. (2003). Towards automated creation of image interpretation systems. (In T.D. Gedeon & L.C.C. Fung (Eds.). *AI 2003: Advances in artificial intelligence*. Lecture Notes in Computer Science 2903. (pp. 653–665) Berlin: Springer)

Levy, J. K., Gopalakrishnan, C. & Lin, Z. (2005). Advances in decision support systems *Development*, 21(4), 593–612

Li, L., Ustin, S. L. & Lay, M. (2005). Application of AVIRIS data in detection of oil- induced vegetation stress at Jornada, New Mexico. *Remote Sensing of Environment*, 94, 1–16

Lichtenthaler, H. K. (1996). Vegetation stress: An introduction to the stress concept in plants. *Journal of Plant Physiology*, 148, 148–154

Liu, X. & Herold, M. (2007). Population estimation and interpolation using remote sensing. (In Q. Weng & D. A. Quattrochi (Eds.), *Urban remote sensing* (pp. 269–290) Boca Raton, FL: CRC)

Lo, C. P. (2006). Estimating population and census data. (In M. K. Ridd & J. D. Hipple (Eds.), *Remote sensing of humans settlements, manual of remote sensing*, Third Edition, Vol. 5 (pp. 337–377) Bethesda, MD: American Society for Photogrammetry and Remote Sensing)

Lu, Y. H., Trinder, J. C. & Kubik, K. (2006). Automatic building detection using the Dempster-Shafer algorithm. *Photogrammetric Engineering & Remote Sensing*, 72(4), 395–403

Mackenzie, D. (2007). Billions at risk from wheat super-blight. *New Scientist*, 2598, 6–7. Retrieved April 16, 2007 from http://environment.newscientist.com/channel/earth/mg19425983.700-billions-at-risk-from-wheat-superblight.html

MacKenzie, D. C. (2003). Collaborative tasking of tightly constrained multi-robot missions. (In Schultz et al. (Eds.), *Multi-robotsSystems: From swarms to intelligent automata, Proc. second international workshop on multi-robot systems* (pp. 39–50). Washington, DC: Kluwer)

Matsuoka, M., Vu, T. T. & Yamazaki, F. (2004). Automated damage detection and visualization of the 2003 Bam, Iran, earthquake using high-resolution satellite images. (In *Proceedings of the 25th Asian conference on remote sensing* (pp. 841–845))

McMaster, K. J. (2005). Forest blowdown prediction: A correlation of remotely sensed contributing factors. *Northern Journal of Applied Forestry*, 22(1), 48–53

Mikkelsen, T., Larsen, S. E. & Thykier-Nielsen, S. (1984). Description of the Risø puff diffusion model. *Nuclear Technology*, 67, 56–65

Mielke, H. W., Powell, E. T., Gonzales, C. R. & Mielke, P. W., Jr. (2006). Hurricane Katrina's impact on New Orleans soils treated with low lead Mississippi River Alluvium. *Environmental Science & Technology*, 40(24), 7623–7628

Mirik, M., Michels, G. J., Jr., Kassymzhanova-Mirik, Elliott, N. C. & Bowling, R. (2006). Hyperspectral spectrometry as a means to differentiate uninfested and infested winter wheat by greenbug (Hemiptera: Aphididae). *Journal of Economic Entomology*, 99(5), 1682–1690

Miura, H. & Midorikawa, S. (2006). Updating GIS building inventory data using high-resolution satellite images for earthquake damage assessment: Application to metro Manila, Philippines. *Earthquake Spectra*, 22(1), 151–168

Montoya, L. (2003). Geo-data acquisition through mobile GIS and digital video: An urban disaster management perspective. *Environmental Modelling & Software*, 18, 869–876

Moshou, D., Bravo, C., Oberti, R., West, J., Bodria, L., McCartney, A. & Ramon, H. (2005). Plant disease detection based on data fusion of hyper-spectral and multi-spectral fluorescence imaging using Kohonen maps. *Real-Time Imaging*, 11, 75–83

Moskal, L. M. & Franklin, S. E. (2004). Relationship between airborne multispectral image texture and aspen defoliation. *International Journal of Remote Sensing*, 25, 2701–2711

Moulin, C., Rougeault, S., Hamon, D. & Mauchien, P. (1993). Uranium determination by remote time-resolved laser-induced fluorescence. *Applied Spectroscopy*, 47(12), 2007–2012

Munger, F. (2006). Thwarting threats: ORNL works to detect terrorist activity, speed flow of info spreads across U.S. map. *The Knoxville News Sentinel*, 27 February 2006. Retrieved March 19, 2007 from http://www3.knoxnews.com/kns/news_columnists/article/0,1406,KNS_359_4498879, 00.html

Murtha, P. A. (1976). Vegetation damage and remote sensing: Principal problems and some recommendations. *Photogrammetria*, 32, 147–156

Myint S. W., Lam, N. S. -N. & Tyler, J. M. (2004). Wavelets for urban spatial feature discrimination: Comparisons with fractal, spatial autocorrelation, and spatial co-occurrence approaches. *Photogrammetric Engineering & Remote Sensing*, 70(7), 803–812

Nourbakhsh, I., Sargent, R., Wright, A., Cramer, K., McClendon, B. & Jones, M. (2006). Mapping disaster zones. *Nature*, 439, 787–788

Office of Technology Assessment (1980). *The effects of nuclear war*. (London: Croom Helm)

Orr, M. & Spohn, H. (1967). A multivariate statistical model for indirect bomb damage assessment. *Operations Research*, 15(4), 706–719

Osteraas, J. D. (2006). Murrah Building bombing revisited: A qualitative assessment of blast damage and collapse patterns. *Journal of Performance of Constructed Facilities*, 20(4), 330–335

Papastefanou, C. (2006). Residence time of tropospheric aerosols in association with radioactive nuclides. *Applied Radiation and Isotopes*, 64, 93–100

Pardue, J. H., Moe, W. M., Mcinnis, D., Thibodeaux, J., Valsaraj, K. T., Maciasz, E., Van Heerden, I., Korevec, N. & Yuan, Q. Z. (2005). Chemical and microbiological parameters in New Orleans floodwater following Hurricane Katrina. *Environmental Science & Technology*, 39(22), 8591–8599

Paul, R., Subramanian, S. & Bharadwaj, S. (2006). Creation of digital city model using single high resolution satellite image. *Photogrammetric Engineering & Remote Sensing*, 72(4), 341–342

Perissin, D. & Rocca, F. (2006). High-accuracy urban DEM using permanent scatters. *IEEE Transactions on Geoscience and Remote Sensing*, 44(11), 3338–3347

Presley, S. M., Rainwater, T. R., Austin, G. P., Platt, S. G., Zak, J. C., Cobb, G. P., Marsland, E. J., Tian, K., Zhang, B., Anderson, T. A., Cox, S. B., Abel, M. T., Leftwich, B. D., Huddleston, J. R., Jeter, R. M. & Kendall, R. J. (2006). Assessment of pathogens and toxicants in New Orleans, LA following hurricane Katrina. *Environmental Science & Technology*, 40(2), 468–474

Ramabhushanam, E. & Lynch, M. (1994). Structural assessment of bomb damage for World Trade Center. *Journal of Performance of Constructed Facilities*, 8(14), 229–242

Ranson, K. J., Kovacs, K., Sun, G. & Kharuk, V. I. (2003). Disturbance recognition in the boreal forest using radar and Landsat-7. *Canadian Journal of Remote Sensing*, 29(2), 271–285

Rauch, J. T., Jr. (2004). *Assessing airpower's effects: Capabilities and limitations of real-time battle damage assessment*. Thesis, The School of Advanced Airpower Studies, Maxwell Air Force Base, Alabama: Air University Press, 56 p

Rejaie, A. & Shinozuka, M. (2004). Reconnaissance of Golcuk 1999 earthquake damage using satellite images. *Journal of Aerospace Engineering*, 17(1), 20–25

Ridd, M. K., Bjorgo, E., Camp, L. E., Card, D. H., Chung, J. M., Dudley-Murphy, E. A., Gillies, R. R., Hipple, J. D., Hernandez, M. W., Hofmann, P., Jurgens, C., Lam, N. N. S.-N., Massasati, A. S., Merola, J. A., Mesev, V., Nash, G. D., Quattrochi, D. A. & Wheeler, D. J. (2006). Documenting dynamics of human settlements. (In M. K. Ridd & J. D. Hipple (Eds.), *Remote sensing of human settlements. Manual of remote sensing*, Third Edition, Vol. 5 (pp. 521–670) Bethesda, MD: American Society for Photogrammetry and Remote Sensing)

Rizzo, V. & Iodice, A. (2006). Satellite differential SAR interferometry for the evaluation of effects of hydrogeological disasters: Definition of a scale for damage evaluation. *Annals of Geophysics*, 49(1), 253–260

Rogan, J., Franklin, J. & Roberts, D. A. (2002). A comparison of methods for monitoring multitemporal vegetation change using Thematic Mapper imagery. *Remote Sensing of Environment*, 80, 143–156

Rogge, D. M., Rivard, B., Zhang, J. & Feng, J. (2006). Iterative spectral unmixing for optimizing per-pixel endmember sets. *IEEE Transactions on Geoscience and Remote Sensing*, 44(12), 3725–3736

Roper, W. E. (2005). Spatial multi-database integration for emergency operations support. *International Journal of Technology Transfer and Commercialisation*, 4(1), 3–30

Ross, T. & Lott, N. (2003). *A climatology of 1980–2003 extreme weather and climate events*. (Asheville: US Department of Commerce NOAA/NESDIS)

Rosso, P. H., Pushnik, J. C., Lay, M. & Ustin, S. L. (2005). Reflectance properties and physiological responses of *Salicornia virginica* to heavy metal and petroleum contamination. *Environmental Pollution*, 137, 241–252

Rycroft, M. J. (2000). From environmental risks, as problems, to earth observations as solutions to those problems. *Surveys in Geophysics*, 21, 115–125

Sadowski, F. G. & Covington, S. J. (1987). *Processing and analysis of commercial satellite image data of the nuclear accident near Chernobyl, U.S.S.R.* U.S. Geological Survey Bulletin 1785, Reston, VA: Department of the Interior, U.S. Geological Survey, U.S. G. P.O., 19 p

Saito, K., Spence, R. J. S., Going, C. & Markus, M. (2004). Using high-resolution satellite images for post-earthquake building damage assessment: A study following the 26 January 2001 Gujarat earthquake. *Earthquake Spectra*, 20, 1–25

Sakamoto, M., Takasago, Y., Uto, K., Kakumoto, S. & Kosugi, Y. (2004). Automatic detection of damaged area of Iran earthquake by high-resolution satellite imagery. *Proceedings of IEEE IGARSS*, 2, 1418–1421

Sanderson, D. C. W., Cresswell, A. J., Hardeman, F. & Debauche, A. (2004). An airborne gamma-ray spectrometry survey of nuclear sites in Belgium. *Journal of Environmental Radioactivity*, 72, 213–224

San Miguel-Ayanz, J., Vogt, J., De Roo, A. & Schmuck, G. (2000). Natural hazards monitoring: Forest fires, droughts, and floods—The example of European pilot projects. *Surveys in Geophysics*, 21, 291–305

Schneider, W. & Steinwender, J. (1999). Landcover mapping by interrelated segmentation and classification of satellite images. *International Archives of Photogrammetry and Remote Sensing*, 32(Part 7–4–3 W6), Valladolid

Schowengerdt, R. A. (1997). *Remote sensing: Models and methods for image processing*. Second Edition. (San Diego, CA: Academic)

Schwartz, P. & Randall, D. (2003). *An abrupt climate changes scenario and its implications for United States national security*. Global Business Network, Emeryville, CA, October 2003. [Electronic version]. Retrieved April 16, 2007 from http://www.gbn.com/ArticleDisplayServlet.srv?aid = 26231)

Schwarz, G. F., Rybach, L. & Klingelé, E. E. (1997). Design, calibration, and application of an airborne gamma spectrometer system in Switzerland. *Geophysics*, 62(5), 1369–1378

Schweitzer, B. & McLeod, B. (1997). Marketing technology that is changing at the speed of light. *Earth Observation Magazine*, 6(7), 22–24

Seaman, N. L. (2003). Future directions of meteorology related to air-quality research. *Environment International*, 29, 245–252

Shackelford, A. K. & Davis, C. H. (2003). A hierarchical fuzzy classification approach for high-resolution multispectral data over urban areas. *IEEE Transactions on Geoscience and Remote Sensing*, 41(9), 1920–1932

Shearer, A. W. (2005). Whether the weather: Comments on 'An abrupt climate change scenario and its implications for United States national security'. *Futures*, 37, 445–463

Shippert, P. (2004). Why use hyperspectral imagery? *Photogrammetric Engineering & Remote Sensing*, 70(4), 377–80

Showalter, P. S. (2001). Remote sensing in disaster research: A review. *Disaster Prevention and Management*, 10(1), 21–29

Silleos, N., Perakis, K. & Petsanis, G. (2002). Assessment of crop damage using space remote sensing and GIS. *International Journal of Remote Sensing*, 23(3), 417–427

Skuterud, L., Goltsova, N. I., Naeumann, R., Sikkeland, T. & Lindmo, T. (1994). Histological changes in *Pinus sylvestris* L. in the proximal-zone around the Chernobyl power plant. *The Science of the Total Environment*, 157, 387–397

Skyttner, L. (2002). Monitoring and early warning systems—A design for human survival. *Kybernetes*, 31(2), 220–245

Smith, M. A. J., Dekker, R., Kos, J. & Hontelez, J. A. M. (2001). The availability of unmanned air vehicles: A post-case study. *Journal of the Operational Research Society*, 52, 161–168

Song, L.-P., Yu, C. & Liu, Q. H. (2005). Through-wall imaging (TWI) by radar: 2-D tomographic results and analyses, *IEEE Transactions on Geoscience and Remote Sensing*, 43, 2793–2798

Soukhova, N. V., Fesenko, S. V., Klein, D., Spiridonov, S. I., Sanzharova, N. I. & Badot, P. M. (2003). ^{137}Cs distribution among annual rings of different tree species contaminated after the Chernobyl accident. *Journal of Environmental Radioactivity*, 65, 9–28

Stefanov, W. L., Ramsey, M. S. & Christensen, P. R. (2003). Identification of fugitive dust generation, transport, and deposition areas using remote sensing. *Environmental & Engineering Geoscience*, 9(2), 151–165

Stilla, U. & Soergel, U. (2007). Reconstruction of buildings in SAR imagery of urban areas. (In Q. Weng & D. A. Quattrochi (Eds.), *Urban remote sensing* (pp. 47–67). Boca Raton, FL: CRC)

Stramondo, S., Bignami, C., Chini, M., Pierdicca, N. & Tertulliani, A. (2006). Satellite radar and optical remote sensing for earthquake damage detection: Results from different case studies. *International Journal of Remote Sensing*, 27(20), 4433–4447

Stull, R. B. (1991). *An introduction to boundary layer meteorology*. (Dordrecht, The Netherlands: Kluwer)

Sutton, P. C., Elvidge, C. & Obremski, T. (2003). Building and evaluating models to estimate ambient population density. *Photogrammetric Engineering & Remote Sensing*, 69(5), 545–553

Suzuki, R., Masuda, K. & Dye, D. G. (2007). Interannual covariability between actual evapotranspiration and PAL and GIMMS NDVIs of northern Asia. *Remote Sensing of Environment*, 106, 387–398

Tobler, W. (1969). Satellite confirmation of settlement size coefficients. *Area*, 1, 30–34

Tomás, R., Márquez, Y., Lopez-Sanchez, J. M., Delgado, J., Blanco, P., Mallorquí, J. J., Martínez, M., Herrera, G. & Mulas, J. (2005). Mapping ground subsidence induced by aquifer overexploitation using advanced differential SAR interferometery: Vega Media of the Segura River (SE Spain) case study. *Remote Sensing of Environment*, 98, 269–283

Tralli, D. M., Blom, R. G., Zlotnicki, V., Donnellan, A. & Evans, D. L. (2005). Satellite remote sensing of earthquake, volcano, flood, landslide and coastal inundation hazards. *ISPRS Journal of Photogrammetry & Remote Sensing*, 59, 185–198

Tucker, C. J. (1979). Red and photographic infrared linear combinations for monitoring vegetation. *Remote Sensing of Environment*, 8, 127–150

Tucker, M. & San, B. T. (2003). SPOT HRV data analysis for detecting earthquake- induced changes in Izmit, Turkey. *International Journal of Remote Sensing*, 24(12), 2439–2450

Turker, M. & San, B. T. (2004). Detection of collapsed buildings caused by the 1999 Izmit, Turkey, earthquake through digital analysis of post-event aerial photographs. *International Journal of Remote Sensing*, 25, 4701–4714

United States, Congress, Senate, Committee on Agriculture, Nutrition, & Forestry (1986). *Possible impact on agriculture of the explosion of the Soviet nuclear plant at Chernobyl: Hearing before the Committee on Agriculture, Nutrition, and Forestry, United States Senate, Ninety-ninth Congress, second session, May 15, 1986*. S. HRG. 99–718, Washington, DC: U.S. G.P.O., 59 p

United States, Congress, Joint Economic Committee, Subcommittee on Agriculture and Transportation (1987). *The Chernobyl disaster: Implications for world food security and the U.S. farm economy: Hearing before the Subcommittee on Agriculture and Transportation of the Joint Economic Committee, Congress of the United States, Ninety-ninth Congress, second session, May 5, 1986*. S. HRG. 99–1029, Washington, DC: U.S. G.P.O., 68 p

Van der Meer, F. D. (2004). Pixel-based, stratified and contextual analysis of hyperspectral imagery. (In S. M. deJong & F. D. van der Meer (Eds.), *Remote sensing image analysis: Including the spatial domain*. (pp. 153–180). Dordrecht, The Netherlands: Kluwer)

Van de Voorde, T., De Genst, W. & Canters, F. (2007). Improving pixel-based VHR land-cover classifications of urban areas with post-classification techniques. *Photogrammetric Engineering & Remote Sensing*, 73(9), 1017–1027

Vincent, P., Larsen, S., Galloway, D., Laczniak, R. J., Walter, W. R., Foxall, W. & Zucca, J. J. (2003). New signatures of underground nuclear tests revealed by satellite interferometry. *Geophysical Research Letters*, 30(22), 2141. doi: 10.1029/2003GL018179, 2003

Vincente-Serrano, S. (2007). Evaluating the impact of drought using remote sensing in a Mediterranean, semi-arid region. *Natural Hazards*, 40, 173–208

Volcani, A., Karnieli, A. & Svoray, T. (2005). The use of remote sensing and GIS for spatio-temporal analysis of the physiological state of a semi-arid forest with respect to drought years. *Forest Ecology and Management*, 215, 239–250

Vosselman, G. & Dijkman, S. (2001). 3D building model reconstruction from point clouds and ground plans. *International Archives of Photogrammetry and Remote Sensing* XXXIV-3/W4, Annapolis, MD, 22–24 October 2001, pp. 37–43

Warner, T. (2005). Hyperspherical direction cosine change vector analysis. *International Journal of Remote Sensing*, 26(6), 1201–1215

Wightman, J. M. & Gladish, S. L. (2001). Explosions and blast injuries. *Annals of Emergency Medicine*, 37(6), 664–678

Wilhite, D. A. (2000). Drought as a natural hazard: Concepts and definitions. (In D. A. Wilhite (Ed.), *Drought: A global assessment*, Natural Hazards and Disasters Series (pp. 3–18). London: Routledge)

Williams, J. H. & Ashenden, T. W. (1992). Differences in the spectral characteristics of white clover exposed to gaseous pollutants and acid mist. *New Phytologist*, 120, 69–75

Winkelmann, I., Strobl, C. & Thomas, M. (2004). Aerial measurements of artificial radionuclides in Germany in case of a nuclear accident. *Journal of Environmental Radioactivity*, 72, 225–231

Winter, M. E. (1999). N-FINDR: An algorithm for fast autonomous spectral end-member determination in hyperspectral data. (In M. R. Descour & S. S. Shen (Eds.), *Imaging spectrometry V*, Proc. SPIE, Vol. 3753 (pp. 266–275))

Yang, Y. & Fathy, A. E. (2005). See-through-wall imaging using ultra wideband short- pulse radar system. In *IEEE Intl. Antennas and Propagation Symposium*, Vol. 3B, pp. 334–337, 3–8 July 2005

Yonezawa, C. & Takeuchi, S. (2001). Decorrelation of SAR data by urban damages caused by the 1995 Hyogoken-nanbu earthquake. *International Journal of Remote Sensing*, 22(8), 1585–1600

Yost, K. A. & Washburn, A. R. (2000). Optimizing assignment of air-to-ground assets and BDA sensors. *Military Operations Research*, 5(2), 77–91

Chapter 8
Estimating Flood Damage in Texas Using GIS: Predictors, Consequences, and Policy Implications

Samuel D. Brody[1] **and Sammy Zahran**[2]

Abstract Floods continue to pose the greatest threat to the property and safety of human communities among all natural hazards in the United States. While the link between urbanization and flooding is established, the degree to which specific characteristics of the built environment affect the level of damage sustained by a community has never been thoroughly investigated at the regional scale. Our study addresses this lack of research by examining the relationship between the built environment and flood impacts in Texas, which consistently sustains the most damage from flooding more than any other state in the country. Specifically, we calculate property damage resulting from 423 flood events over a five year period between 1997 and 2001 at the county level. We identify the impact of several built environment measures, including wetland alteration, impervious surface, and dams on reported property damage while controlling for biophysical and socioeconomic characteristics. Statistical results suggest that naturally occurring wetlands play a particularly important role in mitigating flood damage. These findings provide guidance to homeland security experts and flood managers on how to most effectively mitigate the costly impacts of floods at the community level.

Keywords Built environment, flood damage, Texas, wetland alteration

8.1 Introduction

Floods can be considered one of the most important homeland security issues in the United States (US) as they pose the greatest threat among all natural hazards to the safety and economic well-being of local communities. The economic impact from floods is estimated in the billions of dollars annually (Peilke 1996; ASFPM 2000). According to data from the Spatial Hazard Events and Losses Database for the United States (SHELDUS), the average annual flood count has increased six-fold

[1]Texas A&M University

[2]Colorado State University

from 394 floods per year in the 1960s to 2,444 flood events a year in the 1990s. Kunreuther and Roth (1998) estimated $22 billion in flood losses from 1949–1988 in the US compared to $80 billion from 1989–1997. This evident spike in flood-related property damage cannot be explained away by inflation in monetary systems, population growth along the coast, or increases in annual mean precipitation. The rising cost of floods is, instead, driven by the way humans plan for and subsequently develop their communities. Individual and community-based decisions pertaining to where buildings and impervious surfaces are distributed, and the degree to which hydrological systems are altered, may be exacerbating losses from repetitive floods. Increasing development associated with residential, commercial, and tourism activities, particularly in coastal and low-lying areas, has diminished the capacity of hydrological systems (e.g. watersheds) to naturally hold and store surface water run-off. As a result, private property, households, and the economic well-being of coastal communities are increasingly vulnerable to the risks of repetitive flooding events (Mileti 1999).

While the importance of maintaining the integrity of hydrological systems is well understood, the degree to which the human built environment affects the level of damage sustained by a community has never been thoroughly investigated at the regional scale (Peilke 2000). Aside from small-scale case studies conducted within a single watershed or jurisdiction, no study to date has thoroughly tested the impact of the human built environment based on multiple flood events over time, at large spatial scales, and controlling for multiple biophysical and socioeconomic characteristics.

We address this lack of research and the general homeland security issue by using multiple sources of geospatial data. Traditionally, social science research has not made widespread use of geospatial data sources or geographic information systems (GIS) when measuring variables and conducting analyses. However, using this technology to map policy and strategy across landscapes provides a far greater degree of locational specificity when attempting to understand the factors affecting flood damage. Using geospatial techniques can thus produce research results that help decision makers be more spatially precise and effective when adopting flood mitigation policies.

We apply these geospatial techniques to empirically examine the relationship between the built environment and flood impacts in the eastern portion of Texas. Specifically, we calculate property damage resulting from 423 flood events over a five year period between 1997 and 2001 at the county level and then use statistical modeling to isolate the impact of several built environment measures including wetland alteration, impervious surface, and dams on reported property damage. Texas is an ideal study area since it consistently has the most deaths and damages from flooding of any state. Of the 42 flood events listed as causing more than a billion dollars in damage between 1980 and 1998, 4 were in Texas (NCDC 2000). According to Federal Emergency Management Agency (FEMA) statistics on flood insurance payments from 1978 to 2001, Texas suffered $2.25 billion dollars in property loss, more than California, New York and Florida combined (NFIP: *http://www.fema.gov/nfip/pcstat.htm*).

Results from our study provide important information to environmental planners and flood managers on how development and modifications of natural landscapes adversely impact flood outcomes. Such information is critical given the continued development of coastal areas and the increasing vulnerability of human populations to inland coastal flooding. Our findings may thus provide guidance on how to build more sustainable, resilient communities over the long term.

8.2 Adverse Impacts of the Built Environment on Flooding

8.2.1 Impervious Surfaces

In the US, rapid growth and sprawling development patterns have contributed to a marked increase in urbanization and built-up land uses. As of 2000, almost 80 percent of the population resided in urban areas (US Census Bureau 2000). Between 1982 and 1997, there was a 34 percent increase in the amount of land devoted to urban or built-up uses; this area is projected to increase by 79 percent in the next 25 years, raising the proportion of the total land base that is developed from 5.2 to 9.2 percent (Alig et al. 2004). Conversion of agricultural and forest lands to urban areas can diminish a hydrological system's ability to store and slowly release water, resulting in increased severity of flooding (Carter 1961; Tourbier and Westmacott 1981).

A major component of urbanization and contributor to flood occurrence is the increase in impervious surfaces. The link between impervious surface coverage and floods has been established since the late 1960s (Leopold 1968; Seaburn 1969). As the area of impervious surface coverage increases, there is a corresponding decrease in infiltration and an increase in surface runoff (Dunne and Leopold 1978; Paul and Meyer 2001). According to Arnold and Gibbons (1996), as the percent catchment (i.e. drainage basin) impervious surface cover increases to 10–20 percent, runoff increases twofold. Most recently, White and Greer (2006) found that as urbanization in the Peñasquitos Creek watershed in southern California increased from 9–37 percent, total runoff increased by an average of 4 percent per year. This yearly figure represents a rise of over 200 percent based on the authors' study period from 1973 to 2000. Greater surface runoff volume often results in increased frequency and severity of flooding from streams. Finally, Brody et al. (2007) note that an increase in impervious surfaces (as measured through remote sensing imagery) correlated with a significant increase in stream flow over a twelve year period across 85 coastal watersheds in Texas and Florida.

The urban built environment has also been linked to increased peak discharges (Leopold 1994; Burges et al. 1998; Brezonik and Stadelmann 2002). In this instance, the lag time (time difference between the center of precipitation volume to the center of runoff volume) is compressed, resulting in floods that peak more rapidly (Hirsch et al. 1990). For example, Rose and Peters (2001) report peak discharge increases of approximately 80 percent in urban catchments with 50 percent impervious area. Flood discharge in proportion to impervious surface cover were at least 250 percent higher in urban compared to forested catchments in Texas and New York after similar storms (Espey et al. 1965; Seaburn 1969; Paul and Meyer 2001). Examining mean peak discharges for 27 storms in the Croton River Basin in New York, Burns et al. (2005) observed a 300 percent increase in a catchment with an impervious area of only 11.1 percent. These studies demonstrate that urbanization not only increases runoff volume, but also peak discharges and associated flood magnitudes.

8.2.2 Wetland Alteration

One of the most significant ways impervious surface cover can exacerbate flooding is through the alteration or elimination of naturally occurring wetlands. The development of wetlands is considered central to the loss of natural water retention within watershed units and an increase in flood hazards for local communities. Wetlands not only provide the ecological infrastructure for watershed systems, but are also believed to provide natural flood mitigation by maintaining a properly functioning water cycle (Mitch and Gosselink 2000; Lewis 2001).

Early research on wetlands and flooding focused on the differences between drained and natural wetlands as a basis for assessment. The results from these studies showed that undrained peat bogs reduce low-return period flood flow and reduce overall storm flows when compared to their drained counterparts (Verry and Boelter 1978; Heikuranen 1976; Daniel 1981). Additional work relying primarily on linear regression analysis yielded similar results. For example, Conger (1971) demonstrated that the ability of wetlands to store water significantly reduced peak flows for recurrence intervals up to 100 years. Examining four different types of wetlands, found that each had a negative effect on flood flows. Novitski (1985) also concluded that basins with as little as 5 percent lake and wetland area may result in 40 percent to 60 percent lower flood peaks.

More recent research utilizing simulation models also demonstrates the flood-reducing role of wetlands. Ammon et al. (1981) modeled the effects of wetlands on both water quantity and quality of Chandler Slough Marsh in South Florida. Results indicate that maximum flood peak attenuation is higher with increasing areas of marsh. The authors concluded that Chandler Slough Marsh increases storm water detention times, changes runoff regimes from surface to increased subsurface regimes, and is 'moderately effective as a water quantity control unit' (p. 326). Ogawa and Male (1986) also developed a simulation model to explore the preservation of wetlands as a flood mitigation strategy. Using four scenarios of downstream wetland encroachment ranging from 25–100 percent loss, the authors found that increased encroachment resulted in significant increases in peak flow.

Other studies are not as clear on the benefits of wetland protection and restoration as a tool for flood mitigation. The 1994 Galloway Report concluded that upland wetlands could be effective for smaller floods, but diminish in value as storage capacity is exceeded for larger floods. This report states that the effect of wetlands on peak flows for large floods along main rivers are inconclusive and that additional research is needed. Also, using model simulations, Padmanabhan and Bengston (2001) found that wetland restoration in the Maple River watershed in North Dakota did not have significant effects on high-return period flood events.

Research based on direct observation also supports the notion that wetlands play an important role in reducing the degree of flooding. For example, recent findings demonstrate that wetlands are able to absorb and hold greater amounts of floodwater than previously thought. Based on an experiment that involved constructing wetlands along the Des Plaines River in Illinois, it was found that a marsh of only 5.7 acres could retain the natural run-off of a 410-acre watershed. This study estimated that only 13 million acres of wetlands (3 percent of the upper Mississippi watershed) would have been needed to prevent the catastrophic flood of 1993 (Godschalk et al. 1999). Other observational research suggests there is a critical threshold for the effects of wetland loss on flood storage. For example, in a study that utilized the record of stream flow data from stream gauge stations, Johnston et al. (1990) found that small wetland losses in watersheds with less than 10 percent of wetlands could have a significant effect on increased flooding.

It is clear from the literature that the value of wetlands for flood mitigation and the intersection between environmental protection and flood management needs further study. A comprehensive review of the literature conducted by Bullock and Acreman (2003) showed that wetlands do play a significant role in modifying the hydrological cycle. The authors find that for 23 of the 28 studies on wetlands and flooding, 'floodplain wetlands reduce or delay floods' (p. 366). Overall, research suggests that wetlands may reduce or slow downstream flooding.

8.2.3 Mitigation: Structural Versus Nonstructural Techniques

Flooding and the reduction of flood damage were initially addressed through community works projects, such as dams, levees, dikes, and channel improvements. Structural approaches to flood mitigation can protect property and lives, particularly for upstream flood events. For example, the US Army Corps of Engineers estimated that flood damages from 1991 to 2000 were approximately $45 billion dollars, yet the organization's flood control measures prevented over $208 billion dollars of additional damage (USACE 2002). Zahran et al. (under review) noted that an increase in the number of dams decreased the odds of a death or injury from a flood by 21.6 percent.

While the benefits of structural mitigation devices are clear, so are their disadvantages.

Structural flood-control projects can encourage development in vulnerable areas that would have otherwise remained undeveloped. Also, when a flood event exceeds the capacity of a flood control structure, the resulting flood damages can be significantly higher (White 1945; Hertzler 1961; Burby et al. 1985; Stein et al. 2000; Larson and Plasencia 2001). Finally, structural measures such as dams are built at a very high price. Since the 1940s, the USACE has spent over $100 billion dollars (1999 dollars) on structural flood protection projects (Stein et al. 2000). From a cost-effectiveness standpoint, structural solutions to flood mitigation such as dams are extremely costly, can exacerbate development in flood prone areas out of a false sense of security (Harding and Parker 1974; Tobin 1995; Pielke 1999), and can present a hazard in themselves in the case of structural failure.

More recent public-sector flood mitigation initiatives have taken a non-structural approach. The most widely implemented alternative is the National Flood Insurance Program (NFIP). Established in 1968 as an attempt to combat rising flood losses, the NFIP has, by many accounts, successfully brought flood insurance and a form of flood mitigation to the forefront of many communities. However, several scholars have raised concern over NFIP's effect on subsidizing and encouraging floodplain development, the overall equitability of the program, and the high financial costs of repetitive losses (Godschalk et al. 1999; Platt 1999; Birkland et al. 2003).

FEMA's Community Rating System (CRS) adopted in the early 1990s encourages communities to go beyond the NFIP's minimum standards for floodplain management by providing discounts of up to 45 percent on flood insurance premiums for residents of participating communities. Credit points are assigned for 18 measures organized into the following 4 broad categories of floodplain planning and management activities: public information, mapping and regulation, flood damage reduction, and flood preparedness. Premium discounts correspond to credit points. Communities with higher CRS scores (on a scale from 9 to 1) have implemented a greater number of the 18 flood mitigation measures and thus receive a higher premium discount for insurance coverage. Discounts range from 5 (class 9) to 45 percent (class 1) depending on the degree to which a community plans for the adverse impacts of floods. Specific activities under the CRS program include: preservation of floodplain areas as open spaces to enhance the storage of flood water; the use of low-density zoning to protect the natural functions of floodplains; and the adoption and implementation of a comprehensive floodplain management plan. In general, proactive local land use policies effectively reduce flood damage because they focus development on areas least likely to be inundated in a storm.

8.3 Research Methods and Data Analysis

8.3.1 Sample Selection

As mentioned above, we selected for analysis 423 damaging flood events across a 37 county study area in eastern Texas between 1997 and 2001. This area is ideal for examining the impact of the built environment on inland flooding (excluding tidal or surge-based flooding) for the following reasons: 1) Texas suffers significantly more property damage from floods than any other state in the country; 2) these floods tend to be spatially repetitive over time; 3) eastern Texas has been experiencing large increases in impervious surfaces and alteration of wetlands associated with rapid coastal development; and 4) these development patterns vary spatially across counties.

8.3.2 Concept Measurement

8.3.2.1 Dependent Variable

The dependent variable, *flood property damage*, is measured as the total dollar loss (in CPI adjusted 1997 US$) from a flood event. This variable is log-transformed to approximate a Gaussian distribution. Data on flood property damage are collected from the SHELDUS database at the Hazard Research Lab at University of South Carolina, Columbia (Hazards Research Lab 2006). The database consists of a county-level inventory of 18 natural hazard types, including hurricanes, floods, wildfires, and drought. Each hazard event record includes a start and end date, estimated property loss, as well as the number of human injuries and deaths. Our property damage variable ranges from $1,000 to $69 million. It is important to note that dollar estimates of property loss from flooding are often obtained through 'windshield' observation or secondary reports that may under or over-estimate the true loss experienced by a property owner. While this is the best available information collected by the National Weather Service, and we assume any error is unsystematic, the results should be interpreted with some level of caution. Also, our property loss variable does not include indirect financial loss (e.g. loss of wages) from floods, which may prove to be a significant economic burden.

8.3.2.2 Biophysical Variables

To properly estimate the effect of built environment on flood related property damage, one must control for storm intensity and flood duration. In our model, we measure four biophysical predictors of flood damage: precipitation (day of the flood event), precipitation (day before the flood event), flood duration, and floodplain overlap. *Precipitation (day of the flood event)* and *precipitation (day before the flood event)* are measured as the average surface precipitation (in hundredths of an inch) recorded by county weather stations. We recorded precipitation the day before a recorded storm in recognition of the lag time (shortened in urban areas) between runoff and stream peak discharge. The number of weather stations varies across counties and within a county longitudinally. Harris County has the highest number of weather stations with 12. We collected daily surface precipitation data from the NNDC Climate Data Online search engine. Search results include latitude, longitude, and altitude coordinates for weather stations, name and county location, as well as 'quality controlled' data on daily (24-hour observation period) surface precipitation. The highest precipitation total recorded for a flood event in our study was 14.11 inches in Jackson County in November of 1998. We measured *Flood duration* as a dichotomous variable. A flood event was assigned a score of one (1) if it lasted more the one day, and a score of zero (0) if it lasted one full day or less. Duration estimates were derived from SHELDUS records on the start and end dates of a hazard event. *Floodplain overlap* is calculated as the percentage of a county's area within a FEMA defined 100-year floodplain (delineated areas that have a one percent chance of flooding in any one year) using geographic information systems (GIS) analytical techniques. Specifically, we used a simple intersection command for county and floodplain boundaries to derive an estimate for the area of each jurisdiction in the floodplain. Floodplain estimates were derived from FEMA Digital Q3 flood data which we obtained for all of the counties in the sample. Jefferson County has the highest percentage of its land area within the 100-year flood plain in our sample with approximately 60 percent.

8.3.2.3 Built Environment Variables

We measured and analyzed three built environment variables shown to affect the degree of community-wide flood damage. We calculated *impervious surface* as the percentage of land covered by impervious surfaces (i.e. pavement, buildings, etc.) in a county area. Impervious surface was developed using GeoCover satellite imagery from NASA Stennis Space Center. Imagery from 1990 and 2000 were classified by several iterations of an unsupervised classification method. Digital Ortho Quarter Quads (DOQQ) imagery was used to confirm classification accuracy. We summed impervious surface area by county units for 1990 and 2000 and then calculated monthly values for the study period assuming an equal interval rate of change. *Wetland alteration* is measured as the cumulative total of spatially-defined wetland permits the day of a flood event. Wetland permits, required under Section 404 of the Clean Water Act, enable an applicant to alter a naturally occurring wetland for a construction project and were obtained from the US Army Corp of Engineers (USACE) District Office in Galveston, Texas. Permit records include permit type (individual permits; letters of permission; general permits; and nationwide permits), the date of permit issuance, and the latitude/longitude coordinates of the permit approved development activity. Of the 10,921 permit records received from the USACE, 7,957 had sufficient information to be located geographically. The number of permits was then recorded cumulatively by country for each flood event. Finally, the number of *dams* in a county area was tabulated to estimate the extent to which water embankments function to reduce flood property damage. Locations of dams were obtained from the US Army Corps of Engineers and summed by each county unit. Harris County has four dams, the highest number in our sample.

8.3.2.4 Policy and Socioeconomic Control Variables

We measured both a policy-related and socioeconomic variable that act as statistical controls to better isolate the influence of the built environment on flood damage. *FEMA CRS rating* scores are based on the FEMA Community Rating System (CRS). As mentioned above, the CRS promotes mitigation of flood damage through insurance premium discounts and other financial incentives. We calculated the premium discount for each jurisdiction in the sample on a scale from 0 to 45, where 0 is no discount and 45 is the highest percentage of premium discount allowed under the CRS program. Counties with higher discounts engage in more flood mitigation activities. It is important to note that discounts move in 5 percent increments. In Texas, Galveston County has the highest FEMA rating with a 15 percent premium discount.

To control for the economic status of a locality, we obtained income data from the US Census Bureau's 1990 and 2000 Summary Tape Files. *Median household income* is measured as the sum of money income received in calendar year 1999 by all household members 15 years old and over, including household members not related to the householder, people living alone, and other non-family household members. We included household income in the statistical model to control for the degree of wealth, and by proxy, the value of structures in each county. We presume that wealthier communities have the financial capacity to more effectively mitigate flooding but at the same time have greater financial capital that can be lost from damaging floods.

8.3.3 Data Analysis

We analyzed the data in two phases. First, we report descriptive statistics related to the spatial and temporal pattern of flood damage over the five-year study period. Second, we use multiple regression analyses to estimate the effect of the built environment and various control variables on reported flood damage in eastern Texas. Tests for estimate reliability including specification, muliticollinearity, and autocorrelation exhibited no significant violation of regression assumptions. Based on statistical diagnostics, we did however detect heteroskedasticity in the data leading us to analyze regression equations with robust standard errors.

8.4 Results

Over the five year study period, 423 flood events caused over $320 million in reported property damage among Texas coastal counties. The average amount of damage per flood during this time period was $423,765.90. The majority of this damage occurred during a two-day tempest beginning October 17 and 18, 1998.

This storm spread over much of the 37-county study area, from Galveston to Guadalupe County. Guadalupe County incurred the most damage in the sample with approximately $89 million over 20 flood events (Table 8.1). Neighboring Gonzales County experienced a similar degree of flood damage (almost $89 million) over 23 events. In contrast, Hidalgo County reported the lowest amount of

Table 8.1 Floods, property damage, and property damage per flood by Texas County, 1997–2001

Rank	County	Total Floods	Total Property Loss	Average Property Loss
1	Guadalupe	20	89,100,000.00	4,457,400.00
2	Gonzales	23	88,900,000.00	3,867,022.00
3	DeWitt	33	74,200,000.00	2,247,470.00
4	Karnes	19	17,200,000.00	904,157.90
5	Jefferson	12	10,700,000.00	889,166.70
6	Liberty	12	10,600,000.00	882,000.00
7	Bastrop	12	7,255,000.00	604,583.30
8	Lavaca	34	6,653,000.00	195,676.50
9	Fayette	13	4,560,000.00	350,769.20
10	Harris	26	1,628,000.00	62,615.38
11	Newton	6	1,280,000.00	213,333.30
12	Brazoria	13	944,000.00	72,615.38
13	Fort Bend	10	926,000.00	92,600.00
14	Walker	12	882,000.00	73,500.00
15	Orange	17	845,000.00	49,705.88
16	Jasper	7	585,000.00	83,571.43
17	Hardin	6	555,000.00	92,500.00
18	Galveston	17	512,000.00	30,117.65
19	Polk	9	409,000.00	45,444.44
20	Montgomery	17	385,000.00	22,647.06
21	San Jacinto	16	319,000.00	19,937.50
22	Wharton	11	263,000.00	23,909.09
23	Jackson	10	255,000.00	25,500.00
24	Tyler	3	235,000.00	78,333.33
25	Chambers	9	217,000.00	24,111.11
26	Matagorda	4	206,000.00	51,500.00
27	Austin	11	196,000.00	17,818.18
28	Madison	9	180,000.00	20,000.00
29	Trinity	10	141,000.00	14,100.00
30	Colorado	10	117,000.00	11,700.00
31	Nueces	2	100,000.00	50,000.00
32	Starr	3	67,000.00	22,333.33
33	Sabine	1	45,000.00	45,000.00
34	Cameron	1	21,200.00	21,200.00
35	Waller	3	15,000.00	5,000.00
36	Brooks	1	10,000.00	10,000.00
37	Hidalgo	1	2,000.00	2,000.00
		423	**320,508,200.00**	**423,765.90**

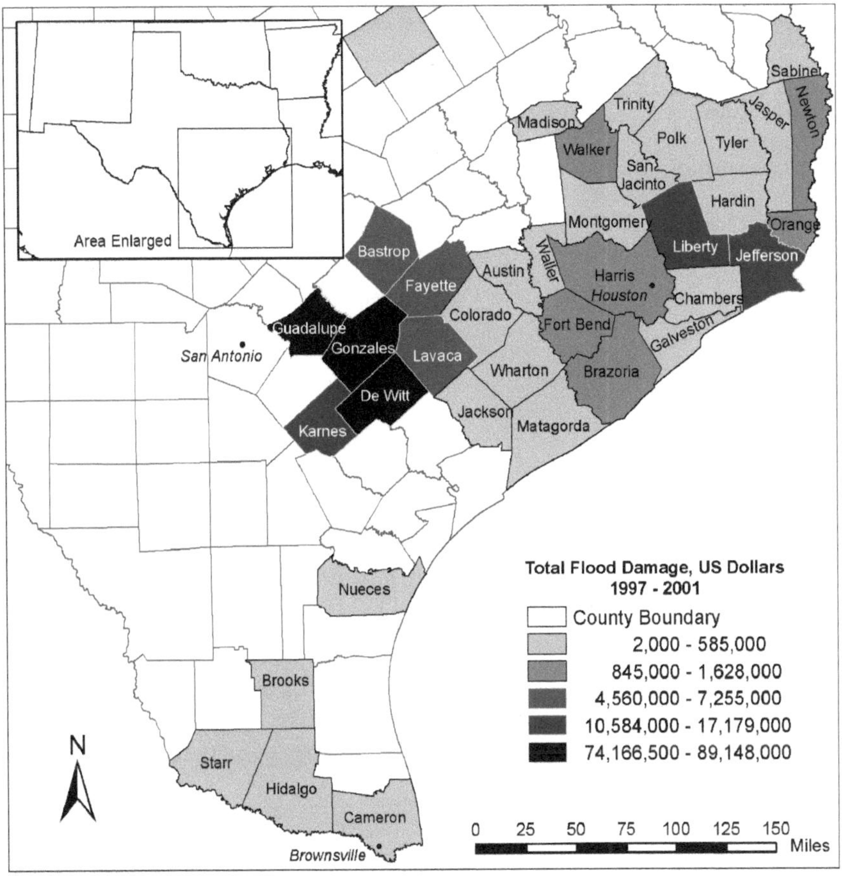

Fig. 8.1 Cumulative flood damage from 1997–2001

damage, \$2,000 for one only event. The number of events does not always correspond to the amount of property damage. For example, while Lavaca County experienced 34 flooding events during the study period, it reported only \$6.6 million in property damage. Similarly, Harris County endured 26 flood events, but its total property damage was an estimated \$1.6 million. Figure 8.1 displays the cumulative property damage from floods over the study period. Flood events caused most of the reported property damage to the west of the study area within Guadalupe, Gonzales, DeWitt, and Karnes counties. A geographic hotspot of damage also occurs to the east center on Jefferson and Liberty counties.

Multivariate regression analysis with standardized coefficients indicates the most influential factors on the degree of flood damage in eastern Texas (Table 8.2). We loaded the following three suites of variables sequentially to the model to test their effects both individually and as a group: biophysical, built, and socioeconomic

Table 8.2 OLS regression models predicting property damage from floods in Texas, 1997–2001

	B	Beta	B	Beta	B	Beta
Natural Environment Variables						
Precipitation (day of event)	.0515* (.0277)	.1636	.0464* (.0281)	.1476	.0473* (.0278)	.1504
Precipitation (day before event)	.1259*** (.0336)	.3083	.1338*** (.0342)	.3277	.1320*** (.0338)	.3234
Floodplain overlap	.4516* (.2565)	.0665	.2177 (.2620)	.0320	.1748 (.2813)	.0257
Duration of flood	.4234*** (.1311)	.1976	.4396*** (.1327)	.2003	.4295*** (.1311)	.2004
Built Environment Variables						
Impervious surface			.0100* (.0052)	.0826	.0085* (.0050)	.0702
Wetland alteration			.0004*** (.0001)	.1161	.0005*** (.0002)	.1581
Dams			−.0723** (.0401)	−.1061	−.0644* (.0392)	−.0944
Socioeconomic Variables						
FEMA rating					−.0184* (.0102)	−.1073
Median household income					6.3e-06 (5.1e-06)	.0604
Constant	3.8428*** (.0769)		3.7502*** (.0857)		3.5867*** (.1676)	
N	423		423		423	
F	20.80		15.96		12.88	
Probability > F	.000		.000		.000	
Adjusted R-squared	0.2984		.3130		.3189	
Root MSE	.71384		.70891		.70756	

Robust standard errors are in parentheses. Null test of coefficient equal to zero, *p < .10, **p < .05; ***p < .01

environments. Biophysical variables as a whole explain the most variance on the dependent variable (over 29 percent). Adjusting for precipitation the day of the flood event, rainfall amount on the day before the actual flood event is the strongest predictor of damage, followed by the duration of a flood (where $p < .05$). Precipitation the day of the flood event and the percentage of a county within the 100-year floodplain are, by comparison, weaker yet still statistically significant predictors of flood damage (where $p < .1$).

With the addition of human-built environmental factors to the model, the floodplain variable is no longer statistically significant at the.1 level. Increasing amounts of wetland alteration (the majority of which are located within the 100-year floodplain), correspond to a significant increase in reported property damage ($p < .01$).

Of the built environment variables examined, wetland alteration is the strongest partial correlate of flood property damage (β =.1161). Increasing amounts of impervious surfaces within each county also contributes to marked increases in flood damage (where $p < .1$). The presence of dams as flood control devices appears to reduce the amount of damage ($p < .1$) almost to the same degree to which damage is exacerbated by wetland alteration (β = –.1061). In effect, what is gained by dams in the mitigation of flood outcomes is statistically offset by development activities in wetlands.

In the fully-specified model containing socioeconomic variables, approximately 32 percent of variation in flood related property damage is explained. Counties with higher reductions in insurance premiums based on their FEMA CRS scores experience lower amounts of flood damage at the.1 level of significance. The effect size of FEMA rating (β = –.1073), summarizing the flood mitigation efforts undertaken by a locality, rivals the effect size of precipitation on the day of a flood event (β = –.1504). Increasing amounts of precipitation on the day before the actual flood event remains the strongest predictor among biophysical variables examined. Wetland alteration continues to have the largest effect on the dependent variable among built environment variables (β = –.1581). The predictive power of the number of dams within a county, representing structural solutions to flood mitigation, decreases (β = –.0944, $p <.1$) with the addition of socioeconomic controls.

8.5 Discussion

Analysis of the data indicates that specific characteristics of the human built environment in eastern Texas have an important influence on property damage resulting from floods, even when controlling for biophysical and socioeconomic factors. These findings provide guidance to hazard planners and homeland security specialists on how to most effectively mitigate the costly impacts of floods at the community level. First, as expected, flood damage is largely governed by the amount and duration of precipitation associated with a given storm. Yet, our data show that the *timing* of precipitation is particularly important in terms of its effect on the amount of resulting property loss. Heavy precipitation the day before the actual flood event is by far the strongest predictor of total property damage. This result may be a function of the delay between initial rainfall and resulting rise in water levels causing damage. In addition, saturated soil from a heavy rainfall can transform even modest amounts of precipitation during subsequent days into damaging flood events. In other words, the amount of rainfall before a flood event weakens the absorption capacity of hydrologic systems, increasing the probability and extent of property damage the day of the flood event. Even in urban areas where this lag time to peak discharge is shortened by increased runoff volumes, it is important for decision makers and the public to understand that heavy precipitation followed by sunny skies can still result in significant flood damage the next day. Preparedness measures are thus necessary that build in a reaction time to accommodate the relatively slow onset of flood waters.

Second, our results show that the alteration of naturally occurring wetlands is the most important built environment indicator of flood damage. Impervious surfaces have long been criticized for their contribution to increased flooding and associated damage. However, the most significant impact may not be based solely on the total amount of imperviousness in a watershed or drainage basin, but rather where exactly these built surfaces are placed. Altering or removing a wetland to construct parking lots, roads, rooftops, etc. effectively eliminates its ability to capture, hold, and store water runoff. In addition to acting as a vehicle for clean water and wildlife habitat, wetlands may be the best natural defense against the adverse impacts of coastal flooding.

This general trend is evident when looking more closely at our data. For example, 3.15 inches of rainfall in October, 1997 caused approximately $15,000 of reported damage in De Witt County where at the time only five wetland altering permits had been granted. Four years later, when there were 17 wetland alteration permits, roughly the same amount of rainfall caused $150,000 of damage. Similarly, 1.5 inches of rainfall in April, 1997 caused approximately $50,000 of reported damage in Wharton County in which 17 wetland alteration permits had been issued up till that point. Four years later, with 26 wetland development activities permitted, roughly the same amount of rainfall caused $100,000 of damage. Finally, Galveston County had 546 wetland permits issued by April, 1997. A rainfall event of 0.09 inches caused $5,000 of property damage. In September 2000 the same amount of precipitation caused $100,000. At this time there were 921 wetland permits issued, according to Army Corps of Engineers records.

Disrupting the natural hydrological system can exacerbate flooding or create flood problems in areas not originally considered vulnerable to this hazard. Thus, developments initially considered safe from flood threats become an unexpected target of expensive flood damage over time. If wetlands serve as a natural flood mitigation device, this positive function should be considered by local land use and zoning ordinances before regional developments take place. The planning goal in this situation is to allow development to proceed without reducing the hydrological function and value of wetland systems. Achieving this goal will involve identifying the location of naturally occurring wetlands and then protecting these critical areas through local land use policies, such as zoning restrictions, land acquisition programs, clustered development, density bonuses, transfer of development rights, etc. (see Brody and Highfield 2005). Such a proactive approach may result in net economic benefits to a locality by reducing costs related to both repair of damaged structures and engineering solutions (e.g. culverts, retention ponds, storm drains, etc.) used to mitigate floods when the natural systems are compromised.

Third, structural solutions to flood mitigation significantly reduce flood damage as evidenced by the performance of our variable measuring the number of dams in each county. However, based on the standardized coefficients in our fully specified model, wetlands may be more effective than dams in mitigating property loss over time. Dams are also extremely costly mitigation alternatives, can exacerbate development in flood prone areas out of a false sense of security (Harding and Parker 1974; Tobin 1995; Pielke 1999), and can present a hazard in themselves in the case of structural failure.

Fourth, our empirical results suggest that mitigation measures under FEMA's CRS program reduce property damage from floods. Communities that engage in mitigation activities related to public information, mapping and regulations, and flood damage reduction in exchange for reduced flood insurance premiums experience significantly lower amounts of flood related property damage at the.1 level of significance. In fact, the effect of CRS participation appears to reduce community-wide flood damage more than dams, which are far more costly. This finding lends support to the implementation of non-structural mitigation strategies to reduce community-wide flood damage. Strong mitigation measures may partly explain why our floodplain measure does not perform as strongly as anticipated in fully specified models. We speculate that counties with a greater percentage of flood prone areas are also better prepared for the damaging effects of floods (there is in fact a significantly positive association between CRS participation and the percentage of floodplains within a county).

8.5.1 The Price of a Permit Versus Other Mitigation Measures

In addition to comparing the relative effects of predictor variables on flood damage (by interpreting standardized coefficients), because our dependent variable is measured in dollar figures, we can address the questions: *What is the price of a wetland permit and what are the economic tradeoffs of various mitigation measures?* Based on our fully-specified model, a single wetland permit translates into an average of $211.88 in additional property damage per flood (we exponentiated the coefficient and multiplied by the observed average value of flood damage to derive the marginal increase associated with a unit change in the number of permits). This translates, on average, into an additional $38,138 per flood from wetland alteration. To fully internalize this external cost, planning organizations ought to consider setting the acquisition costs of a wetland permit at an appropriate level (in our case at $211.88). An increase in the economic burden of acquiring (and maintaining) a permit, may decrease the willingness of an individual to alter wetlands in the first place. By comparison, the presence of a dam results in a $27,290 decrease in average property damage for each flood event in our sample. This means that, on average, only 129 wetland alteration permits offset the flood-reducing effects of dams. Given the expense of building dams, their negative environmental impacts, and the possibility of structural failure, protecting naturally occurring wetlands may be a more rational policy alternative.

The economic gains obtained by non-structural mitigation measures are also substantial for those counties participating in the FEMA CRS program. Based on our results, a unit increase in FEMA rating produces a 1.84 decrease in average flood cost. In dollar terms, this equals approximately $7,797 less in property damage per flood. Because FEMA scores move in 5 percent increments, a real unit increase in FEMA CRS rating corresponds to a $38,989 reduction in average cost per flood. If all localities in our sample achieve the maximum premium discount of 45 percent, the average damage of a flood is reduced to under $100,000, roughly a quarter of that caused by the average flood in our study. Thus, mitigation is an essential component in any flood reduction program aimed at protecting the property and safety of communities.

8.6 Conclusion

Our study provides evidence that flood damage is not solely a function of rainfall, but is also driven by the scale and type of human development. Furthermore, property damage is influenced not so much by how much is built, but precisely where within an ecological system development unfolds. Location-based development decisions thus become critical to mitigating property damage from floods in the future. As stated by a 1966 Federal Task Force, 'floods are an act of God; flood damages result from the acts of [people]' (TFFFCP 1966, p. 14). Assuming that communities have a choice as to where and how they develop, decision makers are wise to build places to live, work, and recreate that simultaneously maintain the functionality of hydrological systems and the flood moderating features of naturally occurring wetlands.

Acknowledgements This chapter is based on research supported in part by the US National Science Foundation Grant No. CMS- 0346673 and the Institute for Science, Technology and Public Policy at Texas A&M University. The findings and opinions reported are those of the authors and are not necessarily endorsed by the funding organizations or those who provided assistance with various aspects of the study. Parts of this chapter will appear in the international journal *Disasters*.

References

Alig, R.J., Kline, J.D. & Lichtenstein, M. (2004). Urbanization on the US landscape: Looking ahead in the 21st century. *Landscape and Urban Planning*, 69, 219–234

Ammon, D.C., Wayne, C. & Hearney, J.P. (1981). Wetlands' use for water management in Florida. *Journal of Water Resources. Planning and Management*, 107(WR2), 315–327

Arnold, L.A. & Gibbons, C.J. (1996). Impervious surface coverage-The emergence of a key environmental indicator. *Journal of the American Planning Association*, 62, 243–258

(ASFPM) Association of State Floodplain Managers (2000). *National flood programs in review – 2000*. (Madison, WI: ASFM)

Birkland, T., Burby, R., Conrad, D., Cortner, H. & Mitchner, W. (2003). River ecology and flood hazard mitigation. *Natural Hazards Review*, 4(1), 46–54

Brezonik, P.L. & Stadelmann, T.H. (2002). Analysis and predictive models of stormwater runoff volumes, loads and pollutant concentrations from watersheds in the twin cities metropolitan area, Minnesota, USA. *Water Resources*, 36, 1743–1757

Brody, S.D. & Highfield, W.E. (2005). Does planning work? Testing the implementation of local environment planning in Florida. *Journal of the American Planning Association*, 71(2), 159–176

Brody, S.D., Highfield, W.E., Ryu, H. & Spanel-Weber, L. (2007). Examining the relationship between wetland alteration and watershed flooding in Texas and Florida. *Natural Hazards*, 40(2), 413–428

Bullock, A. & Acreman, M. (2003). The role of wetlands in the hydrological cycle, *Hydrology and. Earth System Sciences*, 7(3), 358–389

Burby, R.J., French, S.P. & Cigler, B.A. (1985). *Flood plain land use management: A national assessment.* (Boulder, CO: Westview Press)

Burges, S.J., Wigmosta, M.S. & Meena, J.M. (1998). Hydrological effects of land-use change in a zero-order catchment. *Journal of Hydrological Engineering*, 3, 86–97

Burns, D., Vitvar, T., McDonnell, J., Hassett, J., Duncan, J. & Kendall, C. (2005). Effects of suburban development on runoff generation in the Croton River Basin, New York, USA. *Journal of Hydrology*, 311, 266–281

Carter, R.W. (1961). Magnitude and frequency of floods in suburban areas. (In *Short papers in geologic and hydrologic sciences*. USGS Professional Paper 424-B: B9-B11)

Conger, S. (1971). *Estimating magnitude and frequency of floods in Wisconsin, U.S.* Geological Survey open-file report, Madison, WI

Daniel, C. (1981). *Hydrology, geology, and soils of Pocosins: A comparison of natural and altered systems, Pocosin Wetlands.* (Straudsburg, PA: Hutchinson Ross)

Dunne, T. & Leopold, L.B. (1978). *Water in environmental planning.* (New York: Freeman)

Espey, W.H., Morgan, C.W. & Masch, F.D. (1965). *A study of some effects of urbanization on storm runoff from a small watershed. Tech. Rep. 44D 07–6501 CRWR-2.* (Austin, TX: Center for Research in Water Resources. University of Texas)

Godschalk, D.R., Beatley, T., Berke, P., Brower, D.J. & Kaiser, E.J. (1999). *Natural hazard mitigation: Recasting disaster policy and planning.* (Washington, DC: Island Press)

Harding, D. & Parker D. (1974). Flood hazards at Shrewsbury, United Kingdom. (In Gilbert F. White (Ed.), *Natural hazards: Local, national, and global* (pp. 43–52). New York: Oxford University Press)

Hazards Research Lab (2006). The spatial hazard events and losses database for the United States, Version 4.1 [Electronic edition]. (Columbia, SC: University of South Carolina. Retrieved from http://www.sheldus.org)

Heikuranen, L. (1976). Comparison between runoff condition on a virgin peatland and a forest drainage area. *Proceedings of the Fifth International Peat Congress*, 76–86

Hertzler, R.A. (1961). Corps of Engineers experience relating to flood-plain regulation. *Papers on Flood Problems.* (Chicago, IL: University of Chicago Press)

Hirsch, R.M., Walker, J.F., Day, J.C. & Kallio, R. (1990). The influence of main on hydrological systems. (In M.G. Wolman & H.C. Riggs (Eds.), *Surface water hydrology*, vol. 0–1 (pp. 329–359). Boulder, CO: Geological Society of America)

Johnston, C.A., Detenbeck, N.E. & Niemi, G.J. (1990). The cumulative effect of wetlands on stream water quality and quantity. A landscape approach. *Biogeochemistry*, 10(2), 105–141

Kunreuther, H.J. & Roth (1998). *Paying the price: The status and role of insurance against natural disasters in the US.* (Washington, DC: John Henry Press)

Larson, L. & Plasencia, D. (2001). No adverse impact: A new direction in floodplain management policy. *Natural Hazards Review*, 2(4), 167–181

Leopold, L.B. (1968). *Hydrology for urban planning- A guidebook on the hydrologic effects of urban land use.* (USGS Circular 554, 18 pp)

Leopold, L.B. (1994). *A view of the river* (Cambridge, MA: Harvard University Press)

Lewis, W.M. (2001). *Wetlands explained: Wetland science, policy, and politics in America.* (New York: Oxford University Press)

Mileti, D.S. (1999). *Disasters by design: A reassessment of natural hazards in the United States.* (Washington, DC: Joseph-Henry Press)

Mitch, W.J. & Gosselink, J.G. (2000). *Wetlands*, 3rd ed. (New York: Wiley)

NCDC (National Climatic Data Center) (2000). *Billion dollar U.S. weather disasters* [Electronic version]. Retrieved from http://www.ncdc.noaa.gov/ol/reports/billionz.html

Novitski, R.P. (1985). The effects of lakes and wetlands on flood flows and base flows in selected northern and eastern states. *Proceedings of the Conference on Wetlands of the Chesapeake, Easton, Maryland*, Environmental Law Institute, 143–154

Ogawa, H. & Male J.W. (1986). Simulating the flood mitigation role of wetlands. *Journal of Water Resources Planning and Management*, 112(1), 114–128

Padmanabhan, F.G. & Bengston, M.L. (2001). Assessing the influence of wetlands on flooding. *Proceedings of the ASCE Conference on Wetlands Engineering and River Restoration*, Reno, NV, 1–12.

Paul, M.J. & Meyer, J.L. (2001). Streams in the urban landscape. *Annual Review of Ecological Systems*, 32, 333–365

Pielke, R.A. (1996). *Midwest flood of 1993: Weather, climate, and societal impacts*. (Boulder, CO: National Center for Atmospheric Research)

Pielke, R.A. (2000). Flood impacts on society: Damaging floods as a framework for assessment. (In D.J. Parker (Ed.), *Floods* (pp. 133–155). New York/London: Routledge)

Platt, R.H. (1999). *Disasters and democracy: The politics of extreme natural events*. (Washington, DC: Island Press)

Rose, S. & Peters, N. (2001). Effects of urbanization on streamflow in the Atlanta area (Georgia, USA): A comparative hydrological approach. *Hydrological Proceedings*, 14, 1441–1457

Seaburn, G.E. (1969). Effects of urban development on direct runoff to East Meadow Brook, Nassau County, Long Island, New York. USGS Professional Paper 627-B

Stein, J., Moreno, P., Conrad, D. & Ellis, S. (2000). *Troubled waters: Congress, the Corps of Engineers, and wasteful water projects*. (Washington, DC: Taxpayers for Common Sense and National Wildlife Federation)

TFFFCP (Task Force on Federal Flood Control Policy) (1966). *A unified national program for managing flood losses*. Report No. 67-663. U.S. GPO, Washington, DC

Tobin, G.A. (1995). The levee love affair: A stormy relationship. *Water Resources Bulletin*, 31, 359–367

Tourbier, J.T. & Westmacott, R. (1981). *Water resources protection technology: A handbook of measures to protect water resources in land development*. (Washington, DC: The Urban Land Institute)

(USACE) United States Army Corps of Engineers (2002). Services to the public: Flood damage reduction [Electronic version]. Retrieved July 7, 2005 from http://www.usace.army.mil/public.html#Flood

U.S. Census Bureau (2000). http://www.census.gov/main/www/cen2000.html (accessed 26 July 2006)

Verry, E.S. & Boelter, D.H. (1978). *Peatland hydrology, wetland functions and values: The state of our understanding* (pp. 389–402). (Minneapolis, MN: American Water Resources Association)

White, G.F. (1945). *Human adjustment to floods: A geographical approach to the flood problem in the United States*. Research Paper No. 29, University of Chicago, Department of Geography, Chicago, IL

White, M.D. & Greer, K.A. (2006). The effects of watershed urbanization on the stream hydrology and riparian vegetation of Los Peñasquitos Creek, California. *Landscape and Urban Planning*, 74, 125–138

Zahran, S., Peacock, W.G., Grover, H. & Vedlitz, A. (2008). Social vulnerability, natural, and built environments: Flood casualties in Texas, 1997–2001. Environmental Planning and Management (under review)

Chapter 9
Agent-Based Modeling and Evacuation Planning

F. Benjamin Zhan[1] and Xuwei Chen[2]

Abstract In evacuation planning, it is advantageous for community leaders to have a thorough understanding of the human and geophysical characteristics of a community, be able to anticipate possible outcomes of different response and evacuation strategies under different situations, inform the general public, and develop a set of evacuation plans accordingly. In order to achieve this goal, evacuation managers in a community can use computer modeling techniques to simulate different 'what-if' scenarios, use the results from these simulations to inform the public, and generate different evacuation plans under different circumstances. The complexity associated with evacuation planning in an urban environment requires a computer modeling framework that can incorporate a number of factors into the modeling process. These factors include the nature of the disaster in question, the anticipated human behavioral patterns in the evacuation process, the unique geography and transportation infrastructure in a given area, the population distribution in the area, the population dynamics over different time periods, and the special needs of different population groups, to name a few. Agent-Based Modeling (ABM) provides a general approach that can be used to account for these factors in the modeling and simulation process. In this chapter, the authors provide an overview of agent-based modeling and simulation, illustrate how agent-based modeling and simulation were used in estimating the evacuation time for the Florida Keys, and report some preliminary results in planning a hypothetical route for evacuating the elderly from a nursing home on Galveston Island, Texas, based on network dynamics during an evacuation.

Keywords Agent-based modeling, GIS, emergency response, homeland security, hurricane evacuation, spatial analysis

[1]Texas State University

[2]Northern Illinois University

D.Z. Sui (ed.) *Geospatial Technologies and Homeland Security*,

9.1 Introduction

Mass evacuation planning in large geographical areas is a very complex and difficult task. The complexity lies in the fact that a sound evacuation plan would have to take account of a number of concomitant factors. These factors include the nature of the disaster in question, the unique geography and transportation infrastructure in a given area, the anticipated human behavioral patterns in the evacuation process, the population distribution in the area, the population dynamics over different time periods, and the special needs of different population groups, to name a few.

Unfortunately, it is difficult if not impossible for any community to fully understand how these factors would affect each other unless the community had previous experience of a disaster. In order to better prepare for a certain type of disaster for a vulnerable community, it is advantageous for community leaders to have a better understanding of the human and geophysical characteristics of a community, be able to anticipate possible outcomes of different response and evacuation strategies under different situations, inform the general public, and develop a set of evacuation plans accordingly. In order to achieve this goal, evacuation managers in a community can use computer modeling techniques to simulate different 'what-if' scenarios, use the results from these simulations to inform the public, and to generate different evacuation plans under different circumstances.

The complexity associated with evacuation planning in an urban environment requires a computer modeling framework that would incorporate the factors stated above into the modeling process. Agent-based modeling (ABM) provides a general approach that can be used to account for these factors in the modeling and simulation process. An agent-based modeling approach decomposes a complexity system into individual components (agents) and seeks to understand the behaviors of the entire system based on behaviors of individual agents in the system and the interactions of these agents. In the context of evacuation planning in the United States, an agent is an evacuating vehicle whose movement behaviors are determined by the driver of the vehicle. Once a vehicle enters the road system to evacuate, there are only a few possibilities that it can do. These possibilities represent the possible behaviors of the agent. The system level behavior of an evacuation is the overall traffic resulting from the interactions of the evacuating vehicles.

In the rest of this chapter, Section 2 will review related work in agent-based modeling for evacuation planning and disaster preparedness. Section 3 presents a brief description of a modeling framework for evacuation planning. Section 4 describes two case studies. The first case study illustrates how agent-based modeling can be used to estimate the evacuation time for the Florida Keys. The second case study shows how agent-based modeling can help us better understand evaluation dynamics and how these dynamics would affect evacuation planning and disaster preparedness in evacuating certain populations groups in Galveston Island, Texas. Section 5 provides some concluding remarks.

9.2 Related Work

Agent-based modeling is a type of microscopic modeling technique that addresses disaggregate characteristics of constituent units in a complex system and the dynamic interactions between these individual autonomous entities (Bonabeau 2002). As traffic is dynamic and disaggregate in nature, agent-based modeling is well suited for modeling and simulating traffic and transportation systems.

In contrast to macrosimulation, agent-based modeling has advantages in better representing realistic situations during evacuations. It also provides greater flexibilities and capabilities to assess different scenarios under emergency situations. Therefore, agent-based modeling can help us better understand the outcomes of different evacuation options in evacuation planning and disaster management and gain insights about the dynamics of an evacuation that usually cannot be achieved through traditional macroscopic modeling approaches. In addition, the capability of agent-based modeling in capturing the fine details of traffic movements can also help us better communicate with a general audience about the complexity of an evacuation.

Vehicles are represented as individual entities in agent-based modeling of traffic dynamics. The movements of individual vehicles are captured based on the characteristics of individual drivers and certain rules that govern the interactions between individual drivers. Typical rules include car-following and lane-changing rules. Because evacuations, particularly large-scale evacuations, usually involve a large number of vehicles over large transportation networks, it is a computationally very intensive task to model and simulate traffic at the individual vehicle level. The advancement of computer technology and computing power since the 1990s has made agent-based modeling and simulation of large scale evacuations possible.

9.2.1 Agent-Based Modeling of Evacuation Prior to 1990

Because micro-simulation of large-scale evacuations is a computationally very demanding task, and because of insufficient computing technologies before 1990, very few studies of emergency evacuation at the microscopic level were documented in the research literature before 1990. Some earlier examples of agent-based micro-simulation modeling included the work by Peat and colleagues (1973), Moeller, Urbanik, and Desrosiers (1981), Tweedie and colleagues (1986), and the study reported by Stern and Sinuany-Stern (1989).

Among these early studies, Peat's research team (1973) applied NETSIM, a microscopic-based urban traffic simulation system, to study urban evacuation planning. Moeller, Urbanik, and Desrosiers (1981) developed the CLEAR model, a generic transportation network evacuation model which calculated evacuation times for nuclear incidents, while Tweedie and colleagues (1986) introduced a probabilistic model for simulating evacuation with pre-selected evacuation routes. Finally, Stern and Sinuany-Stern (1989) developed a behavior-based SLAM Network

Evacuation Model (SNEM) for network evacuation. In order to reduce the computational burden, most early microsimulation models were designed only to model small-scale evacuations on primary road networks and were in fact quite simple.

9.2.2 Agent-Based Modeling of Evacuation After 1990

Since the early 1990s, with the development of newer and more computer technologies as well as more advanced software systems, there has been a surge of interest in using agent-based microscopic modeling and simulation to investigate various issues in evacuations.

In the early 1990s, some researchers developed their own models for different case studies. For example, based on the SLAM model, Sinuany-Stern and Stern (1993) and Stern and colleagues (1996) examined the sensitivity of network clearance time with respect to various traffic factors and route choice mechanisms for radiological emergency situations. Their studies suggest that the simulated evacuation time is closer to reality when interactions with pedestrians are taken into consideration. When agent-based transportation modeling software packages became available, some researchers conducted their studies using over the shelf software packages. Farahmand (1997) gave an example of using WITNESS to estimate evacuation time and identify optimal routes in a hurricane evacuation in the coastal area of the Rio Grande Valley in southeast Texas. Rontiris and Crous (undated) used EMME/2 to model evacuation after a nuclear accident.

In one notable study, Batty et al. (2003) constructed an agent-based model to simulate the effects of route change during an annual Carnival event over a period of two days. They examined how the system level behaviors of the carnival emerged from the accumulated interactions between individual agents and demonstrated how congestion and safety concerns could be resolved by introducing traffic controls. In another study, researchers at the MIT Intelligent Transportation Systems (ITS) Program developed a simulation package called MITSIMLab. These researchers produced five scenarios that were used to analyze issues and circumstances involved in the evacuation of the Los Alamos National Laboratory and its surrounding communities (Jha et al. 2004).

In a study about earthquake evacuations, Kagaya and his colleagues (2005) developed a multi-agent model to simulate human behavior. In yet another study, Radwan and other researchers (2005) used INTEGRATION to study hurricane evacuations in Central Florida and tested a proposed framework for modeling emergency evacuation. More recently, Mysore et al. (2006) applied agent-based modeling techniques to study the effects of a potential Sarin gas attack in Manhattan.

What is missing in the literature is the development of a comprehensive and cohesive modeling framework that can account for all necessary factors that should be considered in evacuation planning, and how these factors affect each other. The discussions presented in this chapter serve the purpose of making some of the initial steps toward the ambitious goal of developing a sound and comprehensive agent-based modeling framework for evacuation planning.

9.2.3 *Software Packages for Microsimulation of Transportation Systems*

To date, the most widely used microsimulation software packages for modeling traffic flows in emergency situations are: (1) DYNASMART-P, (2) TSIS/CORSIM, (3) Paramics, and, (4) VISSIM. The DYNASMART-P model (Dynamic Network Assignment-Simulation Model for Advanced Roadway Telematics-Planning version), developed by researchers at the University of Maryland, has been used to investigate the effects of household activity chains on evacuation time estimates (Murray-Tuite and Mahmassani 2004). This research included the examination of traffic scheduling and assignment in a hypothetical area-wide evacuation of a toxic airborne hazardous material release incident in the city of El Paso, Texas (Chiu 2004) and an evaluation of traffic operation strategies of evacuating downtown Minneapolis (Kwon and Pitt 2005). Han and other researchers (2006) used DYNASMART-P to test a one-destination evacuation approach to optimize destination and route choices in a hypothetical country-wide evacuation case study. They found that the one-destination model could help reduce evacuation time substantially in contrast to the conventional fixed destinations or routes approaches.

TSIS/CORSIM, developed by the US Federal Highway Administration in the 1970s (FHWA 2005), has been widely used in evacuation studies. CORSIM combines the capabilities of two simulation models: NETSIM, for surface streets, and FRESIM, for freeways. At the neighborhood scale, Gu (2004) used CORSIM to simulate the evacuation of the Virginia Tech main campus in Blacksburg, Virginia. In studies of contra-flow practice in evacuations, CORSIM has been used to examine whether the contra-flow strategy could help expedite the evacuation process in the City of New Orleans, Louisiana (Theodoulou and Wolshon 2004) and assess the performance of contra-flow evacuation termination points (Lim and Wolshon 2005).

In a study examining major traffic corridors in Birmingham, Alabama, Sisiopiku and colleagues (2004) used CORSIM to develop a regional transportation model to test emergency response plans for several hypothetical emergency situations. Hamza-Lup and researchers (2005) used CORSIM to extract road link speed at each pre-defined time step for a Smart Traffic Evacuation Management System (STEMS) in a hypothetical evacuation event in downtown Orlando, Florida. In studies of hypothetical evacuations in Washington, D.C., CORSIM served as the simulation environment in a real-time emergency evacuation system (ATTAP 2006) and was used to test signal timing operations (Chen 2005). Liu, Lai, and Chang (2006) also used CORSIM to investigate a two-level integrated optimization system for planning an emergency evacuation of Ocean City, Maryland.

The third software package, Paramics, was developed by Quadstone Limited in Britain, and is suitable for simulating traffic at the individual level on local arterials as well as regional freeway networks (Quadstone 2002). Using Paramics, Church and Sexton (2002) investigated how evacuation times may be affected under different evacuation scenarios, such as opening an alternative exit, invoking traffic control, and changing the number of vehicles leaving a household, and so forth, at the neighborhood and community levels. In another study by Cova and Johnson (2002) Paramics was used to test neighborhood evacuation plans in an urban environment and assess the effects of a proposed second access road on household evacuation time. Paramics was also chosen for use in a study for evacuating downtown Salt Lake City, Utah. Paramics has also been used to compare the performance of a network flow evacuation model (Cova and Johnson 2003), for applications in corridor transportation analysis for evacuation planning and management (Radwan and Ramasamy 2003), as well as, for examining the relative performance of stage evacuation strategies and simultaneous evacuation strategies (Chen and Zhan 2006).

Finally, VISSIM, a simulation system developed by *Planung Transport Verkehr* (PTV) in Germany, offers the capacity for researchers to simulate traffic flows in multi-modal transportation systems, including cars, trucks, buses, heavy rail, trams, LRT, bicycles and pedestrians (PTV 2003). VISSIM has been used to analyze the effectiveness of lane reversal in large-scale evacuations (Williams et al. 2007; Tagliaferri 2005), to examine traffic operations in an evacuation related to a large-scale nuclear power plant accident (Han 2005), and to investigate evacuation procedures for the Florida Keys (Chen et al. 2006).

In addition to the software systems described above, a software development effort has been under way at Oak Ridge National Laboratory (ORNL) (Franzese and Han 2001). This effort aims to develop a computer-based incident management decision aid system (IMDAS) at the microscopic-level. The system can be used to estimate evacuation time and test various traffic operations and management strategies.

9.3 A Modeling Framework for Evacuation Planning

Based on the complexity of evacuation planning, we propose a modeling framework for evacuation planning for any given area in question. This framework aims to account for four domains and the interactions amongst them: the geophysical domain, the social domain, the cognitive and behavioral domain, and the (geospatial) information domain. We describe each of these four domains and the general modeling framework in this section.

9.3.1 The Geophysical Domain and Vulnerable Communities

The geophysical domain is concerned with the vulnerability of the community in question based on the geographical characteristics of the community from a geographical location perspective. For example, communities like New Orleans, the Florida Keys, and Galveston Island, Texas have unique geographical characteristics because of their geographical locations. These communities are also examples of some of the most vulnerable communities to the effects of a hurricane strike.

The Florida Keys, a chain of low-lying islands from the southern tip of the Florida Peninsula, stretch for more than 190 miles. The Keys had over 76,000 residents in 2005 based on US Census estimates. The Florida Keys present both high risk for hurricane landfalls and exceptional vulnerability to the effects of hurricanes. Galveston Island is a barrier island located on the Texas Gulf coast, about 80 kilometers south of Houston. The island is about 43 kilometers long and no more than 5 kilometers wide at its widest point. The island had a population of over 57,000 people in 2005 based on US Census estimates. Its geographic location makes the island very vulnerable to hurricane landfalls.

In addition to their unique geographic locations that are vulnerable to hurricane landfalls, evacuation of citizens to safer locations from the Florida Keys and Galveston Island is further complicated because of limited road capacities in both communities. The only road that connects the Florida Keys to the Florida mainland is US Highway 1, which is the only viable mass evacuation route in the event of a hurricane. In addition, an 18-mile stretch of this highway passes through a remote, low lying swamp before entering the highway system on the Florida mainland. Similarly for Galveston Island, the only access road to the mainland is Interstate Highway 45. The unique transportation infrastructure makes timely evacuation more difficult for residents in these two communities.

9.3.2 The Social Domain and Vulnerability of Different Population Groups

The social domain addresses the issue of how to best serve disadvantaged population groups during evacuation planning (Cutter 2006; Laska and Morrow 2006). In a given community, certain population groups may be at a disadvantage when they have to evacuate in anticipation of, for instance, a hurricane strike. These population groups include, but are not limited to, the elderly, people who do not have access to an automobile, and the physically, mentally, or visually impaired. These population groups are most vulnerable to the effects of a disaster. Any community with a sound evacuation plan ought to know the geographical distribution of these population groups and prepare in advance how to evacuate these population groups when a disaster strikes.

9.3.3 The Cognitive and Behavioral Domain and Evacuation Planning

One of the most difficult issues in evacuation planning is perhaps the unpredictability of a number of factors that influence human behaviors in anticipation of a forecast disaster such as a hurricane (Lindell and Prater 2007; Kang et al. 2007). In addition, the unpredictability lies in the fact that human behaviors are experiential in nature—meaning people behave differently based on their own experience. One good example of this type of experiential behavior and its unpredictability is what we witnessed during the evacuation in the greater Houston area while Hurricane Rita was approaching the Texas Gulf coast in September of 2005. Because citizens in Texas watched the devastating damage in New Orleans after Hurricane Katrina hit in August 2005, this experience encouraged many citizens in the greater Houston area to evacuate when it was forecast that Hurricane Rita was approaching. This high rate of evacuation participation undoubtedly resulted in heavy traffic congestion in areas around Houston, leading to significant frustration when people were stuck in traffic for many hours.

The fact that Hurricane Rita missed the Houston area, coupled with the mounting frustration during the evacuation, will definitely impact future evacuation decisions of Houston area households and citizens in the event of another hurricane in the area. What these people will do in the future when a hurricane is forecast to hit becomes very difficult to predict. A sound evacuation plan will have to account for the unpredictability of the factors that affect people's decision-making processes as well as human cognition and behavioral patterns during an evacuation.

9.3.4 The Information Domain and Collaborative Evacuation Planning

For evacuation planning, it is almost inescapable that we have to involve several agencies, different organizations, citizen groups, and individual citizens. A key issue then is how these different groups can work collaboratively to share information from different sources. The information, in many cases, is geospatial in nature. Therefore, it is very critical for all parties involved in the planning process to have a platform to share information, and perhaps more importantly to collaborate and coordinate all efforts in planning an evacuation in a community.

Another aspect of information sharing is that individual citizens often either miss the picture of evacuation in the area or have no access to information about the overall context of the evacuation. This lack of information may lead to deadly results when a disaster strikes. For example, when information about the overall traffic congestion is not available to individuals, people may make a decision in choosing an evacuation route based on their personal knowledge of the area without information about the overall traffic situation in the area. They may end up stuck in traffic for a very long time because too many citizens are heading to the same road segments.

9.3.5 Agent-Based Modeling and Simulation as a means to Understand the Complexity and Dynamics of Evacuation Planning

The four domains discussed above constitute the four important aspects that should be taken into consideration when planning for an evacuation in a community. The difficulty lies in the fact that we have very limited knowledge about how these domains would interact during an actual evacuation. Because it is very costly and almost impossible to test-run an evacuation in a given area, one possible approach for community disaster managers to understand the complexity and dynamics related to an evacuation is through modeling and simulation. An agent-based modeling approach, focusing on the behaviors of individual participants in an evacuation and the collective system level behaviors resulting from the interactions of the individuals, can be used to model and simulate an evacuation in a community and help disaster managers better prepare for potential tragedies. As Chen and Zhan (2006) provided examples of how an evacuation process can be parameterized and simulated, we do not intend to provide a similar detailed discussion in this chapter.

The real power of agent-based modeling and simulation is to provide answers to a set of 'what-if' scenarios so that community leaders can better prepare for possible outcomes under different circumstances during an evacuation. Community leaders can use simulation results to inform the public about the safest actions with respect to different ground situations during an evacuation. The real power of agent-based modeling and simulation is its potential to reveal some counterintuitive collective system level behaviors when all individuals appear to make rational decisions at the individual level. This type of counterintuitive system level behavior may differ from one location to another, and vary from one culture to another.

9.4 Case Studies

Within the framework discussed above, we have conducted modeling and simulation studies for the Florida Keys and Galveston Island, Texas. The Florida Keys case study aimed to estimate the total evacuation time under different scenarios (Chen et al. 2006). The Galveston Island case study attempted to find the best evacuation route for senior citizens in a nursing home while an evacuation is in progress. Before we describe the two case studies, we need to review the basic modeling and simulation procedures.

9.4.1 Modeling and Simulation Procedures

9.4.1.1 Network Preparation and Coding

Because the modeling and simulations are conducted at the individual vehicle level, we had to code the road networks and traffic control systems in the study areas so that the movement of individual vehicles in the simulations reflects the ground truth as closely as possible. Network preparation includes the delineation of different network lanes, speed limits associated with different road segments, and turning restrictions at each intersection.

9.4.1.2 Trip Generation

For evacuation planning, local authorities have usually divided the area of anticipated impact into a number of evacuation zones. An important issue in conducting any evacuation modeling and simulation is to estimate the number of evacuating vehicles in a given zone—a procedure called trip generation in the transportation analysis literature. This is a difficult task, and there are no widely accepted methods that can be used to accurately calculate trips generated in each zone during an evacuation (Lindell and Prater 2007; Kang et al. 2007). To determine the number of evacuating vehicles in each zone, both non-seasonal and seasonal populations should be taken into account. We used the formula proposed by Nelson and his colleagues to determine the number of non-seasonal evacuating vehicles (N_v) in a zone (Nelson et al. 1989). This formula is provided below:

$$N_v = N_h \times N_{vh} \times R_p \times P_{vu}$$

where, N_v is the number of evacuating vehicles;
N_h is the number of households;
N_{vh} is the number of vehicles per household;
R_p is the percentage of people participating in an evacuation;
P_{vu} is the percentage of vehicle usage.

The parameters in this formula are usually determined based on evacuation behavior studies in a study area. Seasonal evacuating vehicles can be determined according to the number of hotel/motel rooms, the occupancy rate of hotel/motel rooms during hurricane seasons, and the participating rate of tourists in an evacuation. These numbers can often be obtained through behavior studies of hurricane evacuations as well.

9.4.1.3 Departure Timing

Once the number of evacuating vehicles in a zone is determined, the next step is to estimate the number of vehicles that is assumed to leave the zone for a given time interval (e.g., one hour) during the evacuation. This number can either be estimated using a stochastic process or based on empirical data. In this study, we used empirical data from behavior surveys for both the Florida Keys and Galveston Island (Baker 2000; Lindell et al. 2001). Typically, there are three types of response rates that define departure timing: rapid (early), medium (normal), and slow (late). Rapid or early response rates correspond to the assumption that people start to evacuate shortly after an evacuation order is issued, whereas slow or late response rates are assigned to a situation when people are assumed to evacuate during a much longer time frame. It should be noted that response rates provide data about the number of people that start to evacuate during a given time interval. Within each time interval, the exact time for a vehicle to leave its origin is typically determined randomly.

9.4.1.4 Destination and Route Choice

The area of impact (risk area) of an approaching hurricane depends on the strength of the hurricane (e.g., Category 3 or 5). Theoretically, any place outside the risk area can be a destination choice. In practice, evacuation destinations can be either pre-specified (e.g. shelters), or anywhere that is outside the risk area. It is important to note that destination choice for an individual is closely related to the evacuation routes connecting the origin of this individual in the risk area to a location in areas that are considered to be safe. If there is only one evacuating roadway going out of the risk area, then destination choice can be very limited. Otherwise, there are more options for an individual to choose a destination. When there are two or more destinations for an evacuating vehicle to choose from, we can either set a pre-specified destination or any location that would take the evacuating vehicle the least amount of time to reach.

There are different routes between an origin–destination pair, and therefore route choice is another important issue in modeling and simulating an evacuation. Similar to destination choice, some states in the US have state-mandated evacuation routes. In other states, although there are specified official evacuating routes, evacuating vehicles are not required to take these routes. In reality, drivers' behaviors in choosing a particular route are hard to predict, and it is typical for us to assume that drivers tend to take the fastest route (least driving time) to reach their destinations. In the simulations reported in this chapter, we used a dynamic routing method to assign evacuation routes to vehicles, meaning drivers are assumed to know the overall traffic situation in the area, and evacuating vehicles en route would adjust their routes according to real time traffic conditions.

9.4.2 *Time Needed to Evacuate People in the Florida Keys*

The Florida Keys has a unique geography that presents both high risk of hurricane landfall and exceptional vulnerability to the effects of a hurricane strike. The entire Keys need to be evacuated if a category 3 or higher hurricane strikes this area, and the State of Florida has a mandated evacuation clearance time of 24 hours in such an event. We conducted agent-based modeling and simulation of hurricane evacuation for the Keys (Chen et al. 2006). One of the questions that we attempted to answer through the simulation was: What is the minimum clearance time needed to evacuate 92,596 people from the Keys? This is population size calculated based on the 2000 US Census population data, census under counts, and the number of tourists estimated to be in the area.

9.4.2.1 Preparation of Road Networks

The evacuating roadways for the Florida Keys, starting at the northern boundary of Key West, stretch to the intersection of US Highway 1 and the Homestead Extension of the Florida Turnpike. In the simulations reported in this chapter, the attributes of the links, such as link type, number of lanes, lane width, and speed limit, were set the same as those used in the Miller report (Miller Consulting Inc. 2001).

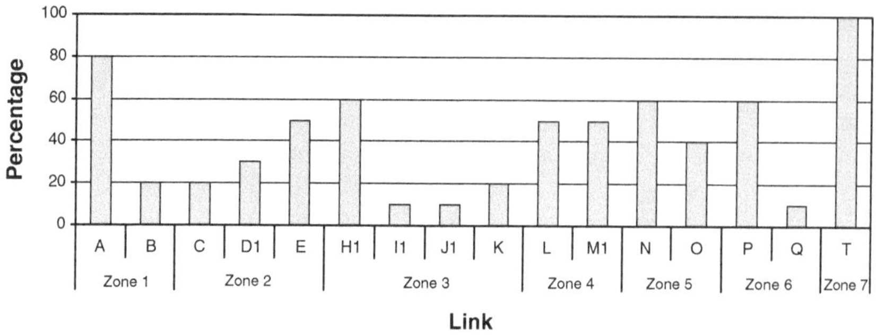

Fig. 9.1 Percentages of vehicles in each zone leaving from the links (Note: Key West is located in Zone 1 and the zones are numbered sequentially, starting from Key West) (Miller Consulting 2001)

9.4.2.2 Evacuation Zones

The Miller Study divided the Florida Keys into seven evacuation zones from Key West to the Upper Keys. We used the same seven evacuation zones in the simulation. The evacuating traffic from the seven zones was loaded onto US Highway 1 through sixteen links. Based on the trip generation formula described above, it was estimated that a total of 41,016 evacuating vehicles were to be evacuated for the 92,596 people. The number of vehicles leaving from each of the sixteen links at a given time interval was adopted from the numbers used in the Miller study (Fig. 9.1).

9.4.2.3 Evacuation Timing

To determine the minimum clearance time for the Keys, we used the late response curve based on the study by Baker (2000). Baker's study assumes that all evacuating vehicles leave within 16 hours after an evacuation order is issued (Fig. 9.2). In this response curve, the time period from the ninth hour to the tenth hour after the order is issued has the highest percentage of people leaving their residences.

9.4.2.4 Destination and Route Choice

We defined two types of destinations in the Keys simulation. It is assumed that a small proportion of evacuating vehicles would go to motels or homes of their friends in Monroe County. The rest would go outside of the county. For this second group of people, we set the destination of the evacuating vehicles as the intersection of US Highway 1 and the Homestead Extension of the Florida Turnpike. For the Keys, besides US Highway 1, Card Sound road in the Upper Keys can also partially serve as an evacuation route. Therefore, we assume that drivers are free to choose one of the two routes that reduces destination drive times.

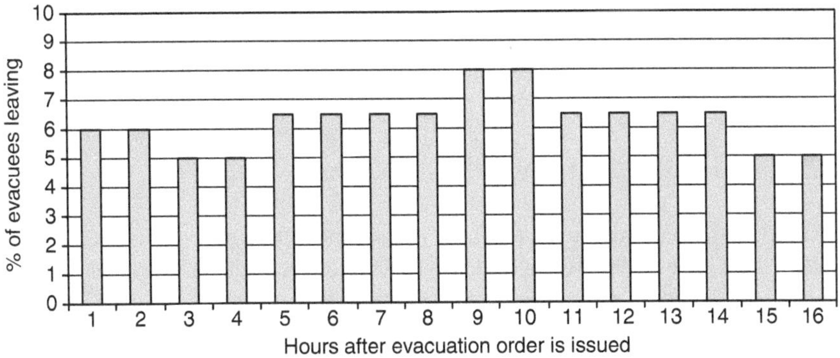

Fig. 9.2 Late response curve of evacuation in the Florida Keys (Miller Consulting 2001)

9.4.2.5 Simulation Results

In order to eliminate randomness in microsimulation, we performed ten simulation runs and calculated the average evacuation time. Simulation results suggest that it would take from 20 hours and 11 minutes to 20 hours and 14 minutes to evacuate 92,596 people from the Florida Keys. This clearance time is less than the Florida state mandated 24-hour clearance time limit.

9.4.3 Dynamic Evacuation Route Planning for Galveston Island

Galveston Island is one of the most vulnerable regions for hurricanes in Texas because of its geographical location. Evacuation is necessary for the whole island in response to a Category 2 or stronger hurricane (TXDPS 2005). No shelters will be open in the island when a hurricane approaches this area. In addition, there is only one way—the IH-45 Bridge—connecting the island with inland areas. Furthermore, the evacuation of Galveston Island is further complicated by the fact that traffic in the rest of Galveston County and the greater Houston area is already congested under normal circumstances, and traffic congestion in the area will become significantly worse during an evacuation of Galveston County and Houston.

9.4.3.1 Preparation of Road Networks

While preparing Galveston Island road networks for the simulation, we incorporated the official evacuation routes by the State of Texas. However, the official evacuation route system of Galveston Island only includes four routes (PSB&J 2004).

Because it is very time consuming to code all local roadways in the simulation, we constructed a road network skeleton that included only major roads of the island for the simulations reported in this chapter.

9.4.3.2 Evacuation Zones and Trip Generation

We used traffic analysis zones (TAZs) to represent trip origins. It is assumed that evacuating vehicles from every TAZ leave the centroid of each TAZ and then get onto the nearest evacuation route. The number of evacuating vehicles from each TAZ was determined using the formula by Nelson et al. (1989). The number of seasonal evacuating vehicles was calculated based on the estimations from the PBS&J study (2004). We used a population size of 35,219 for the Galveston Island simulation.

9.4.3.3 Evacuation Response Rate and Evacuation Timing

We conducted the Galveston Island simulation assuming that people would follow the rapid response curve suggested by Lindell et al. (2001). This response curve assumes that the entire population on the island would evacuate during a five-hour time period with different proportions of the population leaving during each of the five hours (Fig. 9.3). The response curve also suggests that about 70% of the evacuees leave within the first three hours.

9.4.3.4 Destination and Route Choice

Because we were only interested in the evacuation dynamics of the island, we assumed that an evacuating vehicle reaches safe area once it crosses the IH-45 Bridge. Given that the State of Texas does not require evacuating vehicles to take

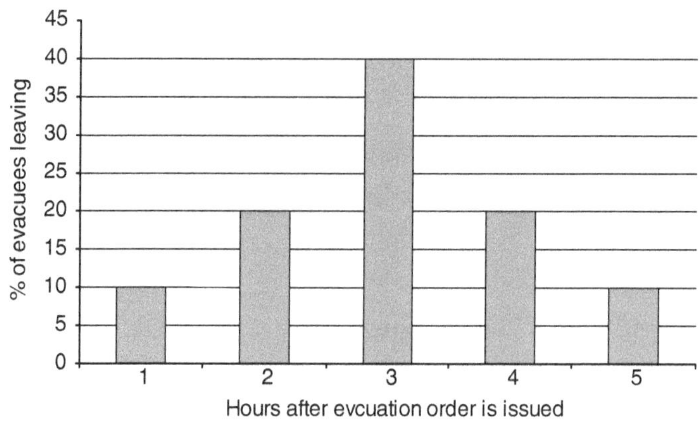

Fig. 9.3 Rapid response curve of evacuation in Galveston Island, Texas (PBS&J 2004)

the official evacuation routes specified for a community, we used dynamic routing in the simulation, meaning drivers have the liberty to adjust their routes based on changing traffic conditions while en route.

9.4.3.5 Results of Route Planning for Evacuating Senior Citizens from a Nursing Home

We report simulation results related to a 'what–if' scenario in Galveston Island—what if we have to evacuate a group of senior citizens while the evacuation is in progress. As discussed in Section 3, it is important for community leaders and disaster managers to prepare how to evacuate disadvantaged population groups. This 'what–if' scenario is one of the situations that community leaders need to prepare for during evacuation planning.

There is a nursing home located at 2302 Avenue E on Galveston Island. If for any reason we have to evacuate the senior citizens from this nursing home during an evacuation, the route for this evacuation would depend on the network dynamics corresponding to the evacuation and the network states (congestion level of each road segment). Figure 9.4(a) shows the speed of each road segment when an evacuation is

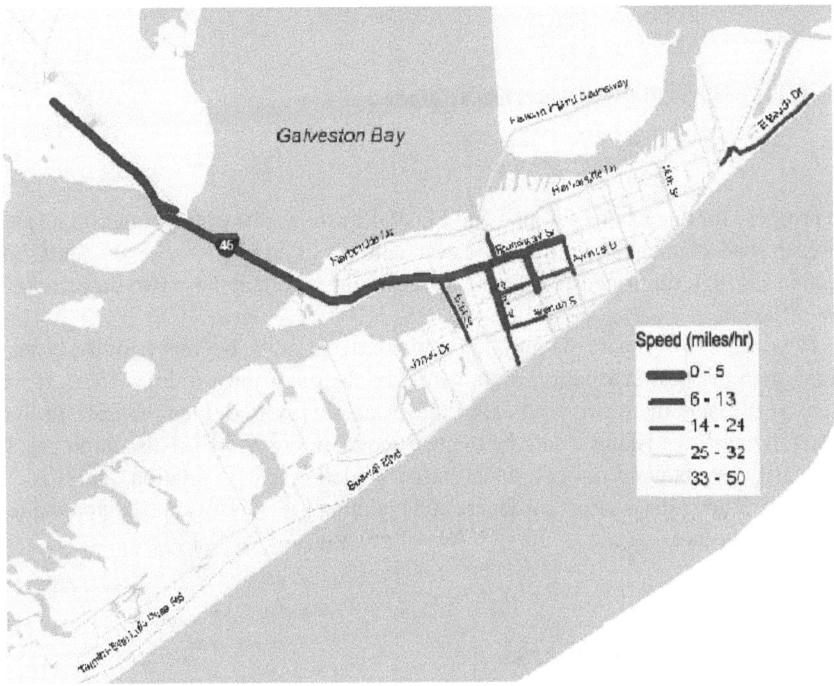

a

Fig. 9.4 Traffic situations at different times during a simulated evacuation: (a) One and half hours into the evacuation;

(continued)

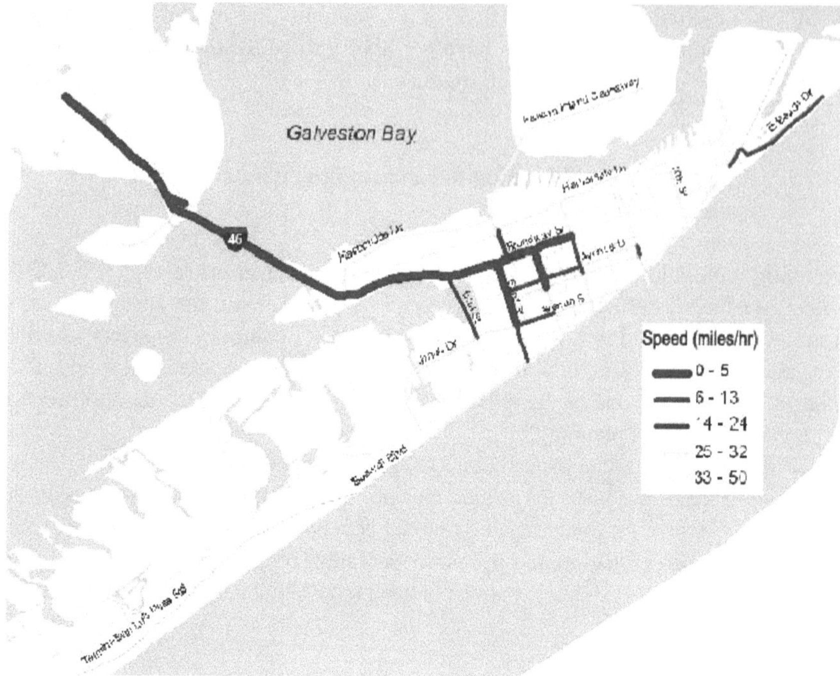

b

Fig. 9.4 (continued) (b) Seven hours into the evacuation

in progress for one and half hours, and Fig. 9.4(b) shows the traffic congestion level of each road segment when the same evacuation is in progress for seven hours. As seen in Fig. 9.4, the congestion levels on the road network at these two time periods are different.

If we were to evacuate the group of senior citizens at the beginning of the evacuation, then the best evacuation route would be the one shown in Fig. 9.5(a). If we were to evacuate them two hours after the evacuation started, then the fastest route to get them off the island would be the one shown in Fig. 9.5(b). This simple result can easily be obtained through agent-based modeling and simulation, and it is an example of good disaster preparedness and planning for a disadvantaged population group in a vulnerable community.

9.5 Concluding Remarks

Evacuation planning remains a challenging task for community leaders and disaster managers. One of the difficulties is that community leaders often have limited knowledge about what to expect from the evacuation of a large area with a large

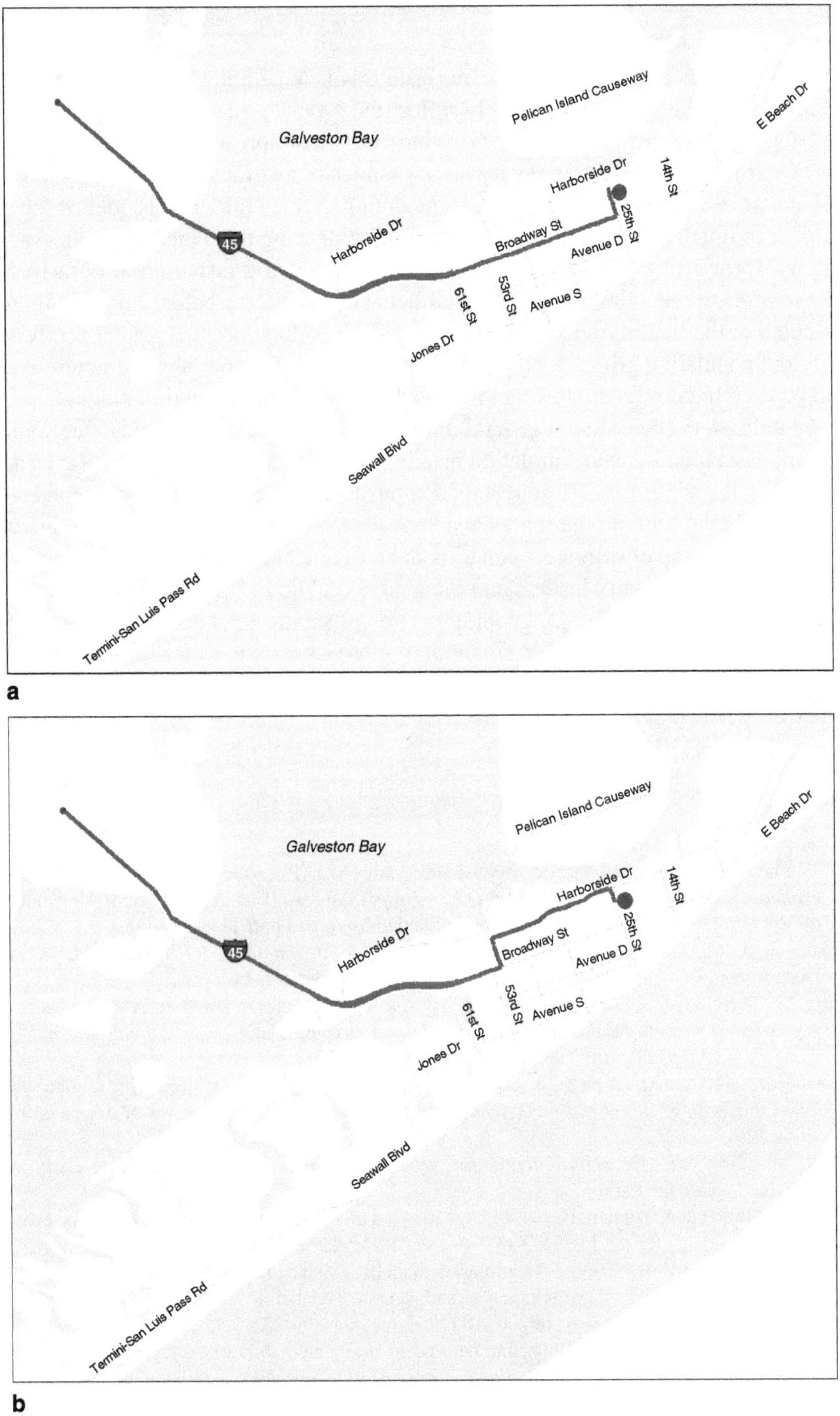

Fig. 9.5 Network dynamics during evacuation and evacuation route planning: (a) Evacuation route for senior citizens in a nursing home during the first five minutes of the evacuation; (b) Evacuation route for senior citizens in a nursing home when the evacuation is in progress for two hours

population size. A practical way to mitigate this lack of knowledge is to conduct computer simulations that accurately reflect the ground situation of an evacuation in a given community. Traditional modeling and simulation approaches do not have the power to capture the complexity and dynamics of an evacuation in large metropolitan areas. We propose a general modeling framework that accounts for a number of important and concomitant factors influencing the dynamics of an evacuation. These factors include the unique geography and transportation infrastructure in a given area, the different human behavioral patterns before and during an evacuation, the underlying social structure of the population, the special needs of different population groups, and an integrated geospatial information structure that can be used to coordinate different parties involved in the evacuation process.

In addition to presenting a general modeling framework, we also describe some preliminary modeling and simulation results of evacuations in the Florida Keys and Galveston Island, Texas. These results demonstrate that it is possible to use agent-based modeling and simulation as a means to gain a better understanding about possible outcomes of different scenarios in an evacuation. This better understanding will help community leaders and the general public to better prepare for a certain type of disaster in a given geographic area. It should be stated that this is only a start in developing computer simulation approaches that will help community leaders and the general public better prepare for a potential disaster.

References

An Applied Technology and Traffic Analysis Program (ATTAP) (2006). *A real-time emergency evacuation system for the Washington, D.C. metropolitan area* [Electronic version]. University of Maryland. Retrieved from http://attap.umd.edu/bbs/zboard.php?id = highlights

Baker, E. J. (2000). *Hurricane evacuation behavioral assumptions for the Florida Keys*. Department of Geography, Florida State University, Tallahassee, FL

Batty, M., DeSyllas, J. & Duxbury, E. (2003). The discrete dynamics of small-scale spatial events: Agent-based models of mobility in carnivals and streetpParades. *International Journal of Geographical Information Science*, 17, 673–697

Bonabeau, E. (2002). Agent-based modelling: Methods and techniques for simulating human systems. *Proceedings of the National Academy of Sciences of the United States of America*, 99, 7280–7287

Chen, M. (2005). *Traffic signal timing for urban evacuation*. Master thesis, University of Maryland – College Park

Chen, X., Meaker J. & Zhan, F. B. (2006). Agent-based modelling and analysis of hurricane evacuation procedures for the Florida Keys. *Natural Hazards*, 38, 321–338

Chen, X. & Zhan, F. B. (2006). Agent-based modelling and simulation of urban evacuation: Relative effectiveness of simultaneous and staged evacuation strategies. *Journal of the Operational Research Society*, doi: 10.1057/palgrave.jors.2602321)

Chiu, Y. (2004). Traffic scheduling simulation and assignment for area-wide evacuation. (In *Proceedings of the 2004 IEEE Intelligent Transportation Systems Conference* (pp. 537–542))

Church, R. L. & Sexton, R. M. (2002). *Modelling small area evacuation: Can existing transportation infrastructure impede public safety?* Vehicle Intelligence & Transportation Analysis Laboratory, University of California, Santa Barbara, CA

Cova, T. J. & Johnson, J. P. (2002). Microsimulation of neighborhood evacuations in the urban-wildland interface. *Environment and Planning A*, 34, 2211–2229

Cova, T. J. & Johnson, J. P. (2003). A network flow model for lane-based evacuation routing. *Transportation Research Part A – Policy and Practice*, 37, 579–604

Cutter, S. (2006). *The geography of social vulnerability: Race, class, and catastrophe* [Electronic version]. Retrieved September 25, 2007 from http://understandingkatrina.ssrc.org/Cutter/

Farahmand, K. (1997). Application of simulation modelling to emergency population evacuation. (In S. Andradcittir, K. J. Healy, D. H. Withers, & B. L. Nelson (Eds.), *Proceedings of the 1997 winter simulation conference* (pp. 1181–1188))

Federal Highway Administration (FHWA) (2005) *Corridor simulation* (*CORSIM/TSIS*) [Electronic version]. Retrieved December 20, 2005, from http://ops.fhwa.dot.gov/trafficanalysistools/corsim.htm

Franzese, O. & Han, L. (2001). Traffic modelling framework for hurricane evacuation. (In *Proceedings of the Transportation Research Board 80th annual meeting*. Washington, DC)

Gu, Y. (2004). *Integrating a regional planning model (TRANSIMS) with an operational model (CORSIM)*. Master thesis, Virginia Polytechnic Institute and State University

Hamza-Lup, G. L., Hua, K. A., Peng, R. & Ho, A. H. (2005). A maximum flow approach to dynamic handling of multiple incidents in traffic evacuation management. (In *Proceedings of the 2005 IEEE Intelligent Transportation Systems conference* (pp. 1147–1152))

Han, L. D. (2005). Evacuation modelling and operations using dynamic traffic assignment and most desirable destination approaches. (In *Proceedings of the Transportation Research Board 84th annual meeting*. Washington, DC)

Han, L., Yuan, F., Chin, S. & Hwang, H. (2006). Global optimization of emergency evacuation assignments. *Interfaces*, 36, 502 – 513

Jha, M., Moore, K., & Pashaie, B. (2004). Emergency evacuation planning with microscopic traffic simulation. *Transportation Research Record*, 1886, 40–48

Kagaya, S., Uchida, K., Hagiwara, T. & Negishi, A. (2005). An application of multi-agent simulation to traffic behavior for evacuation in earthquake disaster. *Journal of the Eastern Asia Society for Transportation Studies*, 6, 4224–4236

Kang, J. E., Lindell, M. K. & Prater, C. S. (2007). Hurricane evacuation expectations and actual behavior in hurricane Lili. *Journal of Applied Social Psychology*, 37, 881–897

Kwon E. & Pitt, S. (2005). Evaluation of emergency evacuation strategies for downtown event traffic using a dynamic network model. *Transportation Research Record*, 1922, 149–155

Laska, S. & Morrow, B. H. (2006). Social vulnerabilities and hurricane Katrina: An unnatural disaster in New Orleans. *Marine Technology Society Journal*, 40, 16–26

Lim, E. & Wolshon, B. (2005). Modeling and performance assessment of contraflow evacuation termination points. *Transportation Research Record*, 1922, 118–128

Lindell, M.K. & Prater, C.S. (2007). Critical behavioral assumptions in evacuation analysis for private vehicles: Examples from hurricane research and planning. *Journal of Urban Planning and Development*, 133, 18–29

Lindell, M. K., Prater, C. S., Sanderson, W. G., Lee, H. M., Yang, Z., Mohite, A. & Hwang, S. N. (2001). *Texas gulf coast residents' expectations and intentions regarding hurricane evacuation*. Hazard Reduction & Recovery Center, Texas A&M University. College Station, TX

Liu, Y., Lai, X. R. & Chang, G. L. (2006). Two-level integrated optimization system for planning of emergency evacuation. *Journal of Transportation Engineering*, 132, 800–807

Miller Consulting Inc. (2001). *Florida Keys hurricane evacuation report*. Contract No. C7391, Florida Department of Transportation. Miami, FL

Moeller, M., Urbanik, T. & Desrosiers, A. (1981). *CLEAR (Calculated Logical Evacuation and Response): A generic transportation network evacuation model for the calculation of evacuation time estimates*. Prepared for the Nuclear Regulatory Commission by Pacific Northwest Laboratory. Washington, DC

Murray-Tuite, P. M. & Mahmassani, H. S. (2004). Transportation network evacuation planning with household activity interactions. *Transportation Research Record*, 1894, 150–159

Mysore, V., Narzisi, G. & Mishra, B. (2006). *Emergency response planning for a potential sarin gas attack in Manhattan using agent-based models*. Agent Technology for Disaster Management. Hakodate, Japan

Nelson, C. E., Coovert, M. D., Kurtz, A., Fritzche, B., Crumley, C. & Powell, A. (1989). *Models of hurricane evacuation behavior*. Department Of Psychology, University of South Florida, Tampa, FL

Peat, Marwick, Mitchell, & Company. (1973). *Network flow simulation for urban traffic control system–Phase II. v. 1*. Prepared for the Federal Highway Administration. Washington, DC

Planung Transport Verkehr AG (PTV) (2003). *VISSIM3.7 user manual*

Planung Transport Verkehr AG (PTV) (2004). *VISSIM3.7 user manual*

Post, Buckley, Schuh, & Jernigan Inc. (PBS&J) (2004). *Galveston region hurricane transportation analysis 2004. Draft report*. Prepared for US Army Corps of Engineers, Galveston District. Galveston, TX

Quadstone (2002). *Quadstone Paramics V4.0 Modeller user guide*

Radwan, E. & Ramasamy, S. (2003). *I-4 corridor traffic simulation and visualization*. Center for Advanced Transportation Systems Simulation, University of Central Florida, FL

Radwan, E., Mollaghasemi, M., Mitchell, S. & Yildirim, G. (2005). *Framework for modelling emergency evacuation*. Center for Advanced Transportation Systems Simulation, University of Central Florida, Orlando, FL

Rontiris, K. & Crous, W. (Undated). *Emergency evacuation modelling for the Koeberg nuclear powers station* [Electronic version]. Retrieved December 19, 2005, from http://www.inro.ca/en/pres_pap/asian/asi00/EMME2Asian.pdf

Sinuany-stern, Z. & Stern, E. (1993). Simulating the evacuation of a small city – The effects of traffic factors. *Socio-Economic Planning Sciences*, 27, 97–108

Sisiopiku, V. P., Jones, S., Sullivan, A. J., Patharkar, S. & Tang, X. (2004). *Regional traffic simulation for emergency preparedness* [Electronic version]. University Transportation Center for Alabama, The University of Alabama, The University of Alabama in Birmingham, and The University of Alabama in Huntsville. Retrieved December 20, 2005, from http://utca.eng.ua.edu/projects/final_reports/03226fnl.pdf

Stern, E. & Sinuany-Stern, Z. (1989). A behavioral-based simulation model for urban evacuation. *Papers in Regional Science*, 66, 87–103

Stern, E., Sinuany-Stern, Z. & Holm, Z. S. (1996). Congestion-related information and road network performance. *Journal of Transport Geography*, 4, 169–178

Tagliaferri, A. P. (2005). *Use and comparison of traffic simulation models in the analysis of emergency evacuation conditions*. Master thesis, North Carolina State University

Texas Department of Public Safety (TXDPS) (2005). *Evacuation maps for planning and public information: Houston-Galveston study area* [Electronic version]. Retrieved December 19, 2005, from http://www.txdps.state.tx.us/dem/Hurricanemaps/GalvestonStudyAreaMap.pdf

Theodoulou, G. & Wolshon, B. (2004). Alternative methods to increase the effectiveness of freeway contraflow evacuation. *Transportation Research Record*, 1865, 48–56

Tweedie, S. W., Rowland, J. R., Walsh, S., Rhoten, R. R. & Hagle, P. L. (1986). A methodology for estimating emergency evacuation times. *The Social Science Journal*, 21, 189–204

Williams, B. M., Tagliaferri, A. P., Meinhold, S. S., Hummer, J. E. & Rouphail, N. M. (2007) Simulation and annalysis of freeway lane reversal for coastal hurricane evacuation. *Journal of Urban Planning and Development*, 133, 61–72

Chapter 10
Building Evacuation in Emergencies: A Review and Interpretation of Software for Simulating Pedestrian Egress

Christian J. E. Castle[1] **and Paul A. Longley**[2]

Abstract This chapter begins with an assessment of the reasons for the current surge in interest in developing building evacuation analysis using models that focus explicitly upon the human individual's locus of behavior. We then present an extended overview of available software for modelling and simulating pedestrian evacuation from enclosed spaces, and use this to develop guidelines for evaluating the wide range of pedestrian evacuation software that is currently available. We then develop a sequential conceptual framework of software and model specific considerations to take into account when contemplating application development. This conceptual framework is then applied to a hypothetical building evacuation setting. Our conclusions address not only the efficiency and effectiveness of the software solutions that are currently available, but also the degree of confidence (broadly defined) that underpins their application.

Keywords Application development, building evacuation analysis, framework of assessment criteria, models, simulated pedestrian egress, software

10.1 Introduction

The objectives of this chapter are twofold. First, it develops criteria for evaluating pedestrian evacuation software by outlining features and functions of different software that should be considered when developing an application (Sect. 10.5). These criteria also provide a useful framework for understanding the key principles and techniques that underpin the broader class of pedestrian evacuation models, and their evolution over time. The second, related, objective of this chapter is to review and interpret software designed for simulating the egress of pedestrians from inside a building (Sect. 10.6). Here we will review 27 software programs and identify key literature and reviews associated with each. A hypothetical setting of a subway/

[1] Transport for London

[2] University College London (UCL)

D.Z. Sui (ed.) *Geospatial Technologies and Homeland Security*,

underground station is used to assess features and functionality of these software programs against the requirements of an application development. We begin, however, by justifying the development and use of computer-based models for the evaluation of pedestrian egress from buildings, particularly in relation to emergency management objectives and building design considerations.

10.2 The Growth of Interest in Pedestrian Evacuation Modelling

There are several reasons why computer-based building evacuation analysis has recently become more prominent. Recent world events such as the attacks on the World Trade Center in 2001, the Indian Ocean tsunami in 2004, London bombings in 2005, and Hurricane Katrina in 2005, to name but a few, have brought the issue of emergency management to the fore. Emergency management has historically focused on the immediate and urgent aspects of an incident (i.e. response and post-incident recovery). However, there is a growing awareness that emergency management is much more complex and comprehensive than traditionally perceived (Gunes and Kolel 2000). Although the primary function of emergency services is to protect life and property, a comprehensive approach to emergency management involves more than just reactive responses to incidents as they unfold. It also entails development of methods to avoid incidents in the first place and preparing for those that will unavoidably occur at some point in the future. A recent National Research Council (2007) report identifies the need for practitioners to assimilate model predictions and the best available data as an aid to emergency management. For example, extensive pedestrian evacuation modelling has retrospectively been conducted to investigate the evacuation of the World Trade Center (Averill et al. 2005), in an attempt to learn more about the events which followed the attacks and in a bid to understand how the evacuation process could have been improved.

Previously, the most significant developments in computer-based building evacuation analysis have occurred in the building industry (Galea et al. 2003). The demand for more innovative building designs has increasingly left architects with the dilemma of demonstrating the safety of their concepts (that is, demonstrating that occupants will be able to efficiently exit a structure in the event of an emergency). Building designs involving the construction, renovation, or modification of spaces that exceed the design constraints specified by statutory building regulations prove particularly problematic for architects (Thompson 1994). Intriguing examples include stadiums, trains, underground (subway) or coach stations, airports, large stores and shopping malls, and high-rise buildings. Traditionally, a full-scale evacuation demonstration or adherence to prescriptive building regulations/fire codes has been the main method of meeting these demands (e.g. a building must provide a certain number of exits based on its total occupancy capacity).

For buildings that have already been constructed, it is usual for an evacuation demonstration or fire drill involving a representative target population to be conducted for the structure. However, according to Gwynne and Galea (1997), ethical, practical, and financial limitations can undermine the viability of fire drills. Ethically, a conflict for organizers exists between the threat of participant injury and the lack of realism. Volunteers cannot be subjected to the psychological (e.g. trauma, panic, etc.) or physical (e.g. fire, smoke, debris, etc.) consequences of a real emergency; and thus an exercise can provide only limited information about the evacuation efficiency of a building. Regardless of the success of an evacuation demonstration, the lack of realism reduces confidence about the structure's evacuation capability. In practice, full scale evacuation drills are performed infrequently, especially for buildings that are publicly accessible, such as train stations, airports, museums, etc. Evacuation drills can also be expensive to perform. Finally, if a building has already been constructed, alterations to its layout can be very expensive.

It follows that, in order to fully understand the potential evacuation efficiency of a building, it is essential to assess the relationships between several complex and interrelated facets to the evacuation process. Ventre et al. (1981) proposed a hypothetical equation where the behavior of a crowd could be determined as a function of 'crowd characteristics' (i.e. demographic data, total number of occupants, etc.), 'facility design and layout' (e.g. a building's overall configuration, its predicted capacity for ingress and egress, etc.), and 'management practices' (e.g. staffing, communication facilities, security, admissions, furnishings, etc.). In a similar vein, Gwynne and Galea (1997) have identified four criteria that they consider as necessary to calculate the evacuation efficiency of a building:

1. **Configurational**: These considerations are generally covered by traditional building codes or regulations, and involve building layout, number of exits, exit width, etc.;
2. **Environmental:** Different emergency incidents effect environmental considerations of an enclosure in a variety of ways. For example, heat and toxic/irritant gases can effect an occupant's walking speed and way-finding abilities;
3. **Procedural:** Aspects related to staff actions, levels of occupant evacuation training, occupant prior knowledge of the enclosure, emergency signage, etc.;
4. **Behavioral:** The likely behavioral responses of occupants (e.g. initial response to evacuation cues, maximum walking speed, presence of family or friends, etc.).

Traditional methods of building design fail to consider these factors in a quantitative manner, relying almost totally on judgement and prescriptive building rules or regulations (Gwynne et al. 1999b). Building regulations rely almost totally on the configuration of a structure (e.g. maximum travel-distance to nearest exit) in order to specify the number and location of exits. Thus, building regulations and fire codes can prove very restrictive when designing new buildings. These traditional methods do not consider human movement and behavior or the effect that this can have upon the evacuation process. Computer based evacuation models offer the potential to overcome some of these limitations. State of the art pedestrian evacuation software consider many of the factors identified as necessary for evaluating a building; addressing the needs of both the architect and legislators by permitting the assessment and demonstration of a building design's evacuation capacity.

10.3 The Evolution of Pedestrian Modelling: From Aggregate to Individual

Hypothetically, it is possible to model any system using a set of rules about the behavior of the system's constituent elements. For example, the behavior of a crowd can be modelled through rules likely to characterise the behavior of every individual. However, depending on the size of the crowd and the purpose of the model, this may not be practical or even useful. Continuous-field models address this problem by replacing individual objects with continuously varying estimates of abstracted properties, for example the density of people in a crowd (Goodchild 2005). Alternatively, individual objects can be aggregated into larger wholes, modelling the behavior of the system through these aggregates. However, aggregate systems subsume variation and processes that fall below the implied spatial resolution of the representation. However, the scale and configuration of the constituent elements of information (i.e. objects located in time and space), can exert critical influence on the outcomes of spatial analysis (see Openshaw (1984) for a discussion in the context of the Modifiable Areal Unit Problem (MAUP) and Bailey and Gatrell (1995) in relation to the ecological fallacy concept). Spatially aggregated or coarse resolution data may have no validity independent of a specific application or indeed any validity at all. Problems associated with aggregating data are compounded when modelling interaction between zones, or when seeking to represent process and dynamics (as, for example, in pedestrian modelling). Spatial analysis at the level of unique individuals represented as mobile point referenced 'events'[1] present the logical endpoint of the drive towards disaggregation.

Progress is clearly being made in the use of disaggregated data. Increased computer power and storage capacity has made individual-level modelling more practicable in recent times. An example of this can clearly be seen in the evolution of pedestrian modelling, where there has been a concerted movement from aggregate to individual-level modelling. The evolution of pedestrian evacuation models has been separated into the five stages shown in Table 10.1 by Galea and Gwynne 2006). Specific details of these modelling approaches are discussed further in Sect. 10.4.1.

Essential to the progression of individual-level modelling has been the development of automata approaches, which have been at the forefront of computer modelling research, particularly pedestrian modelling. Benenson and Torrens (2004) define an automaton as a processing mechanism with characteristics that change over time based on its internal characteristics, rules, and external input. Therefore, automata process information received by them from their surroundings, and their characteristics are altered according to the rules that govern their reaction to these stimuli. Two classes of automata tools, CA and ABM have been particularly popular; their use has dominated the pedestrian modelling research literature. It is beyond the scope of this chapter to explore the underlying principles and concepts behind these modelling strategies, and the reader is referred to Castle and Crooks (2006) for a thorough treatment of the subject.

[1] Humans and their activities are depicted in GIS as mobile point-referenced 'events' (Martin 1996).

Table 10.1 Description of pedestrian modelling generations (Adapted from Galea and Gwynne 2006)

<table>
<tr><th colspan="2">Generation</th><th>Description</th></tr>
<tr><td>1</td><td>Hand-based flow/ hydraulic models</td><td rowspan="2">Pedestrian movement is approximated by an equation derived from an analogy to fluid flow in channels. The rate of flow equates to the average density of all occupants within a room or passageway as a function of average pedestrian walking speed. The total evacuation time of a building is thus calculated as the sum of the time taken for each building sector to empty under these conditions.</td></tr>
<tr><td>2</td><td>Computer-based flow/ hydraulic models</td></tr>
<tr><td>3</td><td>Ball bearing models</td><td>This generation of pedestrian evacuation model was the first to consider occupants as individuals. Occupants are completely homogeneous, with exactly the same characteristics and equation(s)/rule(s) determining each pedestrian's movement and behavior. The name of this generation of model relates to the visualisation of simulation output: when moving pedestrians are represented as circles, they look like ball bearings draining through a network of pipes.</td></tr>
<tr><td>4</td><td>Deterministic or Stochastic Rule-based models</td><td>This type of model simulates pedestrians as individual entities whose heterogeneous characteristics determine behavior and govern movement. Movement towards a building exit is thus a response to surrounding environmental conditions (e.g. density) and information availability (e.g. emergency exit signage).</td></tr>
<tr><td>5</td><td>Adaptive pedestrian models</td><td>Here, occupant decision making is adaptive and sensitive to local conditions, because pedestrians have memories and make decisions based on past events in addition to surrounding environment conditions and information made available to them. For example, pedestrians can choose a building exit based on prior knowledge, but may change their decision and decide to use another exit (known or unknown) based on conditions of the surrounding environment and information made available (e.g. congestion). This generation of models is the current state of the art; no model yet incorporates highly adaptive behavior of pedestrians during an evacuation.</td></tr>
</table>

10.4 Pedestrian Evacuation Software

There is a range of software designed to assess emergency egress of occupants from buildings, and this will be our focus here—as opposed to software to simulate evacuation from aircraft, ships, street environments, etc. It is helpful to distinguish between pedestrian evacuation 'software' (e.g. buildingEXODUS, Legion, etc.) that incorporate one or more pedestrian evacuation 'models' (e.g. Okazaki and Matsushita's (1993) magnetic model, Kerridge et al.'s (2001) PEDFLOW, etc.). The models that drive any given software are usually generic in terms of the types and scales of buildings that they can represent, and flexible with regard to the range of emergency scenarios to which they may be applied. On the other hand, pedestrian evacuation models have been specifically developed for particular buildings (e.g. the evacuation of patients from a hospital in London, or a school in Los Angeles) or a class of problems characteristic of particular enclosed settings (e.g. experimenting with room configurations to relieve congestion at pinch points, such as exits, or to increase bi-directional flow within a corridor). Consequently, some pedestrian evacuation models may be unsuitable for purposes beyond their original remit.

Pedestrian evacuation software, by contrast, invariably includes features or functionality beyond that of a pedestrian evacuation model, such as: the ability to import and interpret a building's floor plan in different file formats (e.g. CAD, GIS, image file, etc.); two- and possibly three-dimensional visualisation of the building and simulation; and dynamic presentation of output in multiple forms (e.g. charts, graphs, etc.). Furthermore, some software programs claim to permit the simulation of fire and smoke dispersal as well as the egress of pedestrians. Invariably, this functionality is provided via coupling with another model purposely designed to simulate such phenomena, or the user is required to specify the location and spread of fire manually (i.e. at certain time intervals). Any such fire dispersal model required by the user will need to be considered independently of a pedestrian evacuation model, and is therefore not the focus of this discussion. In this respect, the majority of pedestrian evacuation models are solely designed to simulate pedestrian egress from a building in the event of an evacuation or fire drill. Nevertheless, current software does permit the exploration of scenarios involving different building layouts, and can therefore also explore the effect of parts of a building being blocked by different emergency incidents, for example.

10.5 Guidelines for Assessing Pedestrian Evacuation Software

Pedestrian evacuation software programs adopt various modelling approaches to simulate the egress of pedestrians from buildings. For instance: it is possible to represent an enclosure as either continuous or discrete space; human agents may be represented as anything from a homogeneous ensemble to an assemblage of individuals with unique characteristics; and movement and behavior may be represented deterministically, probabilistically or using a combination of both. Generally, as the level of detail encapsulated within the simulation model increases, the effort required by the user to initialise the software increases, as well as the time required to run the computer simulation. Furthermore, software reflects the purpose for which it was originally designed, the predilections of the software developer (e.g. engineer, psychologist, architect), and the computer power available to the developer at the time. A wide range of software designed to simulate the evacuation of pedestrians from buildings has been developed. Software can be distinguished apart by their choice of development strategy, and the features and functionality that they include.

Nelson and Mowrer (2002) and Kuligowksi and Gwynne (2005) identify and justify a number of pertinent considerations for potential users when choosing software to build an application for assessing pedestrian evacuation from buildings. Castle (2007) expands upon the original list of considerations, and provides a comprehensive discussion of criteria pertaining to each consideration. Castle's guide to assessing pedestrian evacuation software is an invaluable companion to this book chapter, especially for a reader without general knowledge of pedestrian evacuation modelling intricacies.

The criteria that define pedestrian evacuation models fall within nine broad areas, which can be split into two types (Fig. 10.1): 1) model specific criteria about the nature of the simulation model within the software; and 2) software specific criteria concerned with the general features and functionality of the software. The model specific criteria are concerned with how the simulation model works, and how the software can be sensitive to data input and variables defined by the user. The software specific criteria are more general in nature, defining the overall approach and functionality incorporated within different software: they are nevertheless essential to the evaluation process.

10.5.1 Key Questions to Consider

The logical ordering of the nine topics in Fig. 10.1 follows the general sequence in which a user should consider them. The ordering should help minimize redundant effort in choosing a software program (e.g. considering the availability and access of the software before any model specific considerations). The criteria identified in Fig. 10.1 in turn raise the following questions (Table 10.2). The reader should note that each question may subsume sub-topics, and therefore may need to be addressed concurrently.

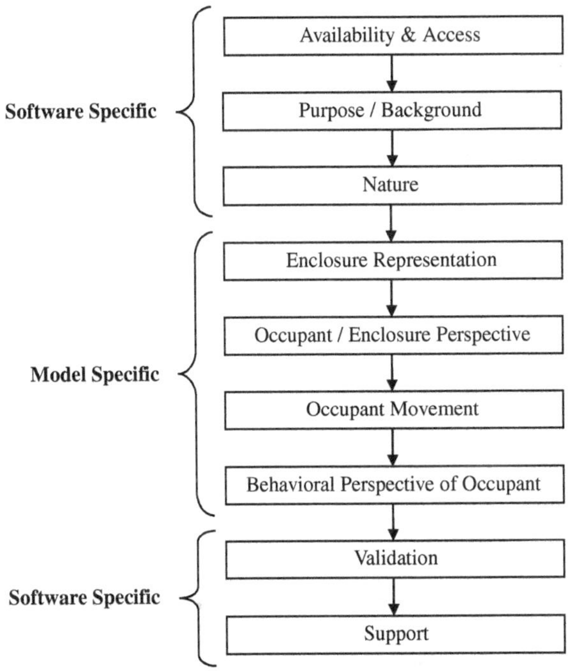

Fig. 10.1 Software and model specific topics of criteria to consider when developing a building evacuation application

Table 10.2 Key questions to consider when developing a building evacuation application

Topic	Question
Availability and Access:	How is the application available (off-the-shelf or on a consultancy basis through the developer or a third-party)?
	What minimum computer hardware specification is required (RAM, central processing unit, etc.)?
	What operating system is required (Windows, Linux, Mac OS X, etc.)?
Purpose/Background:	Is the purpose (e.g. building, aviation, maritime, etc.) of the application suitable for the research investigation?
	Is the focus of an application (e.g. residential buildings, high-rise residential tower blocks, low-rise buildings) suitable for the research endeavour?
	What is the origin of the application (e.g. development environment, expertise of developer/development team)?
Nature:	What is the general nature of the application: movement; optimisation-movement; movement and behavioral; or partial-behavioral?
Enclosure: Representation:	At what scale is the structure represented (e.g. coarse scale network, regular lattice, or continuous space)?
	How, and in what format (e.g. CAD, GIS, image file, etc.) can data be imported into the application in order to represent the enclosure and network connections?
Occupant/Enclosure Perspective:	Does the application have a global or an individual perspective of occupants? – If the perspective is global, what characteristics of the population are represented, and how are they defined? – If the perspective is individual, what individual characteristics of the population are represented, and how are they defined? Do the occupants have a global or individual perspective of the enclosure? – If the perspective is global, what information is available to the occu pants, and how is this information defined? – If the perspective is individual, what information is available to the occupants, and how is this information defined?
Occupant Movement:	How is pedestrian walking speed specified? Are default values provided by the application, or does the user need to initialise this parameter?
	What is the origin and validity of walking speed values input into the application, and are they plausible in the scenarios to be developed (e.g. are non-evacuation walking speeds used, or are values extrapolated by the application in order to simulate evacuation movement and/or walking speeds)?
	How is the direction of pedestrian movement simulated (e.g. using a flow/hydraulic equation, a cell-based structure, a velocity based vector, etc.)?
Behavioral Perspective of Occupants:	What behavioral approach does the application employ (none, implicit, deterministic or stochastic rule-based, artificial intelligence)?
	If the application seeks to simulate the behavior of occupants, what behavioral considerations does it consider, and how does this affect the movement and decision choices of each pedestrian?

(continued)

Table 10.2 (continued)

Topic	Question
Validation:	In terms of both quality and reliability, to what extent has the application been validated?
Support:	Is the application maintained?
	Are developments still being made to the application?
	Is the application actively supported by the developer (training courses, software tutorials, phone or online help, bug reporting/fixing, etc.)?

10.6 An Evaluation of Pedestrian Evacuation Software

In order to move beyond the general criteria that should be considered when contemplating software capable of simulating the egress of occupants from inside a building during an emergency, it is logical to assess each software program in relation to these criteria, and then benchmark these software programs against the requirements of a hypothetical investigation (Sect. 10.6.1). In total, 27 software programs were identified for closer examination. Table 10.3 lists these software programs, identifying key references used to assess the features and functionality of each. In addition to the information obtained from these references, several reviews were used to assist us in evaluating information regarding some of the software (Friedman 1992; Thompson 1994; Paulsen et al. 1995; Gwynne and Galea 1997; Gwynne et al. 1999a, b; Nelson and Mowrer 2002; Galea et al. 2003; Kuligowski 2003; Olenick and Carpenter 2003; RSSB 2003; Sharp et al. 2003; Santos and Aguirre 2004; Averill et al. 2005; Kuligowski 2005; Kuligowski and Peacock 2005).

Availability and accessibility can vary considerably between different software programs. Software programs are available under different financial terms (e.g. free of charge, consultancy basis, one-off fee, annual licence and support fee, or a combination of these agreements), and can have unique computer hardware or operating system requirements. Furthermore, whilst documentation has ostensibly been published about some software programs, it is in practice unavailable to the public. Some software programs are restricted to in-house company use, others are incomplete (perhaps in terms of full functionality, validation, etc.), and still others have been withdrawn from the market. For purposes of this review, the availability and access criterion identified within the guidelines has been used to discriminate between software deemed 'unavailable/withdrawn from the market' rather than 'available', based on the feature/functionality summary shown in Table 10.4. Table 10.5 provides a list of the abbreviated terms used to describe features and functionality pertaining to the identified software. The reader is reminded that full definitions and corresponding explanations of each criteria are provided by Castle (2007).

The information contained within Table 10.4 permits the identification of potential candidate software programs for a particular building evacuation application. For example, the subsequent discussion assesses 'available' software for their appropriateness in

Table 10.3 Pedestrian evacuation software for building analysis

	Application	Availability	Key References
1	ALLSAFE	Free or Fee	Heskestad and Meland (1998)
2	ASERI	Free or Fee	Schneider (2001); Schneider and Könnecke (2001)
3	BFIRES	Withdrawn	Stahl (1978, 1980, 1982)
4	BGRAF	Unavailable	Ozel (1985, 1991)
5	building EXODUS	Free or Fee	Galea et al. (2003, 2006); Gwynne and Galea (1997); Gwynne et al. (1999a, 2001, 2005); FSEG (2004)
6	CRISP	Consultancy	Boyce et al. (1998); Fraser-Mitchell (1999); Björkman and Mikkola (2001); Building Research Establishment (2006)
7	EESCAPE	Consultancy	Kendik (1983, 1986, 1988)
8	EGRESS	Consultancy	Ketchell et al. (1996); Ketchell (2002).
9	EgressPro	Withdrawn	Combustion Science & Engineering (2002)
10	EVACNET4	Free or Fee	Kisko and Francis (1985); Kisko et al. (1998)
11	EvacSim	Unavailable	Poon (1994); Poon and Beck (1995)
12	EXIT89	Unavailable	Fahy (1991, 1994, 1996, 2000)
13	EXITT	Free or Fee	Levin (1987); Weinroth (1988); Kostreva and Lancaster (1998)
14	FPEtool	Free or Fee	Deal (1995)
15	GridFlow	Free or Fee	Bensilum and Purser (2002); Building Research Establishment (2006)
16	Legion	Free or Fee	Connor and Hammer (2004); Connor (2004); Williams (2004); Legion Limited (2006)
17	Myriad	Consultancy	Crowd Dynamics (2006a, b, d, e)
18	Pathfinder	Consultancy	Cappucio (2000)
19	PAXPORT/ PEDROUTE	Unavailable	Halcrow (2006a, b)
20	PedGo	Free or Fee	Klüpfel (2003); Klüpfel and Meyer-König (2003); TraffGO (2006)
21	SGEM	Unavailable	Lo and Fang (2000); Lo et al. (2004)
22	Simulex	Free or Fee	Thompson (1994, 2005); Thompson and Marchant (1995a, b); Thompson et al. (1996); IES (2006)
23	SimWalk	Free or Fee	Savannah Simulations AG (2006a, b, 2007)
24	STEPs	Free or Fee	Hoffmann and Henson (1997); Hoffmann et al. (1998); Rhodes and Hoffmann (1999); Wall and Waterson (2002); Mott MacDonald (2006)
25	TIMTEX	Free or Fee	Harrington (1996)
26	VEgAS	Withdrawn	Still (1993); Crowd Dynamics (2006f)
27	WAYOUT	Free or Fee	Shestopal and Grubits (1994)

Term	Description
Free/Fee[a]:	The model is available for free, a one-off fee, or an annual licence
Consultancy:	The model is available on a consultancy basis only
Unavailable:	The model is not publicly available or is still being developed
Withdrawn:	The model has been withdrawn from the market

[a]Free and fee applications are not distinguished apart, as fees change regularly and can vary depending on the user's circumstances (e.g. the fee can often be reduced or waived if the application is not used for commercial purposes/gain)

Table 10.4 Features and functionality of pedestrian evacuation software

'Available' Software	Purpose	Nature	Enclosure	Enclosure/Occupant	Movement		Behavior	Validation
					Speed/Flow	Direction		
ALLSAFE	One-route	Partial	Coarse	Global/Global	Application (S)	Flow-Equation	Implicit	Other models
ASERI	Any-type	Behave	Continuous	Individual/Individual	Application (P & S)	V-B Velocity	R-B Deter & Stoc	Drill
building EXODUS	Any-type	Behave	R-Lattice	Individual/Individual	Application (S)	Cell-Based	R-B Deter & Stoc	Drill
CRISP	Any-type	Behave	R-Lattice	Individual/Individual	User	Cell-Based	R-B Deter & Stoc	Drill
EESCAPE	One-route	Movement	Coarse	Global/Global	Application (S)	Flow-Equation	None	Drill
EGRESS	Any-type	Behave	R-Lattice	Individual/Individual	Application (S)	Cell-Based	R-B Deter & Stoc	Drill
EVACNET4	Any-type	Move/Opt	Coarse	Global/Global	User	Flow-Equation	None	Drill
EXITT	Residence	Behave	Coarse	Individual/Individual	User	Flow-Equation	R-B Deter & Stoc	None
GridFlow	Any-type	Partial	R-Lattice	Individual/Individual	User (D)	Cell-Based	Implicit	Drill & Literature
Legion	Any-type	Behave	Continuous	Individual/Individual	Application (P)	V-B Velocity	AI	Drill & Other models
Pathfinder	Any-type	Move	Coarse	Global/Global	Application (S)	Flow-Equation	None	None
PedGo	Any-type	Move/ Partial	R-Lattice	Individual/Individual	User (D)	Cell-Based	Implicit	Drill
Simulex	Any-type	Partial	R-Lattice	Individual/Individual	Application (P & S)	Cell-Based	Implicit	Drill & Literature
SimWalk	Any-type	Partial	R-Lattice	Individual/Individual	User	Cell-Based	R-B Deter & Stoc	None
STEPS	Any-type	Move/ Partial	R-Lattice	Individual/Individual	User	Cell-Based	None/Implicit	Reg/Codes
TIMTEX	Low-rise	Move	Coarse	Individual/Global	Application (S)	Flow-Equation	None	Literature
WAYOUT	One-route	Move	Coarse	Global/Global	Application (S)	Flow-Equation	None	Drill

(continued)

Table 10.4 (continued)

'Unavailable' Software	Purpose	Nature	Enclosure	Enclosure/Occupant	Movement		Behavior	Validation
					Speed	Direction		
BFIRES	Low-rise	Behave	Fine	Individual/Individual	User	Cell-Based	R-B Deter & Stoc	None
BGRAF	Any-type	Behave	Fine	Individual/Individual	User	Cell-Based	R-B Deter & Stoc	Drill
EgressPro	One-route	Move	Coarse	Global/Global	Application (S)	Flow-Equation	None	None
EvacSim	Any-type	Behave	Fine	Individual/Individual	Application (S)	Cell-Based	R-B Deter & Stoc	None
EXIT89	Any-type	Partial	Coarse	Individual/Individual	Application (S)	Flow-Equation	Implicit & R-B Deter	Drill
FPEtool	Any-type	Move	Coarse	Global/Global	User	Flow-Equation	None	None
Myriad	Any-type	Move	N/A	Individual/Individual	N/A	N/A	None	Third-party
PAXPORT/ PEDROUTE	Public-T	Partial	Coarse	Global/Global	Application (P & S)	Flow-Equation	Implicit	None
SGEM	Any-type	Move/ Partial	Fine	Individual/Individual	Application (P)	Cell-Based	Implicit	Drill & Other models
VEgAS	Any-type	Behave	Fine	Individual/Individual	Unknown	Cell-Based	AI	None

Table 10.5 Key of abbreviated terms used to describe features and functionality within pedestrian evacuation software

Category	Abbreviation	Feature Description
Purpose	Any Type	Capable of evaluating any building type
	Residence	Specialises in the evaluation of places of residence
	Public-T	Specialises in the evaluation of public transport stations
	Low-rise	Specialises in the evaluation of low-rise buildings
	One-route	Specialises in the evaluation of one exit from a building
Nature	Move	Movement model
	Move-Opt	Movement/optimization model
	Partial	Partial behavior model
	Behave	Behavioral model
Enclosure Representation	Coarse	Coarse scale network
	R-Lattice	Regular lattice
	Continuous	Continuous space
Occupant/ Enclosure Perspective	Global	Global perspective
	Individual	Individual perspective
Movement – Speed & Direction	Application (P and/or S)	Specifies pedestrian walking speeds; values based on primary (P) observations and/or secondary (S) data
	User (D)	User is required to specify pedestrian walking speeds values; default (D) values are provided based on secondary data
	Flow – Equation	Flow/hydraulic equation
	Cell-based	Cell-based movement
	V-B Velocity	Vector-based velocity
Behavioral Perspective	None	No behavior
	Implicit	Implicit behavior
	R-B Deter.	Rule-based deterministic behavior
	R-B Stoc.	Rule-based stochastic behavior
	AI	Artificial Intelligence
Validation	Reg./Code	Validation against fire regulations/codes
	Drill	Validation against fire drills or experiments
	Literature	Validation against published literature on past experiments
	Models	Validation against other Software/model
	Third-Party	Validation by an independent third-party
	None	No validation work could be found for this software

relation to a hypothetical application. Given the references provided within Table 10.3 and the aforementioned software reviews, it is therefore possible for the reader to identify the most suitable software program from these potential candidates.

Before proceeding to the interpretation, a few caveats must be provided. Table 10.4 was populated in early 2007. Modelling approaches or the features/functionality of software may have changed in subsequent releases, and it is possible that we may have misinterpreted the available literature. Furthermore, although FPETool was deemed 'available', this software was not included within our subsequent evaluation because it is primarily designed to enable the analysis of fire related phenomena (i.e. modelling of fire and smoke dispersal: Deal 1995). The software can predict total evacuation time required for occupants to vacate an enclosure, but the developers note the output can be grossly inaccurate in such applications (they cite two to three orders of magnitude of errors in these circumstances). Additionally, it should be noted that Myriad has been designed as a spatial analysis tool for assessing traversable areas of a building based on travel distance to an objective (e.g. a building exit). The software does not simulate the movement or behavior of occupants within an enclosure, rather it analyses the floor plan of a building in order to identify potential areas of congestion based on Fruin's Level of Service Concept (see Fruin 1971). For this reason, Myriad has been coupled with the Simulex software (which is included with the evaluation) to accommodate this deficiency (Crowd Dynamics 2006c).

10.6.1 Hypothetical Example: Interpretation of Suitable Software

In order to discriminate between pedestrian evacuation software programs, it is necessary to identify the features and functionality most desirable in a given application context. Following the structure of the guidelines, and after establishing the availability of different software, it is necessary to identify the purpose of the user's investigation. In order to illustrate these ideas, we will assess the suitability of the identified software in relation to a (hypothetical) evaluation of a complex multi-level subway/underground station that serves as an international transport hub within a major city. The stated aim of our investigation will be to assess the evacuation process of the structure (i.e. total evacuation time, usage of evacuation routes, conditions experienced by passenger such as crowd density, etc.), for two different incident scenarios (e.g. a fire on board a train located at one of the station's platforms and a fire within the ticket hall of the station). Specifically, it is necessary to investigate the effect of variable occupant movement and behavioral characteristics upon the evacuation process. For example, irregular or first time users of the station (e.g. tourists) may be much less familiar with the station layout and its exits than regular commuters, and could be expected to have different movement (e.g. walking speed) and behavioral characteristics (e.g. speed of response to an evacuation cue(s); investigating the alarm to determine if the evacuation is real).

In relation to this hypothetical investigation, software with a purpose limited to the analysis of residential buildings, or buildings with only one exit, would be deemed unsuitable. A station is not a residential structure, and it comprises a complex floor layout serviced by multiple entrances and exits. Since the overarching aim of the investigation is to explore the impact of passengers with different characteristics, it is necessary to identify software capable of simulating variability in pedestrian behavior in addition to movement on an individual basis. Consequently, software incapable of simulating behavior would be deemed unsuitable (e.g. EESCAPE, EVACNET4, etc.). While all of the software programs identified are capable of simulating pedestrian movement (to some degree), software limited to the analysis of movement (typically based on the average flow of an aggregated pedestrian population between rooms or sectors of a building) would also be ruled out. Castle (2007) notes that both the simulation of pedestrian movement and behavior in software is strongly dependent on the scale of enclosure representation, the perspectives of occupants within the enclosure, and the occupants' perspectives of the enclosure itself. Therefore, in relation to the aim of this assessment, software programs that represent an enclosure as a coarse network, occupants as a homogenous ensemble (e.g. EXITT, Pathfinder, TIMTEX, etc.), or define the perspective of occupants and/or the enclosure globally (e.g. ALLSAFE, WAYOUT, etc.) would also be deemed unsuitable.

The final criterion considered pertains to the validation of the software. Software unable to provide any evidence of validation would also be considered less desirable than those that could, depending on reliability and quality of this validation. Unfortunately, information relating to the source(s) of primary data used to specify default pedestrian walking speeds within some software programs could not be identified. Other software programs do not provide default values for pedestrian walking speeds; this parameter must be defined by the user. Consequently, software was not ruled out on this criterion alone. Based on the selection criteria of this hypothetical example, four software programs can be identified as potential candidates for use: ASERI, buildingEXODUS, CRISP, and Legion. Before making a final choice, a user will find it useful to consult publications in relation to each software program (refer to Table 10.3), software reviews (refer to Sect. 11.6), and finally may need to consult with the software developers in order to resolve any unanswered questions.

10.7 Conclusion

The development and implementation of software for evacuation analysis of buildings is an important growth area of modelling and simulation, and is integral to current issues of homeland security analysis. Prevention is better than cure, and it is clear that the drive towards agent-based representations of crowd behavior is providing valuable insights into managing new and existing facilities in ways that are efficient and effective relative to conventional real world evacuation exercises and drills.

This is ever more important, as the range of scenarios and potential threats that need to be considered in creating and maintaining civic infrastructure multiply.

The impetus behind the development of new simulation models, and their embedding in software is leading to an extensive and perplexing array of solutions. However, a central contribution of this chapter has been to illustrate that it is necessary to adopt a 'horses for courses' approach to software development. Moreover, when the detailed characteristics of the (apparently) available options are evaluated, it may become clear that some of the 'horses' are in practice not runners—whether for reasons of failure to update, to tailor to context, or to document.

The last of these reasons is perhaps the most challenging to those seeking to build evacuation applications. In conventional terms of scientific investigation, a key test of model validity is reproducibility, using documentation from the literature that has successfully negotiated the hurdles of peer review. What emerges from our review and interpretation of what is currently available for our hypothetical application, is that documentation is not always as comprehensive, relevant, and confidence-inspiring as might be desired. This should concern the research community. In the domain of homeland security, the apparent efficiency and effectiveness gains of replacing time-consuming and expensive drills with simulations will not (and certainly should not) satisfy the community of potential users if software cannot be certified 'safe to use'.

References

Averill, J. D., Mileti, D. S., Peacock, R. D., Kuligowski, E. D., Groner, N., Proulx, G., et al. (2005). *Federal building and fire safety investigation of the World Trade Center disaster: Occupant behavior, egress, and emergency communications (draft).* (Washington, DC, USA: US Department of Commerce)

Bailey, T. C. & Gatrell, A. C. (1995). *Interactive spatial data analysis* (2nd Edition). (Essex, UK: Longman)

Benenson, I. & Torrens, P. (2004). *Geosimulation: Automata-based modelling of urban pheno mena.* (London, UK: Wiley)

Bensilum, M. & Purser, D. A. (2002). *GridFlow: An object-oriented building evacuation model combining pre-movement and movement behaviours for performance-based design.* (Paper presented at the 7th International Symposium on Fire Safety Science, Worcester Polytechnic Institute, Worcester, MA, USA)

Björkman, J. & Mikkola, E. (2001). Risk assessment of a timber frame building by using CRISP simulation. *Fire and Materials, 25*(5), 185–192

Boyce, K. E., Fraser-Mitchell, J. N. & Shields, T. J. (1998). *Survey, analysis, and modelling of office evacuation using the CRISP model.* (Paper presented at the Human Behavior in Fire, Proceedings of the 1st International Symposium, University of Ulster, Ireland)

Building Research Establishment (2006). *Human behavior in fire and emergency evacuation design* [Electronic version]. Retrieved November 3, 2006, from http://www.bre.co.uk/fire/page.jsp?id=269

Cappucio, P. E. (2000). PathFinder: A computer-based, timed egress simulation. *Fire Protection Engineering, 8*(Fall), 11–12

Castle, C. J. E. (2007). *Guidelines to assess pedestrian evacuation software applications.* (London, UK: Centre for Advanced Spatial Analysis (University College London), Working Paper 115)

Castle, C. J. E. & Crooks, A. T. (2006). *Principles and concepts of agent-based modelling for developing geospatial simulations*. (London, UK: Centre for Advanced Spatial Analysis (University College London), Working Paper 110)

Combustion Science & Engineering (2002). *Computer models forf Fire and smoke: EgressPro* [Electronic version]. Retrieved November 6, 2006, from http://www.firemodelsurvey.com/pdf/EgressPro_2001.pdf

Connor, N. (2004). Simulating pedestrian circulation for olympic planning. *Public Transport International, 53*(2), 28–29

Connor, N. & Hammer, K. (2004). *Crowd pleaser*: Legion Limited

Crowd Dynamics (2006a). *Crowd modelling - Myriad and beyond* [Electronic version]. Retrieved November 6, 2006, from http://www.crowddynamics.com/index.htm

Crowd Dynamics (2006b). *Myriad* [Electronic version]. Retrieved November 6, 2006, from http://www.crowddynamics.com/index.htm

Crowd Dynamics (2006c). *Myriad and Simulex* [Electronic version]. Retrieved November 6, 2006, from http://www.crowddynamics.com/index.htm

Crowd Dynamics (2006d). *Myriad colour conventions* [Electronic version]. Retrieved November 6, 2006, from http://www.crowddynamics.com/index.htm

Crowd Dynamics (2006e). *Station evacuation - Analysis* [Electronic version]. Retrieved November 6, 2006, from http://www.crowddynamics.com/index.htm

Crowd Dynamics (2006f). *VEgAS (Virtual Egress and Analysis System)* [Electronic version]. Retrieved November 6, 2006, from http://www.crowddynamics.com/index.htm

Deal, S. (1995). *Technical reference guide for FPEtool Version 3.2*. (Washington, DC, USA: U.S. Department of Commerce, NISTIR 5486–1)

Fahy, R. F. (1991, July). *EXIT89: An evacuation model for high-rise buildings*. (Paper presented at the Proceedings of the 3rd International Symposium on Fire Safety Science)

Fahy, R. F. (1994, July). *EXIT89: An evacuation model for high-rise buildings - Model description and example applications*. (Paper presented at the Proceedings of the 4th International Symposium on Fire Safety Science, Ottawa, Ontario, Canada)

Fahy, R. F. (1996, March). *EXIT89: High-rise evacuation model - Recent enhancements and example applications*. (Paper presented at the Proceedings of the 7th Interflam 1996 International Conference, Cambridge, England, UK)

Fahy, R. F. (2000, March). *New developments in EXIT89*. (Paper presented at the Proceedings of the 15th Meeting of the UJNR Panel on Fire Research and Safety, San Antonio, TX, USA)

Fraser-Mitchell, J. N. (1999). Modelling human behavior within the fire risk assessment tool CRISP. *Fire and Materials, 23*(6), 349–355

Friedman, R. (1992). An international survey of computer models for fire and smoke. *Society of Fire Protection Engineers Journal of Fire Protection Engineering, 4*(3), 81–92

Fruin, J. J. (1971). *Pedestrian planning and design*. (New York: Metropolitan Association of Urban Designers and Environmental Planners)

FSEG (2004). *BuildingEXODUS Capabilities* [Electronic version]. Fire Safety Engineering Group. Retrieved November 4, 2006, from http://fseg.gre.ac.uk/exodus/exodus_products.html

Galea, E. R. & Gwynne, S. (2006). *Principles and practice of evacuation modelling* (7th Edition). (London: Fire Safety Engineering Group - University of Greenwich)

Galea, E. R., Blake, S. & Gwynne, S. (2003). *A methodology and procedure for the introduction of aircraft evacuation simulation to the aircraft certification process*. VERRES (VLTA Emergency Requirements Research Evacuation Study), a consortium of Cranfield and Greenwich Universities, Virgin Atlantic, Airbus, Sofreavia and the Civil Aviation Authority

Goodchild, M. F. (2005). GIS, spatial analysis, and modelling overview. (In D. J. Maguire, M. Batty & M. F. Goodchild (Eds.), *GIS, spatial analysis and modelling*. Redlands, CA, USA: ESRI Press)

Gunes, A. E. & Kolel, J. P. (2000). Using GIS in emergency management operations. *Journal of Urban Planning and Development, 126*(3), 136–149

Gwynne, S. & Galea, E. R. (1997). *A review of the methodologies used in the computer simulation of evacuation from the built environment*. (London, UK: Fire Safety Engineering Group - University of Greenwich)

Gwynne, S., Galea, E. R., Owen, M., Lawerence, P. J. & Filippidis, L. (1999a). A review of the methodologies used in evacuation modelling. *Fire and Materials*, *23*, 383–388

Gwynne, S., Galea, E. R., Owen, M., Lawerence, P. J. & Filippidis, L. (1999b). A review of the methodologies used in the computer simulation of evacuation from the built environment. *Building and Environment*, *34*, 741–749

Gwynne, S., Galea, E. R., Lawrence, P. J. & Filippidis, L. (2001). Modelling occupant interaction with fire conditions using the BuildingEXODUS evacuation model. *Fire Safety Journal*, *36*(4), 327–357

Gwynne, S., Galea, E. R., Owen, M., Lawrence, P. J. & Filippidis, L. (2005). A systematic comparison of BuildingEXODUS predictions with experimental data from the Stapelfeldt Trials and the Milburn House evacuation. *Applied Mathematical Modelling*, *29*(9), 818–851

Halcrow (2006a). PAXPORT: *Pedestrian planning software* [Electronic version]. Retrieved November 3, 2006, from http://www.halcrow.com/html/documents/pdf/fire_safety/paxport.pdf

Halcrow (2006b). *Pedestrian simulation modelling* - PEDROUTE/PAXPORT [Electronic version]. Retrieved November 3, 2006, from http://www.halcrow.com/html/proj_popup/pedroute_paxport.htm

Harrington, S. S. (1996). TIMTEX: *A hydraulic flow model for emergency egress*. Unpublished M.Sc. thesis, University of Maryland, Maryland, MA, USA

Heskestad, A. W. & Meland, O. J. (1998). *Determination of evacuation times as a function of occupant and building characteristics and performance of evacuation measures*. (Paper presented at the Human Behavior in Fire, Proceedings of the 1st International Symposium, University of Ulster, Ireland)

Hoffmann, N. A. & Henson, D. A. (1997). *Simulating transient evacuation and pedestrian movements in stations*. (Paper presented at the International Conference on Mass Transit Management, Kuala Lumpur, Malaysia)

Hoffmann, N. A., Henson, D. A. & Pope, C. W. (1998). *Analysis of the evacuation of a crush loaded train in a tunnel*. (Paper presented at the 3rd International Conference on Safety in Road and Rail Tunnels, Nice, France)

IES (2006). *Simulex – Simulation of occupant evacuation* [Electronic version]. Retrieved November 3, 2006, from http://www.ies4d.com/content/mediaassets/pdf/Simulex.PDF

Kendik, E. (1983). Determination of the evacuation time pertinent to the projected area factor in the event of total evacuation of high-rise office buildings via staircases. *Fire Safety Journal*, *5*(3–4), 223–232

Kendik, E. (1986). Designing escape routes in buildings. *Fire Technology*, *22*(4), 272–294

Kendik, E. (1988). Means of escape - 1988. How close are we to modelling egress from buildings in fire? *Fire and Material*, *13*(1), 156–163

Kerridge, J., Hine, J. & Wigan, M. (2001). Agent-based modelling of pedestrian movements: The questions that need to be asked and answered. *Environment and Planning B: Planning and Design*, *28*, 327–341

Ketchell, N. (2002). *A technical summary of the AEA EGRESS code* [Electronic version]. Retrieved April 14, 2005, from http://www.aeat-safety-and-risk.com/Downloads/Egress%20Technical%20Summary.pdf

Ketchell, N., Webber, D. M. & Cole, S. K. (1996). Assessment and simulation of crowd evacuation issues. (In R. Barham (Ed.), *Fire Engineering and Emergency Planning* (pp. 162–167). London, UK: E & FN Spon)

Kisko, T. M. & Francis, R. L. (1985). EVACNET+: A computer program to determine optimal evacuation plans. *Fire Safety Journal*, *9*(2), 211–220

Kisko, T. M., Francis, R. L. & Nobel, C. R. (1998). *EVACNET4 user's guide* [Electronic version]. Retrieved November 3, 2006, from http://www.ise.ufl.edu/kisko/files/evacnet/EVAC4UG.HTM

Klüpfel, H. & Meyer-König, T. (2003). Characteristics of the PedGo Software for crowd movement and egress simulation. (In E. R. Galea (Ed.), *Pedestrian and evacuation dynamics 2003: Proceedings of the 2nd international conference on pedestrian and evacuation dynamics* (pp. 331–340). London, UK: CMS Press)

Klüpfel, H. L. (2003). *A cellular automaton model for crowd movement and egress simulation.* Unpublished Ph.D. thesis, University of Duisburg-Essen, Germany

Kostreva, M. M. & Lancaster, L. C. (1998). A comparison of two methodologies in HAZARD I fire egress analysis *Fire Technology, 34*(3), 227–246

Kuligowksi, E. & Gwynne, S. (2005). What a user should know when choosing an evacuation model. *Fire Protection*, Fall, 30–40

Kuligowski, E. D. (2003). *The evaluation of a performance-based design process for a hotel building: The comparison of two egress models.* Unpublished M.Sc. thesis, University of Maryland, MD, USA

Kuligowski, E. D. (2005). *Review of 28 egress models.* (Paper presented at the Workshop on Building Occupant Movement During Fire Emergencies)

Kuligowski, E. D. & Peacock, R. D. (2005). *A review of building evacuation models.* (Washington, DC, USA: Fire Research Division Building and Fire Research Laboratory, U.S. Department of Commerce)

Legion Limited (2006). *Case studies* [Electronic version]. Retrieved November 4, 2006, from http://www.legion.biz/customers/casestudies.html

Levin, B. M. (1987). *EXITT - A simulation model of occupant decisions and actions in residential fires: Users' guide and program description.* (Gaithersburg, MD, USA: National Bureau of Standards U.S. Department of Commerce, Report NBSIR 87–3591)

Lo, S. M. & Fang, Z. (2000). A spatial-grid evacuation model for buildings. *Journal of Fire Sciences, 18*, 376–394

Lo, S. M., Fang, Z., Lin, P. & Zhi, G. S. (2004). An evacuation model: The SGEM package. *Fire Safety Journal, 39*(3), 169–190

Martin, D. J. (1996). *Geographic information systems: Socio economic applications.* (London, UK: Routledge)

Mott MacDonald (2006). *STEPS (Simulation of Transient Evacuation and Pedestrian movementS) Version 2.0.0011: User manual.* (Croydon, UK: Mott MacDonald, STEPS - Simulation Group - Transportation)

Nelson, H. E. & Mowrer, F. E. (2002). Emergency movement. (In P. J. DiNenno (Ed.), *SFPE handbook of fire protection engineering* (3rd Edition) (pp. 367–380). Quincy, MA, USA: Society of Fire Protection Engineers and National Fire Protection Association)

NRC (2007). *Successful response starts with a map: Improving geospatial support for disaster management.* (Washington, DC, USA: National Research Council, National Academies Press)

Okazaki, S. & Matsushita, S. (1993, March). *A study of simulation model for pedestrian movement with evacuation and queuing.* (Paper presented at the Engineering for Crowd Safety: Proceedings of the International Conference on Engineering for Crowd Safety, London, UK)

Olenick, S. M. & Carpenter, D. J. (2003). An updated international survey of computer models for fire and smoke. *Society of Fire Protection Engineers Journal of Fire Protection Engineering, 13*(2), 87–110

Openshaw, S. (1984). *The modifiable areal unit problem (Concepts and techniques in modern geography: 38).* (Norwich, England, UK: Geo-Books)

Ozel, F. (1985). *A stochastic computer simulation to model the behavior of people in fires: An environmental cognitive approach.* (Paper presented at the Proceedings of the Building Technology and Safety Conference, National Institute of Building Sciences, Los Angeles, CA, USA)

Ozel, F. (1991, August). *Simulation of processes in buildings as a factor in the object representation of built environments.* (Paper presented at the Proceedings of Building Simulation 1991, Nice, France)

Paulsen, T., Soma, H., Scheider, V., Wiklund, J. & Løvås, G. (1995). *Evaluation of simulation models of evacuation from complex spaces.* (Trondheim, Norway: The Research Council of Norway, SINTEF Report STF75 A95020)

Poon, S. L. (1994, July). *EvacSim - A simulation model of occupants with behavioral attributes in emergency evacuation of high-rise building fires.* (Paper presented at the Proceedings of the 4th International Symposium on Fire Safety Science, Ottawa, Ontario, Canada)

Poon, S. L. & Beck, V. R. (1995, March). *Numerical modelling of human behavior during egress in multi-storey office building fires*. (Paper presented at the ASIAFLAM 1995, Hong Kong)
Rhodes, N. & Hoffmann, N. A. (1999). *Fire safety engineering for the International Centre for Life, Newcastle-upon-Tyne*. (Paper presented at the Interflam 99, Edinburgh, UK)
RSSB (2003). *Managing large events and perturbations at stations: Stakeholder and domain expert interviews*. (London: Railway Safety Standards Board)
Santos, G. & Aguirre, B. E. (2004). *A critical review of emergency evacuation simulation models*. (Newark, DE, USA: Disaster Research Center, University of Delaware)
Savannah Simulations AG (2006a). *SimWalk pedestrian simulation software: User guide version 2.0*. (Switzerland: Savannah Simulations AG)
Savannah Simulations AG (2006b). *SimWalk: The pedestrian simulation software documentation*. (Switzerland: Savannah Simulations AG)
Savannah Simulations AG (2007). *SimWalk pedestrian simulation software: Homepage* [Electronic version]. Retrieved January 15, 2007, from http://www.simwalk.com/index.html
Schneider, V. (2001). *Application of the individual-based evacuation model ASERI in designing safety concepts*. (Paper presented at the 2nd International Symposium on Human Behavior in Fire, Boston, MA, USA)
Schneider, V. & Könnecke, R. (2001). Simulating evacuation processes with ASERI. (In M. Schreckenberg & S. D. Sharma (Eds.), *Pedestrian and evacuation dynamics* (pp. 303–314). New York, USA: Springer)
Sharp, G., Gwynne, S. & Galea, E. R. (2003). *The effects of ship motion on the evacuation process*. (London: Fire Safety Engineering Group - University of Greenwich)
Shestopal, V. O. & Grubits, S. J. (1994, July). *Evacuation model for merging traffic flows in multi-room and multi-storey buildings*. (Paper presented at the Proceedings of the 4th International Symposium on Fire Safety Science, Ottawa, Ontario, Canada)
Stahl, F. I. (1978). *A computer simulation of human behavior in building fires: Interim report*. (Washington, DC, USA: U.S. Department of Commerce, Report NBSIR 78–1514)
Stahl, F. I. (1980, October–November). *Preliminary findings concerning the validity of "BFIRES": A comparison of simulated with actual fire events*. (Paper presented at the Proceedings of the 2nd International Seminar on Human Behavior in Fire Emergencies, 1978)
Stahl, F. I. (1982). BFIRES-II: A behavioral based computer simulation of emergency egress during fires. *Fire Technology*, *18*(1), 46–65
Still, G. K. (1993). New computer system can predict human behavior response to building fires. *Fire*, *84*, 40–41
Thompson, P. A. (1994). *Developing new techniques for modelling crowd movement*. Unpublished Ph.D. thesis, University of Edinburgh, Edinburgh
Thompson, P. A. (2005, June). *Simulex: Simulated people have needs too*. (Paper presented at the Workshop on Building Occupant Movement During Fire Emergencies)
Thompson, P. A. & Marchant, E. W. (1995a). A computer model for the evacuation of large building populations. *Fire Safety Journal*, *24*(2), 131–148
Thompson, P. A. & Marchant, E. W. (1995b). Testing and application of the computer model 'SIMULEX'. *Fire Safety Journal*, *24*(2), 149–166
Thompson, P. A., Wu, J. & Marchant, E. W. (1996). Modelling evacuation in multi-storey buildings with 'Simulex.' *Fire Engineers Journal*, *56*(185), 6–11
TraffGO (2006). *PedGo* [Electronic version]. Retrieved November 3, 2006, from http://www.traffgo-ht.com/en/pedestrians/products/pedgo/index.html
Ventre, F. T., Stahl, F. I. & Turner, G. E. (1981). *Crowd ingress to places of assembly: Summary and proceedings of an experts' workshop*. (Washington, DC, USA: National Bureau of Standards, US Department of Commerce, Report NBSIR 81–2361)
Wall, J. M. & Waterson, N. P. (2002). Predicting evacuation times – A comparison of the STEPS simulation approach with NFPA 130. *Fire Command Studies*, *1*(1)
Weinroth, J. (1988). Software Review - EXITT: A simulation model of occupant decisions and actions in residential fires. *Fire Technology*, *24*(3), 273–278
Williams, W. (2004). Go with the flow: A new method of analysing how people really move around buildings should take the guesswork out of public spaces. *The Architects' Journal, February*, 8–9

Chapter 11
High-Resolution Coastal Elevation Data: The Key to Planning for Storm Surge and Sea Level Rise

Mark Monmonier

Abstract The coastline is a long, narrow hazard zone vulnerable to storm surge or tsunami as well as to the much slower rise in relative sea level associated with climate change and crustal subsidence. Effective planning for coastal hazards depends upon reliable climatological and tidal data, monitoring networks able to detect large rotating storms or distant undersea earthquakes, and refinement of computer simulation models like SLOSH (Sea, Lake, and Overland Surges from Hurricanes), developed in the 1970s and used for three decades by the National Weather Service to identify areas requiring evacuation. SLOSH, which accommodated 1970s computers with a coarse systematic grid unable to capture the effects of irregular shorelines and small features that can accelerate or attenuate sudden inundation by wind-driven seas, epitomizes the need for better models and more refined elevation data. Reliable modeling is important because the time-consuming evacuation of large coastal populations is thwarted by the unnecessary clearing of areas not at risk. Improved planning for severe coastal storms is possible with more computationally efficient models based on high-resolution, unstructured grids and able to account for wave action as well as topographic irregularities. Although these advanced models require refined elevation data—as do credible projections of sea level rise—for many parts of the United States coast the best current elevation data are derived from obsolete topographic maps with a five- or ten-foot contour interval. Topographic lidar, which can provide a half-foot vertical resolution, is a promising solution to the pressing need for better coastal elevation data. Integration with bathymetric lidar, which can image the sea floor in water as deep as 20 to 30 meters, yields a seamless topographic/bathymetric dataset useful for the high-resolution modeling of storm surge.

Keywords Cartographic coastlines, elevation data, numerical modeling, sea level rise, storm surge

---

Syracuse University

D.Z. Sui (ed.) *Geospatial Technologies and Homeland Security*,

11.1 Introduction

The coastline is not only a well-recognized hazard zone but also an essential focus of homeland security efforts, especially after Hurricane Katrina's strong suggestion that natural disasters can rival the destructive impact of unconventional warfare. The most serious coastal threat along the Atlantic and Gulf Coasts is the wind-driven invasion of sea water known as storm surge. Pacific shorelines are vulnerable to rapid inundation of a different sort: the tsunami triggered by an earthquake, landslide, or volcanic eruption under the sea or near the shore. Mapping the likely impact of storm surge and tsunami run-up became a recognized responsibility of emergency management officials in the 1970s, when the electronic computer provided the impetus for modeling physical processes in specific geographic contexts, climatic and seismic data provided at least rudimentary estimates of recurrence intervals, and digital cartographic data for the coastline and seafloor captured local factors that concentrate or attenuate flooding. As this chapter points out, these efforts as well as more recent attempts at modeling the slower but potentially more devastating consequences of sea level rise have been constrained by imprecise and often unreliable coastal elevation data.

11.2 Historical Background

Although the phrase is a recent coinage, homeland security is not only a well-established rationale for mapping America's coastline but also the raison d'être for the country's oldest federal mapping agency. In 1807 Congress authorized the Survey of the Coast to safeguard coastwise shipping as well as provide military intelligence for defending against European invaders. Because political intrigue and unstable funding compounded the difficulties of organizing and orchestrating systematic surveys, the organization's first charts were not published until the mid 1840s (Theberge 1998)—outrageously tardy by today's standards but four decades ahead of the Geological Survey's large-scale, systematic topographic maps of the interior. Coastal cartography proved especially valuable during the Civil War (1861–1865), when published and unpublished charts from the US Coast Survey (as it was renamed in 1836) helped the Union Navy capture New Orleans and blockade Confederate ports (Weddle 2002). Geodetic surveying of the interior emerged as a second significant responsibility, recognized in 1878 with the expanded name Coast and Geodetic Survey, but coastal navigation remained the agency's central concern. A 1923 study by the Institute for Government Research, a forerunner of the Brookings Institution, noted that the Survey's charts 'also serve for coast defense activities, for harbor improvement work, and other similar purposes' but made no mention of coastal flooding or emergency management (Weber 1923: 28). A significant shift occurred in 1970, when the Nixon Administration consolidated the Coast and Geodetic Survey with the Weather Bureau and several smaller agencies to form the National Oceanic and Atmospheric Administration (NOAA), a new department with a mission that included the 'better protection of life and property from natural hazards' (Nixon 1970).

11.3 Coastal Flooding and the Third Coastline

Collaboration between NOAA, the US Army Corps of Engineers (USACE), and the Federal Emergency Management Agency (FEMA) fostered delineation of the coastal flood zone and the emergence of a third cartographic coastline—the inland extent of inundation during a severe coastal storm—to supplement the high-water and low-water shorelines shown on large-scale coastal charts. More fleeting and less predictable than the other two shorelines, this third coastline is an important national priority as well as a significant technical and scientific challenge that extends the coastal cartographer's vision forward in time. The first coastline, represented more precisely on contemporary charts as the mean high water line—or the mean higher high water line, to recognize the semidiurnal mixed tides of the Pacific Coast—is an approximation of the visible shoreline plotted from offshore by explorers during the Age of Discovery. The second coastline, which appeared on marine charts during the late nineteenth century, is the sounding datum, derived from precise measurement and prediction of tidal variations. The third coastline, a product of computer modeling and statistical climatology, describes the likely extent of inundation for either a single storm or a stochastic flood event with a specific return interval, typically the mythical hundred-year flood, which is more appropriately described as the flood event with a one-percent annual probability. Although the first two coastlines migrate over time as land emerges or subsides, seas rise, and shorelines erode and accrete, the third coastline is essentially a suite of coastlines that depend upon a particular exceedance probability, the sophistication of the modeling technique, the reliability of the data, and the diligence and skill of those running the models and making the maps.

The primary reason for mapping a third coastline is storm surge, the numerical modeling of which is similar to atmospheric modeling insofar as both technologies integrate mathematical formulas representing physical processes with an elaborate accounting scheme based on a tessellation of discrete cells. The typical tessellation has been a polar grid positioned with its smaller cells centered near the shore over the area for which greater precision is required. Figure 11.1 shows the grid for Charlotte Harbor, Florida. The principal modeling technique is known as SLOSH, an acronym for Sea, Lake, and Overland Surges from Hurricanes, and the individual grids customized for specific areas are called SLOSH basins (Jelesnianski et al. 1992). SLOSH was developed in the 1970s and 1980s by Chester Jelesnianski, and used by the National Weather Service to forecast inland flooding from hurricanes. A single simulation is based on a hypothetical or historical storm with a given diameter and central pressure that advances along a designated storm track at a specified speed. Wind velocity within a grid cell at a specific time depends on the

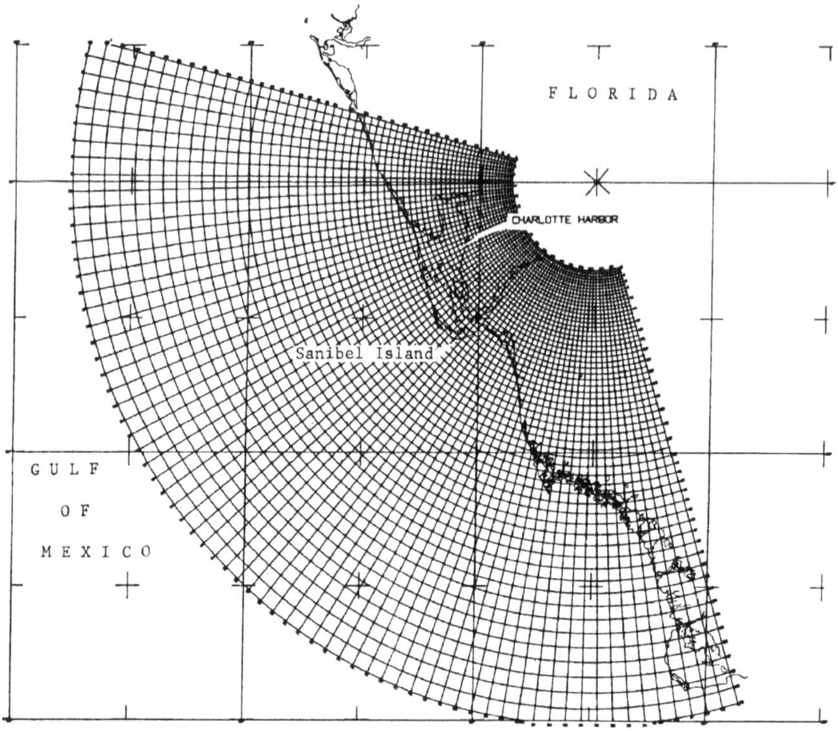

Fig. 11.1 SLOSH basin for Charlotte Harbor, Florida (From Jelesnianski et al. (1992), *SLOSH: Sea, Lake, and Overland Surges from Hurricanes*, 46)

cell's location relative to the current position of the storm center. The model uses this instantaneous velocity to compute a frictional force that transfers water near the surface into neighboring cells immediately downwind. Movement of water across cell boundaries lowers the column of water in a donor cell and raises it in one or more receiver cells, and gravity creates an additional gradient that transfers water into cells with a lower water surface, including onshore cells. SLOSH is a finite-difference model, so-called because it divides the area into discrete parcels and moves forward in time in small increments, typically 100 seconds; before advancing to the next round of wind friction and water transfer, the model settles all accounts, so to speak, and saves for each cell the highest stillwater elevation reached during the most recent or any previous round (Crawford 1979). After validating a SLOSH model by simulating historic storms with known meteorological characteristics and recorded flood levels, meteorologists and emergency managers can experiment with different landfall points, and coastal hydrologists can model the effects of plausible severe storms suggested by climatological data.

As a forecasting tool, SLOSH is only as reliable as its simplifying assumptions, the climatology used to specify model storms, and the accuracy of onshore and offshore elevation data (Houston et al. 1999). Low and problematically variable resolution—cells in a typical SLOSH basin range in width from roughly 0.5 to 7 km—cannot capture the effects of significant irregularities along the shoreline (International Hurricane Research Center 2006), and the model's reliability is especially poor for comparatively large coastal cells well removed from the center of the coarse polar grid, an appropriate accommodation for the markedly slower computers of the 1970s and 1980s. Additionally, SLOSH does not take into account the effect of breaking waves, which can inflict damage at elevations higher than the predicted rise in 'stillwater elevation,' nor does it incorporate the dynamic effects of astronomical tides. In addition, the model does not include the potentially significant contribution of inland flooding, especially significant for a large slow-moving storm with heavy rains. Computationally inefficient by contemporary standards, SLOSH is too slow to support interactive experimentation by storm forecasters. Despite these limitations, NOAA's National Hurricane Center is a strong supporter of SLOSH, which it relied on with mixed success during Hurricane Katrina in 2005 (NOAA 2005).

Newer technology that supports a more reliable mapping of the third shoreline includes an unstructured triangular grid designed to account for small features, above or below the water, that accelerate or retard storm surge (Stone et al. 2005). The example in Fig. 11.2, which focuses on Beaufort Inlet, North Carolina, is a part of a larger grid used with a coastal flooding model to assess the threat of sea level rise along the North Carolina coast (Feyen et al. 2006). The irregular grid provides an efficient, parsimonious integration of bathymetric data registered to a sounding datum with topographic data interpolated from a digital elevation model (DEM). Both data sets are referenced to a common vertical datum. The horizontal and vertical lines in Fig. 11.2 are parallels and meridians identified by decimal latitude or longitude, and the labels affixed to contour lines indicate depth *below* the North American Vertical Datum of 1988, so that positive numbers reflect depth of water while negative numbers represent elevation above the water line. Resolution ranges from 5 km for cells far offshore to 20 m within the narrow, vigilantly dredged channel of the Intracoastal Waterway.

FEMA, which administers the National Flood Insurance Program, draws on a variety of inundation models as well as tidal records and historic shoreline data in developing the Flood Insurance Rate Maps (FIRMs) used in calculating flood insurance premiums and restricting land development near the shoreline (Bellomo et al. 1999; Federal Emergency Management Agency 2006). The flood zone is defined by the estimated hundred-year stillwater elevation plus associated wave effects, and is typically divided into two parts: a V zone, which is deemed susceptible to both erosion and inundation because of breaking waves greater than three feet, and an A zone, farther onshore. As with storm-surge modeling, efforts to forecast the hundred-year flood line are undermined by incomplete models and coarse elevation data, particularly by a reliance on topographic maps with obsolete shorelines and a vertical contour interval of five or ten feet (Crowell et al. 2007). Another serious flaw in coastal flood insurance maps is the inability to disentangle flood damage from coastal erosion, which is not supposed to be covered by federal flood insurance (Crowell et al. 1999).

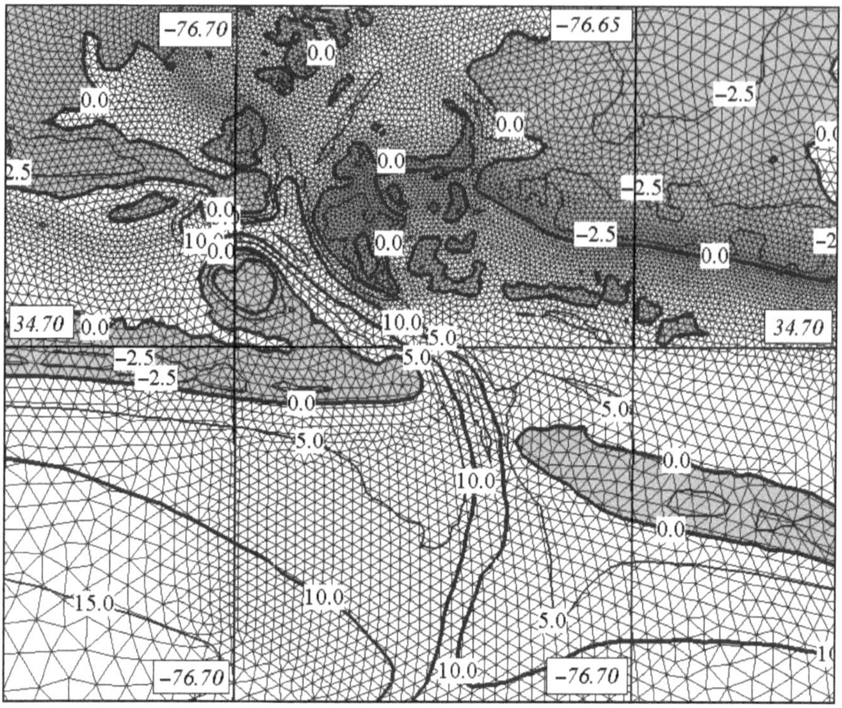

Fig. 11.2 Portion of an unstructured triangular grid for modeling storm surge and coastal flooding (From J. Feyen et al. (2006), Development of a continuous bathymetric/topographic unstructured coastal flooding model to study sea level rise in North Carolina, 354)

Two additional third coastlines complement FIRMs and surge simulations. One is the Coastal High Hazard Area, a regulatory zone typically wider than the flood map's V-zone and considered likely to warrant evacuation during a category-one hurricane (Florida Department of Community Affairs 2006). The other is the coastal evacuation map, translated for convenient use by local officials and the lay public by shifting zone boundaries inland to the nearest arterial roads or other well-known linear features so that the boundary is more conveniently communicated with words and more reliably recognized in the landscape.

11.4 Sea Level Rise and the Fourth Coastline

Growing awareness of sea level rise as a threat to coastal property fosters depiction of a fourth cartographic shoreline, which joined the discourse on climate change as both a planning tool and a propaganda instrument. The earliest primarily rhetorical example appears in a 1971 article in the *Journal of Geography* by geographic climatologist Richard Kopec, who used two maps, one for the world and the other for the United States, to illustrate the astonishing consequences of a 100 m [330 ft] rise, considered 'remote but not impossible' by 2050 (1971: 549). Nine years later an article in the *Annual Review of Energy* by Stephen Schneider and Robert Chen hyped a more subdued forecast with a comparatively detailed map showing the impact of 4.6 and 7.6 m [15 and 25 ft] increases on key government buildings in Washington, D.C. Although their dramatic demonstration that 'one could launch a boat from the west steps of the United States Capitol… and row to the White House South Lawn' (1980: 117) was intentionally rhetorical, they applied the same analysis to hundreds of topographic maps along the Atlantic, Gulf and Pacific coasts, and concluded that inundations this large would displace 5.7 and 7.8 million Americans, respectively. More recent predictions of sea level rise are far less frightening, at least for the present century, but some glaciologists worry that continued increases in atmospheric temperature could lubricate the base of glaciers in Antarctica and Greenland and hasten their downhill movement toward the sea, which could rise drastically, perhaps by 70 m (Alley et al. 2005; Parizek and Alley 2004). These scenarios, which call for acceleration of the gravity-driven, conveyor-belt advance of continental glaciers toward the coast by the downward movement of meltwater through expanding cracks in the ice, are considered plausible but unlikely (Kerr 2006). Even so, a set of maps prepared for a congressional briefing by glaciologist Byron Parizek demonstrate that extreme melting of the Greenland and West Antarctic ice sheets could drastically narrow the Florida peninsula (Bellassen and Chameides 2005: 10). While these examples are admittedly rhetorical, lesser inundations can look nearly as frightening, as demonstrated in the March 26, 2006 issue of *Science*, in which a map describing a 6 m rise showed much of South Florida and Southeastern Louisiana under water (Kerr 2006). Although low-probability cartographic scenarios looking ahead several centuries might be dismissed as scientific fantasy, the catastrophic impact depicted is a persuasive argument for further research.

Despite wariness about an acceleration of global sea level rise resulting from human-induced climate change, scientific consensus predicts much more modest increases in water level for the twenty-first century. Because of multiple sources of uncertainty, these predictions are not single numbers but a range of plausible outcomes based on multiple models and assumptions. The most recent projections, issued in early 2007 by the Intergovernmental Panel on Climate Change (IPCC), forecast a rise between 18 and 59 cm. In many areas, of course, the global rise associated with climate change is affected significantly by local crustal subsidence (or uplift), which can add to (or subtract from) global sea level rise, largely the result of thermal expansion of seawater. Although current rates are lower than the 30 to 110 cm hundred-year rise predicted in 1990, when the IPCC issued its first forecast (Titus and Narayanan 1995: 135), maps of sea level rise have generally tended toward increases of one or two meters over the twenty-first century. For example, the US Environmental Protection Agency (EPA), which in 2000 released a series of small-scale maps for sections of coastline, highlighted coastal elevations below 1.5 m and also delineated a secondary hazard zone with elevations between 1.5 and 3.5 m (Titus and Richman 2001; US Environmental Protection Agency

2000). These limits also reflect a reliance largely on digital data based on existing topographic maps, on which a five- or ten-foot contour interval conveniently places the lowest elevation contour at either 1.52 or 3.05 m above mean sea level. Cautiously avoiding the politically sensitive and scientifically questionable terms 'forecast' or 'prediction,' the EPA maps are appropriately labeled 'Lands Vulnerable to Sea Level Rise.'

In addition to uncertainty about the rate of sea level rise, two factors complicate efforts to map vulnerability at larger scales. One limitation is topographic data based on the National Geodetic Vertical Datum of 1929, when sea level was significantly lower—more than a half foot lower according to the historic rate of 2.7 mm/yr estimated for New York City (Titus and Narayanan 1995: 144–145). The other is the unreliability of linear, straight-line interpolation to estimate a five-foot contour for maps with a contour interval of ten feet. Coastal wetlands typically have an elbow-shaped vertical profile, with a nearly horizontal slope stretching inward from a marshy shoreline and joining a somewhat steeper slope at the dry edge of the wetland (Titus and Richman 2001). Linear interpolation between the shoreline and the ten-foot contour thus yields an estimated five-foot contour much closer to the current shoreline than warranted, thus underestimating the horizontal impact of a rise of 1.5 m or less. Simply put, reliable large-scale planning maps call for markedly better elevation data than found on conventional topographic maps.

A more recent product of the EPA sea level rise program recognizes significant differences in local response to rising seas, which can include defending existing structures and highways with sea walls, revetments, and other engineering strategies (Titus 2004). Whether a shoreline will be aggressively defended depends on the resolve of wealthy property owners and the commitment of local government to tourism as well as state and local regulations designed to protect natural habitats, minimize losses to flooding and erosion, and limit unwise investments of public funds. New EPA maps still under review at this writing recognize the possibility of protective measures by integrating elevation contours of two and five meters with color-coded categories indicating whether defensive measures are almost certain (brown), likely (red), unlikely (blue), or not allowed (light green). The latter two categories reflect a less resistant response for national seashores, state and federal parks, private land in a conservation trust, and most farmland.

11.5 Lidar

Topographic lidar, which became operational in the 1990s, is an efficient and effective tool for creating the high-resolution elevation maps and datasets needed to model reliably the likely effects of storm surge and sea level rise (Brock et al. 2002; Stoker et al. 2006). Lidar is an acronym for **li**ght **d**etection **a**nd **r**anging—still written in all capital letters in some accounts, the term will no doubt lose its capitalization, as happened with *radar*, coined in the 1940s as an acronym for **ra**dio **d**etection **a**nd **r**anging. Like radar, lidar is an active remote sensing system, but instead of using radio-frequency waves, lidar fires pulses of highly coherent, narrowly focused light generated with a laser (another de-capitalized acronym, meaning **l**ight **a**mplification by the **s**timulated **e**mission of **r**adiation). As 'detecting and ranging' implies, both lidar and radar compute distance to a reflective target by measuring the elapsed time between the initial emission of a packet of energy and its return. An airborne platform equipped with a GPS, an inertial measurement unit, and an onboard computer can place the target for each returned pulse in a three-dimensional framework with elevations accurate to within 20 cm or less and horizontal positions accurate to a half meter. In mid-latitude coastal areas, where tree canopy is seldom dense, bare-earth filtering can generate reliable topographic maps with markedly greater accuracy than more conventional photogrammetric approaches (Sithole and Vosselman 2004; Zhang and Whitman 2005).

In addition to topographic lidar systems, bathymetric and atmospheric lidar systems serve hydrographic and meteorological applications. Simultaneous imaging of topography and bathymetry is especially useful in developing seamless topographic/bathymetric datasets for the high-resolution modeling of storm surge. Bathymetric lidar uses laser pulses with two wavelengths, green and red (or very near infrared). In near-shore areas without excessive turbulence, the green pulse reflects from the sea floor while the red light reflects from the water surface. The difference in return time between the green and red pulses (plus a correction for waves) yields the height of the water column, and conflation with sea level at the time of the survey yields the bathymetric elevation. Although turbidity interferes with lidar's ability to image the sea bottom, in clear water the SHOALS (Scanning Hydrographic Operational Airborne Lidar Survey) system developed by the US Army Corps of Engineers was able to detect depths as great as 60 m (Wozencraft and Lillycrop 2003). In more typical situations, bathymetric lidar can penetrate reliably to depths between 20 and 30 m (National Research Council, Committee on National Needs for Coastal Mapping and Charting 2004: 42).

A related remote sensing technology, IFSAR (for **i**nter**f**erometric **s**ynthetic **a**perture **r**adar) can capture elevation data in poor weather but offers lower resolution because of its higher flight path. A typical lidar sensor flown on a fixed-wing aircraft at an altitude between 1,000 and 6,000 feet (300–1,800 m) yields a horizontal resolution of 0.5–1.0 m and a vertical resolution of 15–35 cm, whereas a typical IFSAR unit imaging from an altitude between 20,000 and 30,000 feet (6–9 km) yields a horizontal resolution of roughly 2.5 m and a vertical accuracy of 1.0 m (Mercer 2001). While IFSAR is more efficient in capturing data at low cost over wide areas under challenging conditions, lidar can provide greater detail for a narrower, more focused coastal strip and even greater precision when carried more slowly at a comparatively lower altitude on a helicopter.

Coastal scientists are optimistic about the integration of topographic and bathymetric lidar to support a range of coastal zone studies, including assessments of natural hazards (National Research Council, Committee on National Needs for Coastal Mapping and Charting 2004; Robertson et al. 2004). FEMA is using lidar to improve the base maps used for the National Flood Insurance Program, and several states, notably North Carolina and Florida, have commissioned lidar missions

(National Research Council, Committee on Floodplain Mapping Technologies 2007). Because topographic lidar is expensive, especially when the objective is a high-resolution bare-earth elevation model, state initiatives typically depend upon substantial federal grants. Because the federal government seems reluctant to undertake the massive effort without partnerships or substantially increased appropriations, lawmakers need to recognize reliable flood insurance mapping, dependable evacuation warnings, and competent long-range planning for coastal erosion and sea level rise as indispensable elements in a comprehensive approach to homeland security.

References

Alley, R. B., et al. (2005). Ice-sheet and sea-level changes. *Science*, 310, 456–460

Bellassen, V. & Chameides, B. (2005). *High water blues: The climate science behind sea level rise and its policy implications* [Electronic version]. New York: Environmental Defense. Retrieved March 15, 2006, from http://www.environmentaldefense.org/documents/4846_HighWaterBlues_Update_2005.pdf

Bellomo, D., et al. (1999). Coastal flood hazards and the National Flood Insurance Program. *Journal of Coastal Research*, special issue 28, 21–26

Brock, J. C., et al. (2002). Basis and methods of NASA airborne topographic mapper lidar surveys for coastal studies. *Journal of Coastal Research*, 18, 1–13

Crawford, K. C. (1979). Hurricane surge potentials over southeast Louisiana as revealed by a storm-surge forecast model: A preliminary study. *Bulletin of the American Meteorological Society*, 60, 422–429

Crowell, M., et al. (1999). Evaluation of coastal erosion hazards study: An overview. *Journal of Coastal Research*, special issue 28, 2–9

Crowell, M., et al. (2007). Improving FEMA's coastal risk assessment through the National Flood Insurance Program: An historical overview. *Marine Technology Society Journal*, 41, 18–27

Federal Emergency Management Agency (2006). *Numerical models meeting the minimum requirements of the NFIP* [Electronic version]. Retrieved on February 2, 2007, from http://www.fema.gov/plan/prevent/fhm/en_coast.shtm

Feyen, J., et al. (2006). Development of a continuous bathymetric/topographic unstructured coastal flooding model to study sea level rise in North Carolina. (In M. Spaulding (Ed.), *Proceedings of the Ninth International Conference on Estuarine and Coastal Modeling, Charleston, SC, October 31–September 2, 2005*. (pp. 338–357). Reston, VA: American Society of Civil Engineers). Retrieved February 5, 2007, from www.nauticalcharts.noaa.gov/csdl/Vdatum_pubs/feyen_ECM09.pdf

Florida Department of Community Affairs (2006). *Coastal High Hazard Study Committee final report* [Electronic version]. Retrieved on February 1, 2007, from http://www.dca.state.fl.us/fdcp/dcp/chhsc/final2-1-06.pdf

Houston, S. H., et al. (1999). Comparisons of HRD and SLOSH surface wind fields in hurricanes: Implications for storm surge modeling. *Weather and Forecasting*, 14, 671–686

Intergovernmental Panel on Climate Change (IPCC) (2007). *Climate change 2007: The physical science basis—summary for policymakers* [Electronic version]. Geneva, Switzerland. Retrieved on February 3, 2007, from www.legislative.noaa.gov/Testimony/solomon020807ipccsum.pdf

International Hurricane Research Center. (2006). *Storm surge model evaluation* [Electronic version]. Miami: Florida International University. Retrieved on January 27, 2007, from http://www.ihc.fiu.edu/lcr/research/windstorm_simulation/storm_surge_model_evaluation.htm

Jelesnianski, C. P., et al. (1992). *SLOSH: Sea, lake, and overland surges from hurricanes*. NOAA Technical Report NWS 48. Silver Spring, MD: National Weather Service

Kerr, R. A. (2006). A worrying trend of less ice, higher seas. *Science*, 311, 1698–1701

Kopec, R. J. (1971). Global climate change and the impact of a maximum sea level on coastal settlement. *Journal of Geography*, 70, 541–550

Mercer, B. (2001). *Comparing LIDAR and IFSAR: What can you expect?* Proceedings of Photogrammetric Week 2001, Stuttgart [Electronic version]. Retrieved on August 20, 2007, from http://www.apogee.com.au/pdf/lidar.pdf

National Oceanic and Atmospheric Administration (2005). Storm surge: A rising concern among coastal residents. *NOAA Online Magazine*. Story 178 [Electronic version]. Retrieved on February 10, 2007, from http://www.magazine.noaa.gov/stories/mag178.htm

National Research Council. Committee on Floodplain Mapping Technologies (2007). *Base map inputs for floodplain mapping*. [Electronic Prepublication version]. Washington, DC: National Academies Press. Retrieved on February 19, 2007, from www.nap.edu/catalog/11829.html

National Research Council. Committee on National Needs for Coastal Mapping and Charting (2004). *A geospatial framework for the coastal zone: National needs for coastal mapping and charting*. (Washington, DC: National Academies Press)

Nixon, R. (1970). *Reorganization plan No. 3 of 1970: Message of the President, July 9, 1970* [Electronic version]. Retrieved on January 22, 2007, from http://www.access.gpo.gov/uscode/title5a/5a_4_93_3_.html

Parizek, B. R. & Alley, R. B. (2004). Implications of increased Greenland surface melt under global-warming scenarios: Ice-sheet simulations. *Quaternary Science Reviews*, 23, 1013–1027

Robertson, W., et al. (2004). Mapping shoreline position using airborne laser altimetry. *Journal of Coastal Research*, 20, 884–892

Schneider, S. H. & Chen, R. S. (1980). Carbon dioxide warming and coastline flooding: Physical Factors and climatic impact. *Annual Review of Energy*, 5, 107–140

Sithole, G. & Vosselman, G. (2004). Experimental comparison of filter algorithms for bare-earth extraction from airborne laser scanning point clouds. *ISPRS Journal of Photogrammetry and Remote Sensing*, 59, 85–101

Stoker, J. M., et al. (2006). CLICK: The new USGS center for lidar information coordination and knowledge. *Photogrammetric Engineering and Remote Sensing*, 72, 613–616

Stone, G. W., et al. (2005). The role of barrier islands, muddy shelf and reefs in mitigating the wave field along coastal Louisiana. *Journal of Coastal Research*, special issue 44, 40–55

Theberge, A. E. (1998). *The Coast Survey: 1807–1867*. Vol. 1, *The history of the Commissioned Corps of the National Oceanic and Atmospheric Administration* [Electronic version]. Silver Spring, MD: National Oceanic and Atmospheric Administration. Retrieved on January 12, 2007, from http://www.lib.noaa.gov/edocs/heritage.html

Titus, J. G. (2004, November). *Maps that depict the business-as-usual response to sea level rise in the decentralized United States of America*. [Electronic version] (Paper presented at the OECD [Organisation for Economic Co-operation and Development] Global Forum on Sustainable Development: Development and Climate Change, Paris. Retrieved April 12, 2006 from www.oecd.org/dataoecd/3/23/37815989.pdf

Titus, J. G. & Narayanan, V. (1995). *The probability of sea level rise*. Report EPA 230-R95–008 [Electronic version]. Washington, DC: US Environmental Protection Agency. Retrieved February 15, 2007, from http://yosemite.epa.gov/oar/globalwarming.nsf/content/ResourceCenterPublicationsProbability.html

Titus, J. G. & Richman, C. (2001). Maps of lands vulnerable to sea level rise: Modeled elevations along the US Atlantic and Gulf Coasts. *Climate Research*, 18, 205–228

US Environmental Protection Agency (2000). *Maps of lands vulnerable to sea level rise. July 2000* [Electronic version]. Retrieved April 12, 2006, from http://yosemite.epa.gov/OAR/globalwarming.nsf/content/ResourceCenterPublicationsSLRMaps.html

Weber, G. A. (1923). *The Coast and Geodetic Survey: Its history, activities and organization*. (Baltimore, MD: Johns Hopkins)

Weddle, K. J. (2002). The Blockade Board of 1861 and Union naval strategy. *Civil War History*, 48, 123–142

Wozencraft, J. M. & Lillycrop, W. J. (2003). SHOALS airborne coastal mapping: Past, present, and future. *Journal of Coastal Research*, special issue no. 38, 207–215

Zhang, K. & Whitman, D. (2005). Comparison of three algorithms for filtering airborne lidar data. *Photogrammetric Engineering and Remote Sensing*, 71, 313–324

Chapter 12
Remote Sensing and GIS Applications for Precision Area-Wide Pest Management: Implications for Homeland Security

Yanbo Huang[1], **Yubin Lan**[1], **John K. Westbrook**[1], **and Wesley C. Hoffmann**[1]

Abstract Area-wide pest management essentially represents coordinated adoption of integrated pest management to conduct preventive suppression of a pest species throughout its geographic distribution. Scientists and researchers in area-wide pest management programs have been developing, integrating, and evaluating multiple strategies and technologies into a systems approach for management of field and crop insect pests. Remote sensing, Global Positioning Systems, geographic information systems, and variable rate technology are additional tools that scientists can implement to help farmers maximize the economic and environmental benefits of area-wide pest management through precision agriculture.

Precision area-wide pest management systems were originally developed to reduce the country's daily risk of natural pest introductions. Now the systems are being developed to reduce the daily risk the country faces of pest introductions, both natural and intentional. Aerial application under precision area-wide pest management strategy is one of the most feasible methods to quickly limit the threat of area-wide pest infestations, which increased after the terrorist attacks of September 11, 2001.

Keywords Area-wide pest management, geographic information systems, GPS, remote sensing, variable rate technology

12.1 Introduction

Area-wide Pest Management (APM) is a concept of preventive suppression of a pest species throughout its geographic distribution, rather than reactive field-by-field control. APM essentially represents coordinated adoption of Integrated Pest

[1]Agricultural Research Service, US Department of Agriculture, College Station, TX

Management (IPM), which integrates control tactics to manage one or multiple key pest species in a single field. Scientists and researchers in APM programs have been developing, integrating, and evaluating multiple strategies and technologies into a systems approach for management of field crop insect pests. Remote Sensing, Global Positioning Systems (GPS), geographic information systems (GIS), and variable rate technology (VRT) are additional technologies that scientists can implement to help farmers maximize the economic and environmental benefits of APM through precision agriculture.

Ground and aerial sprayers may be used in implementation of APM. In precision agriculture, ground-based sprayers are effective in reducing pesticide use and production costs, and protecting the environment. When new aerial application technologies become available, research must be conducted before incorporating these aerial application technologies into precision agriculture. A prescription map that indicates within-field differences in crop vigor caused by pest infestations and infections is integrated to direct each aerial application. The map is produced by a remotely sensed image through a fusion with data from field sampling using GPS/GIS technology.

Precision APM systems are being developed to reduce the daily risk the country faces by pest introductions, both natural and intentional. Aerial application under the precision APM strategy is one of the most feasible methods to quickly limit the threat of area-wide pest infestations, which increased after the terrorist attacks of September 11, 2001.

12.2 Precision Agriculture

Precision agriculture is an agricultural management concept dealing with in-field variability. It provides agriculturists with the ability to measure, analyze, and manage variability in a number of production variables. Precision agriculture is not a single technology but a 'smorgasbord' of technologies, which requires acquisition and management of data over a period of time. The more years and better organization of production data, the better one is able to quantify production variability and its causes. The major goals of precision agriculture are to reduce or redistribute input costs, and increase profit by increasing production in quantity and quality while protecting and managing the environment and natural resources. Precision agriculture requires the use of new technologies, such as GPS/GIS, optical sensors, and satellites or aerial images to understand and evaluate the variations. Collected information may be used to more precisely evaluate optimum sowing density, estimate fertilizer and other input needs, and accurately predict crop yields. It seeks to avoid routine practices to a crop, regardless of local soil/climate conditions and may help to better assess local situations of disease or lodging.

12.2.1 IPM and APM

The concept of APM is based on a set of principles that are somewhat different from those of traditional IPM. Definitions of IPM vary widely theoretically. Many agricultural producers implement IPM by using the best available pest management tactic(s) as needed against a key pest(s) on an individual field or farm. APM has evolved from IPM and is currently considered as an effective method to manage pests using an organized and coordinated attack on pest populations over much larger areas. APM is most effective when conducted against a single or small group of pests over large geographical areas that are delineated by biological criteria associated with pest colonization and dispersal potential. Additionally, APM should be coordinated by organizations rather than individuals and should focus on reducing and maintaining a pest population below an economically damaging level.

Numerous APM programs are currently being conducted throughout the world. In the US, several cotton pests (e.g. boll weevil, pink bollworm, sweet potato whitefly, and tobacco budworm) are managed using APM techniques. In 1995, the United States Department of Agriculture, Agricultural Research Service (USDA, ARS) implemented the first formal area-wide pest management program against the codling moth in the Pacific Northwest. That multi-state cooperative program was developed to assess, test, and implement an integrated strategy for the management of the pest on fruit orchards using mating disruption to alleviate the impact of chemical insecticides on natural enemies and to open the opportunity for use of more environmentally friendly control tactics against other pests.

Over the past few years, many discussions have been held by pest management experts to determine the merits of this novel concept. Out of these discussions, six advantages have been identified:

1. Area-wide pest management, when interfaced with IPM programs, offers a long-term solution to the pest problem instead of quick-fix solutions on small acreages
2. When properly implemented, APM can prevent major pest outbreaks and provide a more sustainable management procedure for growers to use
3. APM permits the use of the best and most environmentally friendly management techniques
4. Once fully implemented, area-wide pest management can be more cost effective than managing pests on individual farms
5. Area-wide pest management permits the use of biologically-based management strategies for other pests within the crop
6. The basis of an area-wide pest management system is built upon the development of effective pest monitoring systems and reduction in unnecessary pesticide applications

As IPM expands to area-wide type programs, information management and decision needs become more critical. Pest management using precision farming tools (e.g. GPS/GIS, remote sensing, etc.) will become more important over large geographic management units. Rapid data assessment and precise decision capabilities will be needed by grower/consultants to adequately manage a crop.

The ARS corn rootworm area-wide management program serves as an example of how to use precision farming techniques to manage an important insect pest. The corn rootworm program was implemented in 1996 in response to numerous problems in conventional rootworm management strategies. The program was conducted in Kansas, Illinois, Indiana, Iowa, South Dakota, and Texas on 16-square-mile management units, and targeted western, northern, and Mexican corn rootworms with insecticide-baits. These baits were composed of a feeding stimulant mixed with small amounts of insecticide. Adult rootworms compulsively feed on the baits and die within hours. The baits were applied by airplane or tractor mounted sprayer and used about 95 to 98 percent less insecticide than a typical adult rootworm insecticide application. The baits used to manage these highly mobile insects over broad geographic areas prevented significant numbers of eggs from being laid in corn fields. Reducing egg lay can prevent economic infestations from occurring if corn is planted in the same field the following year. This program allowed decisions to be made by a site manager that reduced unnecessary insecticide inputs (e.g. prophylactic soil insecticide applications) in producer fields. The management and decision making process involved resulted in large quantities of information that could best be evaluated using GPS/GIS techniques.

In implementing APM for effective pest control, improved efficiency of pesticide use, and minimal environmental impact, the management decision should be based on spatial information which characterizes pest distribution. This spatial pest information-based management is the integration of APM and precision agriculture.

12.2.2 Precision APM

In the Midwest, precision agriculture is associated not with sustainable agriculture but with mainstream farmers who are trying to maximize profits by spending money only in areas that need fertilizer. This practice allows the farmers to vary the rate of fertilizing according to the needs identified by GPS-guided grid sampling. Fertilizers that would have been spread in the areas which do not need it can be placed in the areas that do need it, thereby optimizing its use.

Precision agriculture may be used to improve a field or a farm management from several perspectives:

- Agronomical perspective: adjustment of cultural practices to take into account the real needs of the crops (e.g. better fertilization management)

- Technical perspective: better time management at the farm level (e.g. planification of agricultural activity)
- Environmental perspective: reduction of agricultural impacts (better estimation of crop nitrogen needs implying limitation of nitrogen run-off)
- Economic perspective: increase of the output and/or reduction of the input, increase of efficiency (e.g. lower cost of nitrogen fertilization practice)

Other benefits for the farmers may be to help create a history of farm practices and results, which help in decision making and traceability requirements.

Precision APM is a direct application of precision agriculture techniques to APM. In precision APM, management decisions are based on map information which characterizes pest spatial distribution, which are then implemented site-specifically. Implementation of site-specific pest management involves three main steps (Pedigo 2002):

1. Measuring the spatial variability of insect pests within an area
2. Mapping the insect distribution to generate prescription maps with management zones
3. Controlling pests based on the prescription maps

With the strategies of precision APM, the site-specific application varies the rate of pesticide application throughout the field in response to pest variability.

Site-specific management must be approached logically and systematically. Like any major change, it is easier to break the process into manageable action steps that ultimately produce a complete management system. A strategy is offered in contrast to a generalized recipe of tools and practices. A successful strategy considers the farmer's goals and characteristics of the individual farm, farmer, and fields before decisions about tools and practices are made. A strategic approach for site-specific management might include the following:

- Define goals
- List the decisions that must be made to reach the goals
- Determine the data needed to support the decisions
- Determine the tools needed to collect/manage/interpret the data
- Determine requirements for implementation
- Inventory the human, physical, and information resources available
- Make adjustments to meet projected future needs
- Collect and interpret the data needed
- Modify the production plan based on the interpretation of the data collected
- Implement the improved plan
- Repeat the process

Precision APM incorporates remote sensing, GPS, GIS, VRT, and various other electronic technologies that measure and control different components of the management system. In this way precision APM allows crop producers to achieve effective pest control, improved efficiency of pesticide use, minimal environmental impact, and an unprecedented high level of APM in practice.

12.3 Remote Sensing

Remote sensing has increasingly benefited agricultural management during the past several decades. Remote sensing from satellites and aircraft has been widely used for applications in agriculture (Phillips and Johannsen 1969; Johannsen et al. 1977; Jackson 1984; Myers 1983; Owe and D'Urso 2002; Steve and Clark 1990). In particular, passive optical sensors mounted on aircraft, Landsat (NASA), and SPOT (the French Système Probatoire d'Observation de la Terre) have allowed applications such as prediction of crop production and land use change. Precision agriculture is one of the important advances in agricultural applications of remote sensing for sustainable agriculture.

Remote sensing is defined as 'the acquisition of information about an object without being in physical contact with it' (Elachi 1987). The main focus of remote sensing in agriculture is the interaction of electromagnetic energy with plants and soil. The utilized sensors can be grouped into two main categories, photographic and non-photographic. Both provide information about electromagnetic energy and how it interacts with the surface being viewed.

When properly acquired and calibrated, aerial photographs can be used to measure the spectral reflectance of surface features in a manner similar to non-photographic systems. The biggest advantage of non-photographic systems is that they instantly acquire and store the data in digital form immediately. Therefore, calibration of non-photographic systems can be easier. Multispectral scanners, or non-photographic devices, provide the opportunity to sense several wavebands in a wider range of discrete wavelengths than does a photographic sensor. Hyperspectral scanners provide the opportunity to sense many very narrow wavebands over a wide range of wavelengths. However, they do so with a much greater number of sensors. Multispectral systems measure energy in specific, strategically restricted portions of the electromagnetic spectrum. Hyperspectral systems measure several consecutive wavebands across a specified region of the electromagnetic spectrum. These sensors measure energy and store the data in a digital form. The stored digital number of each pixel in the multispectral system represents the energy reflected from some area on the ground. That area is a function of the image array size, instrumental optics, and the altitude of the sensor platform. The ground resolution of the multispectral system is calculated by dividing the ground swath width of the sensor by the image array width. However, non-photographic sensors need to be calibrated periodically if surface reflectance or multi-date analysis is desired, much like that of the photographic sensors. Sensors can be designed to measure the available solar and terrestrial energy, usually from 0.3 micrometer to 14 micrometer. These systems are known as passive sensors because they sense energy originating from an outside source. Active systems generate the energy being sensed. The active systems are generally in the microwave region of 1 mm to 1 m wavelengths, but can be at any wavelength. An example of these types of systems would be radar in the microwave, laser systems in the visible portion of the spectrum, and the electromagnetic sensors that measure electrical conductivity.

A large number of remote sensing products and systems exist for a specific application. To select the most appropriate remote sensing product or develop a remote sensing system for an application, it is necessary to understand:

- Basic principles of remote sensing
- Characteristics of the different information sources
- The nature of the problem
- The value of the information

The large number of satellite remote sensing products available make it difficult to select the appropriate one because each satellite has different revisit times, delivery schedules, ordering requirements, pixel resolutions, sensors, and costs. Some satellites collect on a regular schedule (Landsat), while other satellites (IKONOS, QuickBird, and SPOT) need advance programming (tasking). To obtain high resolution satellite information promptly from QuickBird or SPOT, either High Priority or Rush Tasking may need to be purchased. However, high-resolution (small pixel size) data are not needed for all agricultural problems. In agricultural applications many land managers ask, 'Which satellite remote sensing product is the best?' The answer depends on many factors, including:

- Resolution requirements
- Turn-around and revisit times
- Spectral bands measured by the sensor
- Cost versus value of the information
- Possibility of sharing costs with others within the scene
- Data processing requirements

However, due to the limited availability of fine spatial resolution, near real-time data has slowed the application of satellite remote sensing in the past. Some governmental agencies and companies, such as In-Time, Inc. (Cleveland, Mississippi), provide aircraft-based imagery and are emerging to meet the spatial resolution and temporal requirements for agricultural management and site-specific research. Commercially available satellite imagery such as IKONOS, QuickBird, and SPOT provide additional sources of high-resolution remotely sensed data.

Characterization of pest spatial variability over an area of interest is fundamental for implementing precision APM. Remote sensing is a rapid, effective technology that provides information for spatial variability sampling with GPS/GIS for implementing APM management tactics. Remote sensing is the template that directs both the scouting efforts and the building of spatial, variable-rate pesticide prescriptions. When spatial information is joined with producer and consultant experience, the need for precise, numerically-based decisions is dramatically lessened because the imagery efficiently apportions fields into various habitats relevant to the population biology of the pest. As a result, the spatial and temporal information can be used to improve the timing and placement of pesticide applications to locations only where intervention is necessary. Therefore, if spatial spray applications are correctly timed over a large area, then populations of pests are sufficiently controlled with minimal input of pesticides. However, a precision approach to large-scale pest management requires the establishment of clear channels of communication between managers of adjacent fields and farms.

Advances in precision agriculture technology (GPS, GIS, and VRT) provide the tools needed to apply information from multispectral images to agricultural management problems. However, considerable work is needed before the full benefits of remotely sensed data can be realized.

Pest management is a major component of agriculture that benefits from advances in precision agriculture technology. Lorenzen and Jensen (1989) have shown that powdery mildew is detectable with reflectance measurements in the visible portion of the spectrum. The ability to detect and map insect damage with remotely sensed imagery implies that methods can be developed to focus pesticide applications in the areas of fields most infected, thus conserving beneficial insect populations. Remote sensing technologies were investigated for use in detection of late-season pest infestations and wild host plants of tarnished plant bug in the Mississippi Delta (Sudbrink et al. 2000). Preliminary results indicated that spider mite infestations were discernable from healthy and stressed cotton with aerial videography and spectro-radiometry. Broadleaf wild host plants of tarnished plant bug were distinct from non-host grasses in preliminary imagery and radiometry. Qin et al. (2003) conducted remote sensing analysis of rice disease stresses for farm pest management using wide-band airborne data. Riedell et al. (2005) used remote sensing techniques to reveal that canopy characteristics and spectral reflectance differences between insect infestation damage and disease infection damage can be measured in oat crop canopies. Reisig and Godfrey (2006) explored aerial and satellite remote sensing methods supplemented by ground sampling for their potential to distinguish aphid- and spider mite-infested cotton from uninfested cotton.

12.4 GPS/GIS

GPS receivers provide the means to determine position at locations anywhere on Earth. Developed by the US Department of Defense (DoD) and used for many civilian purposes, GPS has also made precision agriculture a reality. A typical configuration for on-farm applications includes a GPS receiver and antenna, a differential correction receiver and antenna, and cables to interface differentially-corrected (DGPS) data from the receiver to other electronic equipment such as a yield monitor or a variable rate controller. Accurate, automated position tracking with GPS receivers allows farmers and agricultural service providers to record geo-referenced data and to apply variable rates of inputs to smaller areas within larger fields. GPS can provide accurate position data when installed and operated properly, but can produce false readings under poor conditions.

GPS receivers are capable of collecting two kinds of information that can be used for a number of GIS operations. Waypoints are single points containing latitude, longitude, and elevation. Waypoints are used to mark the location of a specific feature. Recreational receivers allow users to represent that location with a symbol and assign it a name. Examples of waypoints are specific weed location, fence posts that need replacement, and location of wild game that must be left over night. Tracks are electronic trajectories (bread crumb trails) of where the receiver has been. Tracks are used to create field boundaries, calculate areas, and create routes or paths to follow with a machine. For agricultural application it is important to pay attention to the number of waypoints the receiver will hold and the number of tracks that can be saved to memory. All data in the GPS receiver can be downloaded with free software and then utilized by GIS software. Applications can be created in the GIS software and uploaded for use in the GPS receiver.

An important part of precision agriculture lies in GIS (Thurston et al. 2003). Generally, GIS is used for analysis, simulation, and model building purposes. GIS software provides a structure for presenting data in the form of maps for visual analysis as points, lines and areas, but the power of GIS goes far beyond the maps. In fact, mapping is a minor part of the use of GIS. The databases associated with GIS, and the tools to manipulate those data sets, are powerful tools for organizing, analyzing, and interpreting data.

GIS software has been used for on-the-go systems, especially in variable rate application research. Larsen et al. (1988) varied the injection sprayer output based on a GIS map and a differential GPS receiver. The injection of the active ingredient was changed uniformly across the boom, based on the GIS map. Al-Gaadi and Ayers (1999) tested the accuracy and lag time of a variable rate, direct injection sprayer based on a GIS map. They found that the reaction time was 2.2 seconds and with an application rate error of less than 1%. The project outlined in this chapter extends the latter project by having each nozzle on the boom controlled separately, although still keeping the spraying rate of the active ingredient the same from each operating nozzle.

Data are basically numbers and/or characters that describe observations or measurements of characteristics of a field. Moving from field-average to site-specific management increases the amount of data available for decisions, but also increases the challenge of storing and organizing the data. In the simplest form, data in a GIS are displayed as a map of the area of interest. Spatial analysis can be used to link data points to individual points on the maps. Statistics, simulations and models are additional analytical tools that can be applied through the GIS to extract more information from the data to support decisions.

The point of entry into precision agriculture for farmers has been the yield monitor. Yield monitors are available for small grains, sugar beets, and potatoes. The resulting GIS map reveals the exact locations of yield variability, the location of items marked by the operator during harvest, and field boundaries. The variability can be analyzed to determine the exact causes and its influence on the yield. The point of entry for a livestock producer is the use of a GPS receiver to determine pasture and field boundaries, location of weed infestations, and the condition of fences. The scale at weaning is one yield monitor, and the other is pounds of forage produced. Remote sensing becomes one of the major tools in identifying the variability in forage production across a pasture. Researchers are developing procedures to estimate forage production based on ground cover and degree of greenness of plants.

Because plants have different light reflectance values (signatures) at certain times in the plant growth cycles, remote sensing can be utilized to determine the exact location of weeds and the acreage of infestations. Ground truthing with the GPS receiver helps validate what the image has illustrated. Once identified and quantified, analyzing the data in a GIS will help determine the cost effectiveness of control, and monitor the effects of the control. Remote sensing technologies are used in the Weed Seeker to identify the presence of green plants. In the Green Seeker, the degree of greenness is used for variable rate top dressing of nitrogen. An example of a direct sensor is Veris Technology, which when pulled through a field, generates a map showing soil electrical conductivity, pH, texture, and moisture.

GPS makes it possible to operate 24 hours a day in all field conditions, and reduce skips and overlaps through the use of navigational guidance. The most complete system, referred to as Autopilot, is capable of steering a tractor or self-propelled equipment along an on-the-go calculated prescribed path. The operator can intervene at any time by taking a hold of the steering wheel. Parallel guidance utilizes a lightbar to guide the operator in steering a machine along an on-the-go determined path. Reduction in the amounts of overlap will improve field efficiency and reduce the amount of materials being applied. A secondary benefit of guidance is the ability to create a map of where the machine has traveled. Application maps can be of benefit when questions arise about where certain applications have occurred.

In precision APM, GPS is used to direct ground sampling of pest distribution and to navigate pesticide application with the control of the spraying system. GIS is used to overlay GPS information, ground sampling, and remote sensing, thus providing necessary data to generate prescription maps for site-specific application of pesticides over the interested area.

12.5 Variable Rate Technology

VRT, directed by a prescription map and synchronized to GPS, functions to automatically vary application rates as the sprayer moves across an area. With VRT, producers are able to apply only the amount needed in any one location. Analysis and processing of remote sensing image data creates variable or 'on–off' field prescriptions. These prescriptions are customarily created in GIS and then converted into a usable format as determined by the application vehicle's guidance system manufacturer. VRT can either be used in sprayer systems mounted on a tractor or on an aircraft (Green 1998; Smith and Thomson 2005). A prescription map is loaded onto an on-board computer in agricultural machinery that is equipped with GPS, and then a variable rate applicator with a rate controller will apply fertilizer, pesticides, water, or will plant seeds at rates that change according to prescription map as the application equipment moves over the area.

Typically, the central component of variable rate application equipment is a computerized controller. This device receives information from several sources, which will be used to control the application equipment. The controller may receive information from the application equipment and other sensors to maintain a database on the actual application rate as a function of field position.

A key component for all precision agricultural operations is the GPS technology to determine the instantaneous position of equipment as it operates in the field, and to store this information in a computer compatible format. Most precision agricultural operations will require real-time differential corrections so that vehicle position information will be accurate when the vehicle is operating in the field. For precision agricultural applications, the GPS positioning technology should be thought of as RT-DGPS, that is, the farmer should always be using real-time differential corrections to minimize position error.

Information contained in the GIS related to a specific field operation is downloaded to the system computer before field operations start. The computerized controller will continuously control variable application rates based upon knowledge gained from the GIS, from field location as provided by RT-DGPS, and perhaps from real time sensors. For example, the desired pesticide application rate may be known as a function of results from soil analysis tests, field location, and crop. The soil analysis test results as a function of field location would be entered into the GIS and downloaded to the computerized controller of the pesticide applicator. If one crop is being grown in the field being treated, then the operator may simply enter the crop from the computerized controller keyboard. However, if two crops are grown in alternating strips, this information would be entered into the GIS as a function of field position, then also downloaded to the variable rate application (VRA) computerized controller. When the equipment is operating in the field, the VRA computerized controller will be receiving RT-DGPS receiver position information and will match required application rate and crop as a function of field location to control the applicator equipment. It may also be possible to have a real time soil sensor, which will provide information on-the-fly about necessary pesticide application rates, rather than using pre-application sampling/analysis techniques.

The application equipment may also have sensors which provide quantitative information on the actual application rates. This information, along with RT-DGPS position, can be recorded to maintain a historical record of application rates. This historical information may allow the farmer to analyze cause and effect in the precision agricultural system, and perhaps influence future decision-making processes implemented in the computerized controller. For example, assuming that sufficient information has been gathered over several years, the farmer may have historical records on the effect of all of the inputs to his system for a specific field, including the crop yield. The GIS would then allow an analysis of cause and effect, based upon many factors, and allow fine-tuning of chemical application rates in subsequent seasons.

Eventually the RT-DGPS system may also be used for vehicle guidance. Most farm vehicle guidance systems today are visual prompting systems for the vehicle operator which can establish accurate vehicle position for application swaths. In the future, the guidance system may automatically guide the application vehicle, both ground and aerial.

Agricultural aircraft can be equipped with GPS/GIS and a multispectral camera system for taking aerial imagery for precision aerial application of crop production and crop protection materials. Ground truth data could be collected for the mission if necessary. Multisensor data fusion technology could be used for analyzing data from air and ground.

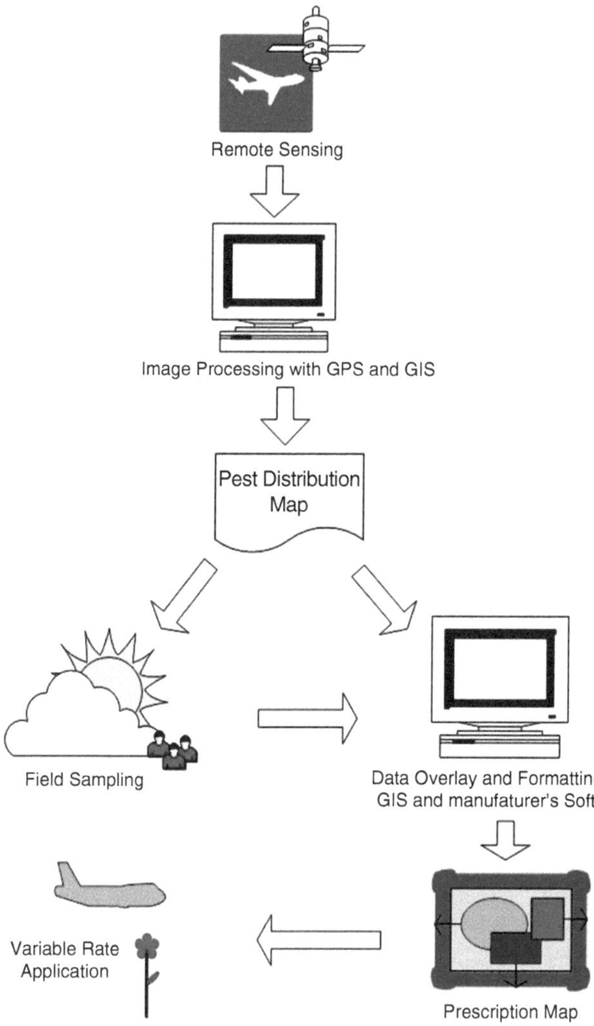

Fig. 12.1 System structure of the pest spraying system by integrating remote sensing, GPS, GIS and VRT technologies

To summarize, general site-specific area pest management or precision APM is implemented as follows:

1. Acquire remotely sensed imagery using multispectral cameras in an agricultural aircraft or multispectral sensors in a high resolution satellite
2. Georeference the acquired imagery with GPS positioning
3. Enhance and process the imagery to generate a pest distribution map with GIS
4. Field sampling referencing to the distribution map
5. Overlay the information of remote sensing, GPS, GIS, and field sampling with GIS to generate a prescription map
6. Pesticide application using VRT according to the prescription map
7. Application log data processing and analysis

Figure 12.1 shows the system structure of the pest spraying system by integrating remote sensing, GPS, GIS and VRT technologies.

12.6 Implications for Homeland Security

Since September 11, 2001, the US has faced an increasing risk of intentional pest introductions. Originally precision APM systems were developed to reduce the risk of recurring natural pest infestations. The components of IPM play an important part of the safeguarding continuum in reducing the risk of pest introduction and establishment. The Center for Plant Health Science and Technology (CPHST) is the scientific support organization for the Plant Protection and Quarantine division (PPQ) of the Animal and Plant Health Inspection Service (APHIS) in USDA. CPHST works to identify and evaluate the pathways used by invasive plant pests and weeds that threaten American agriculture and natural resources, and to assess the risks that these organisms pose to food, fiber, and the environment. In order to reduce the intentional risk, it is straightforward to adopt the developed precision APM systems to quickly limit the threat of area-wide pest infestations.

Area-wide pest management programs are needed to efficiently manage pests that may disperse to infest or infect plants and animals over large agricultural regions (Knipling and Stadelbacher 1983). These pests include insects, pathogens, weeds, and vectors of the aforementioned pests. Whether they are introduced naturally, inadvertently, or maliciously, invasive species are of utmost concern due to agricultural vulnerability and lack of natural (biological and environmental) defense mechanisms. GIS can relate pest distributions to several biotic and abiotic factors via spatial coordinates, and can analyze spatial and temporal patterns in these factors.

Knowledge of pest–host biology and ecology aids in the predictability of pest population dynamics, possible outbreaks, and appropriate mitigation tactics. The survival threshold and development rate of many pest organisms is temperature-dependent. Moisture is another abiotic factor that can limit development of plant pathogens, and extreme moisture values can increase mortality of many insect species depending on the stage of pest development. The successful outcome of area-wide pest management is sustainable production and profitability of the agricultural system, whether it is food, fiber, or natural resources. Consumer confidence in the safety and economic vitality of the agricultural sector can be strengthened by area-wide pest management, which can help avert pest-based domestic and international trade sanctions that can debilitate the US economy. Sanctions may be imposed by as little as a single infection or infestation of a host such as the presence of hoof-and-mouth disease in cattle or the presence of boll weevils or pink bollworms in cotton.

Remote sensing, biological sampling, and biophysical models can contribute to effectively monitor the risk of infestation and subsequent growth and spread of agricultural pests. Many pests are ubiquitous and do not cause serious economic losses or adverse health effects. However, catastrophic conditions occur when the pest, host, and environment are situated so that risks are elevated, such as when soybean rust is aerially transported to soybean production areas (Isard et al. 2005). The proper use of remote sensing, biological sampling, and biophysical models can help target areas that are most susceptible to pest threats and require a quick and thorough mitigation response.

Spatial analysis within GIS can identify several important characteristics of pest infections or infestations. Proximity of pest infestation to relevant physical features such as areas of vigorous crop growth (Willers et al. 2005), bodies of water, native vegetation, and roads can be calculated to determine strategic sampling locations. Patterns of pest dispersion relative to landscape and prevailing wind can be analyzed to determine the spatial extent of the infestation or infection. For example, pest insects may first colonize field borders where precision application of pesticides along only the borders would kill pests and conserve beneficial insects within the field. Analysis of gradients in infestations can be used to determine the area of economic infestation that justifies active mitigation. Scalability is perhaps one of the greatest aspects of GIS application in agriculture because it can be applied to infestations within a single field, large crop production areas, and nationally (Westbrook and Isard 1999).

Homeland security can be immediately fortified and subsequently enhanced based on temporal analysis of pest and host distributions. Historical analysis of infestations can be used for pattern recognition and what–if scenarios. For example, outbreaks of livestock diseases from mosquito-vectored diseases may follow predictable patterns following major floods. The rate of change of infestations can be analyzed to determine if, when, and where to commit mitigation resources.

References

Al-Gaadi, K. A. & Ayers, P. D. (1999). Integrating GIS and GPS into a spatially variable rate herbicide application system. *Appl. Eng. Agric.*, 15(4), 255–262

Elachi, C. (1987). *Introduction to the physics and techniques of remote sensing*. (New York: Wiley)

Green, C. J. (1998). Precision agriculture and cotton production. (In S. K. Esteicher, K. Stefan, & M. Holtz (Eds.), *The Science Corner*. Lubbock, TX: Department of Physics, Texas Tech University)

Isard, S. A., Gage, S. H., Comtois, P. & Russo, J.M. (2005). Principles of the atmospheric pathway for invasive species applied to soybean rust. *BioScience*, 55, 851–861

Jackson, R. D. (1984). Remote sensing of vegetation characteristics for farm management. *SPIE Remote Sens.*, 475, 81–96

Johannsen, C. J., Barr, D. J. & Barney, T. W. (1977). Remote sensing and applications, Remote sensing applications in extension programs conference, April 1977

Knipling, E. F. & Stadelbacher, E. A. (1983). The rationale for areawide management of Heliothis (*Lepidoptera: Noctuidae*) populations. *Bull. Entomol. Soc. Am.*, 29, 29–37

Larsen, W. E., Tyler, D. A. & Neilson, G.A. (1988). Field navigation using global positioning system (GPS). ASAE Paper No. 88–1604. St. Joseph, MI: ASAE

Lorenzen, B. & Jensen, A. (1989). Changes in leaf spectral properties induced in barley by cereal powdery mildew. *Remote Sens. Environ.*, 27, 201–209

Myers, V. I. (1983). Remote sensing applications in agriculture. (In R. N. Colwell (Ed.), *Manual of remote sensing*, 2nd ed., Vol. III (pp. 2111–2228). Falls Church,VA: American Society of Photogrammetry)

Owe, M. & D'Urso, G. (2002). *Remote sensing for agriculture, ecosystems, and hydrology III*. (Bellingham, WA: SPIE)

Pedigo, L. P. (2002). *Entomology and pest management*, 4th ed. (691 pp.) (Englewood Cliffs, NJ: Prentice-Hall)

Phillips, M. W. & Johannsen, C. J. (1969). *Remote sensing in agriculture, agronomy crops and soils notes*, Cooperative Extension Service, Purdue University, No. 130, August 1969

Qin, Z., Zhang, M., Christensen, T., Li, W. & Tang, H. (2003). Remote sensing analysis of rice disease stresses for farm pest management using wide-band airborne data, Proceedings. of IEEE International Geoscience and Remote Sensing Symposium, July 2003. Toulouse, France, pp. 2215–2217

Reisig, D. & Godfrey, L. (2006). Remote sensing for detection of cotton aphid–(*homoptera: aphididae*) and spider mite–(*acari: tetranychidae*) infested cotton in the San Joaquin Valley. *Environ. Entomol.*, 35(6), 1635–1646

Riedell, W. E., Osborne, S. L. & Hesler, L. S. (2005). Insect pest and disease detection using remote sensing techniques. *Meeting Proceedings of 7th International Conference on Precision Agriculture*, July 25–28, 2004, Minneapolis, MN

Smith, L. A. & Thomson, S. J. (2005). Performance of an aerial variable-rate application system with a hydraulically powered chemical pump and spray valve. (Paper No. AA05–009. The 39th annual convention of the National Agricultural Aviation Association, Reno, NV)

Steve, M. D. & Clark, J. A. (1990). *Applications of remote sensing in agriculture*. (London: Butterworths)

Sudbrink, D. L., Jr., Harris, F. A., Robbins, J. T., Snodgrass, G. L. & Thomson, S. J. (2000). Remote sensing of late-season pest damage to cotton & wild host plants of tarnished plant bug in the Mississippi delta. 2000 National Cotton Council Beltwide Cotton Conference, Vol. 2, pp. 1220–1223

Thurston T., Poiker, T. K., & Moore J. P. (2003). *Integrated geospatial technologies: A guide to GPS, GIS and data logging*. (New Jersey: Wiley)

Westbrook, J. K. & Isard, S. A. (1999). Atmospheric scale of motion for dispersing biota. *Agric. Forest Meteorol.*, 97, 263–274

Willers, J. L., Jenkins, J. N., Ladner, W. L, Gerard, P. D., Boykin, D. L., Hood, K. B., McKibben, P. L., Samson, S. A. & Bethel, M. M. (2005). Site-specific approaches to cotton insect control: Sampling, and remote sensing analyses techniques. *Prec. Agri.*, 6, 431–452

Chapter 13
Spatial Epidemiology: Where Have We Come in 150 Years?

Michael Ward

Abstract Modern epidemiology is founded on a tradition of spatial analysis. The genesis of this discipline can be traced to the classic work of John Snow and the Broad Street pump. During the 1850s, cholera outbreaks were an important cause of morbidity and mortality amongst the inhabitants of London. Using simple dot maps and visualization, Snow provided compelling evidence that cases were clustered and that fecal-contaminated drinking water might be the cause of some cholera outbreaks. During the intervening 150 years, spatial epidemiology (alternatively called landscape epidemiology and more broadly, medical geography) has developed into a field within its own right. During the past two decades, advances in geographic information systems and statistical methods for analyzing spatially-referenced health data has allowed epidemiologists to routinely perform spatial analyses. Some of the most beneficial advances in spatial epidemiology have been in the areas of data visualization, detection of disease clusters, identification of spatial risk factors, application of predictive models, and the routine incorporation of GIS into disease surveillance programs. In this chapter, approaches used in spatial epidemiology will be described. Some specific techniques that are currently popular in the discipline will be presented. Several case studies will be used to highlight the application of these techniques within the field of spatial epidemiology and to illustrate the potential value of this discipline to public health and homeland security. The chapter will conclude by considering some of the major obstacles that remain to the consolidation of spatial analysis as a foundation of modern epidemiology, including the availability and quality of spatial disease data, information on the distributions of the populations at-risk, and integration of methods seamlessly into epidemiologic software packages.

Keywords Medical geography, epidemiology, GIS, public health, homeland security

---

Texas A&M University

D.Z. Sui (ed.) *Geospatial Technologies and Homeland Security*,

13.1 Introduction

Epidemiology has been defined as the 'study of the distribution and determinants of health-related states or events in specified populations, and the application of this study to control of health problems' (Last 2001). Epidemiology provides information on risk factors for disease. With knowledge of what risk factors may cause a disease effect, steps can be taken to alter or avoid this factor and thus prevent (or reduce) disease incidence. Thus, the application of epidemiologic knowledge has been called 'preventive medicine,' and information from epidemiologic studies can be incorporated into programs that address public health. Public health has been defined as the state of well-being of a human population, with emphasis on preventing disease, rather than curing it. Thus, in public health practice *the individual is not the patient ... it is the community* (Schwabe 1984).

Understanding the distribution of disease in time and space is a foundation of epidemiology and hence preventive medicine or public health programs. Knowledge of where and when a disease occurs enables the generation of disease causation hypotheses for diseases with unknown or poorly characterized etiology, identification of disease risk factors, and the design of efficient disease surveillance and control programs.

The term 'medical geography' has a long history of use in public health to describe the application of methods borrowed from geography and applied to mapping diseases and investigating the geographic causes of disease occurrence. It is closely associated with the description of disease ecology. More recently, the term 'spatial epidemiology' has been used to describe the application of quantitative epidemiologic methods to disease problems, particular those activities that allow the distribution of disease in geographic space to be described and explained (Elliot et al. 2000). Other terms used include 'geographical epidemiology' and 'landscape epidemiology' (English 1992). These are more general terms that imply simply a focus on those aspects of the landscape (both biotic and abiotic) that might influence the occurrence of disease.

The development of spatial epidemiology as a sub-discipline has been driven particularly by the introduction of spatial statistical and geostatistical methods and the application of geographic information systems. Since most studies currently apply epidemiologic principles and analytical methods to understand why disease occurs within certain regions or landscapes, the term spatial epidemiology is used in this chapter. However, it should be recognized that the results of spatial epidemiologic studies are part of medical geography, which in term informs and guides public health activities.

The drawing of epidemic curves and construction of maps are basic skills used by epidemiologists to investigate disease occurrence, particularly for infectious diseases that might pose a threat to biosecurity. Spatio-temporal statistics and tests have utility in adding precision to qualitative verbal descriptions, facilitating the comparison of distributions, and drawing attention to characteristics unlikely to be noticed by visual inspection. Quantifying spatio-temporal patterns is important for understanding how spatio-temporal phenomena, such as disease occurrence, behave. Statistics quantify patterns, and can bring to attention subtle patterns in disease occurrence that might otherwise be overlooked.

Knowledge of disease determinants and distribution allows the design, application, and monitoring of preventive programs. Many diseases are known to have a strong spatial component of causation. Thus, analysis of disease data that have an implicit spatial component is a foundation of epidemiology and consequently, public health. Such data may be disease outbreaks, data generated by surveillance systems (systematic collection, consolidation, and evaluation of morbidity and mortality reports and other relevant information and the regular dissemination of such data to all that need to know), and specific hypothesis-based epidemiologic research such as surveys, case-control, and cohort studies (studies in which individuals with or without a disease, or exposed or not exposed to a risk factor, are selected and compared). Components of this process include exploratory spatial data analysis (finding interesting patterns), visualization (showing interesting patterns), and spatial modeling (explaining interesting patterns). As part of this process, spatial statistics and tests have utility in adding precision to qualitative verbal descriptions, facilitating the comparison of distributions, and drawing attention to characteristics unlikely to be noticed by visual inspection.

Specific statistical tests and techniques have been available to analyze spatio-temporal disease data for at least 60 years. During the 1950s and 1960s, many techniques were developed and applied, including the autocorrelation statistic (Moran 1950); the nearest neighbor index (Clark and Evans 1954); Geary's *c* statistic (Geary 1954); Ederer-Myers-Mantel disease-clustering procedure (Ederer et al. 1964); Knox's test (Knox 1964); and the temporal scan statistic (Naus 1965, 1966). Within the last 20 years, additional techniques have been developed to meet specific needs, including Cuzick and Edwards' test for inhomogeneous populations (Cuzick and Edwards 1990); population density adjusted autocorrelation (Oden 1995); spatial and spatio-temporal scan statistics (Kulldorff and Nagarwalla 1995; Kulldorff et al. 1998); local neighborhood statistics (Anselin 1995; Getis and Ord 1992); empirical Bayes smoothing; and kriging and other interpolation methods. Despite the development of a range of techniques that can be applied to analyzing spatio-temporal phenomenon, their application in the analysis of disease occurrence has been limited, probably because of a lack of suitable and accessible software (Ward and Carpenter 2000). We face a challenge in implementing these techniques as routine procedures within disease control and prevention programs.

The classic studies of John Snow on cholera outbreaks that occurred in London during the mid-nineteenth century were some of the first quantitative, observational studies that used a comparison group. They are often considered the foundation of modern epidemiology. It is no coincidence that these studies were fundamentally spatial in nature. In this chapter, examples of some techniques currently applied within the field of medical geography are provided to illustrate the potential value of spatial epidemiology to public health and homeland security. Although these examples demonstrate the value of techniques for investigating disease causation and designing public health programs to reduce the impact of disease, they also have an immediate application to homeland security. For example, the use of human and animal pathogens to inflict harm (bioterrorism) would first become apparent as a cluster of disease in time and space. Data would be generated from one or more surveillance systems. Early detection of the deliberate release of pathogens is the key to successfully reducing the impact of such an incident (and also increases the resilience of the homeland against such potential incidents). The analysis of disease surveillance data that has a spatial component facilitates the early detection of bioterrorist disease threats and plays an important role in homeland security.

Although the examples provided in this chapter are veterinary (including zoonotic) examples, the application of these methods are equally applicable to human diseases that do not include animal populations as disease reservoirs. The use of these methods is particularly well-illustrated by studies on vector-borne diseases affecting human populations, and studies on cancer etiology. For example, remotely-sensed data is now used routinely to monitor the risk of malaria (Thomson et al. 1999), and landscape factors that increase the abundance of vectors of Lyme disease (Diuk-Wasser et al. 2006) and trypansomiasis (Vazquez-Prokopec et al. 2005) have been identified. Studies using spatial statistics (Ederer et al. 1964; Bithell 1995; Bithell et al. 1994) have helped our understanding the etiology of leukemia. The case studies presented are also at different spatial scales. However, the geospatial methods illustrated are equally applicable to the different scales that might be encountered in medical geography.

Finally in this chapter, some of the obstacles to implementing spatial epidemiologic techniques and some of the inherent disadvantages with these techniques for analysis of disease data will be discussed.

13.2 Epidemic Equine West Nile Virus

West Nile virus (WNV) is a mosquito-borne disease. In the United States, it was first detected in 1999 as a cause of neurological disease in humans, horses, and birds in the vicinity of New York City. Horses are particularly susceptible to WNV infection, and may present with acute clinical signs of encephalomyelitis (Salazar et al. 2004; Ward et al. 2004a), an inflammation of the central nervous system. Although most affected horses recover in 3–4 weeks with supportive treatment, a small proportion may have a persistent neurological deficiency (Porter et al. 2003). Since the mid-1990s, the number of severe WNV disease outbreaks in equine populations has increased (Castillo-Olivares and Wood 2004). Recent outbreaks include Morocco (1996, 2003), Israel (1998–2000), Italy (1998), and France (2000, 2006). The North American outbreak of equine WNV encephalomyelitis exploded during 2002, with more than 15,000 laboratory confirmed cases in 44 American states, five Canadian provinces and three Mexican states.

The outbreak of WNV disease in horses in Texas during 2002 is one of the largest outbreaks of infectious equine encephalomyelitis ever reported. Despite an active surveillance program (sentinel flocks, mosquito testing, human and equine case reporting) and detection in neighboring Louisiana and Oklahoma, WNV was not detected in the state of Texas until this year. A total of 1,698 equine cases, confirmed by serological testing or other diagnostic procedures, were reported from 204 of 254 Texas counties (Ward et al. 2006). The first cases were reported on June 27 from eastern coastal Texas. The epidemic peak occurred on October 5, and 50 percent of cases were reported during a six-week period, September 3 to October 17. The epidemic (Fig. 13.1) lasted 25 weeks and appeared to consist of three phases: 27 June–July 25 (44 [2.6 percent] cases); July 26–September 27 (633 [37.3 percent] cases); and September 28–December 17 (1,021 [60.1 percent] cases).

County-of-origin was recorded for all reported cases. In addition, estimates of the number of equine present in each Texas county during 2002 was available from the National Agricultural Statistics Service.[1] The number of reported cases were summed by county and county attack rates (cases per 1,000 horses at-risk) were estimated (Fig. 13.2). An attack rate is the proportion of a population developing a specific disease during a finite period of time. It is the most commonly used metric in common source disease outbreaks to measure the average risk of disease occurrence (Office International des Epizooties 1999).

When visually assessing choropleth maps of disease rates and proportions, such as the map shown in Fig. 13.2, an issue that must be addressed is the estimation of these metrics for areas with small denominator information. Because polygons

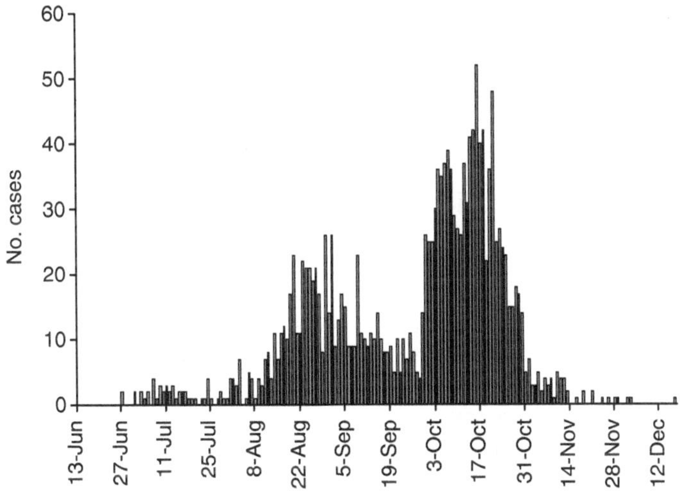

Fig. 13.1 Epidemic curve of cases of equine WNV encephalomyelitis reported in Texas during 2002

[1] NASS; http://www.nass.usda.gov/census/census02/profiles/tx/index.htm. Accessed 23 May, 2005.

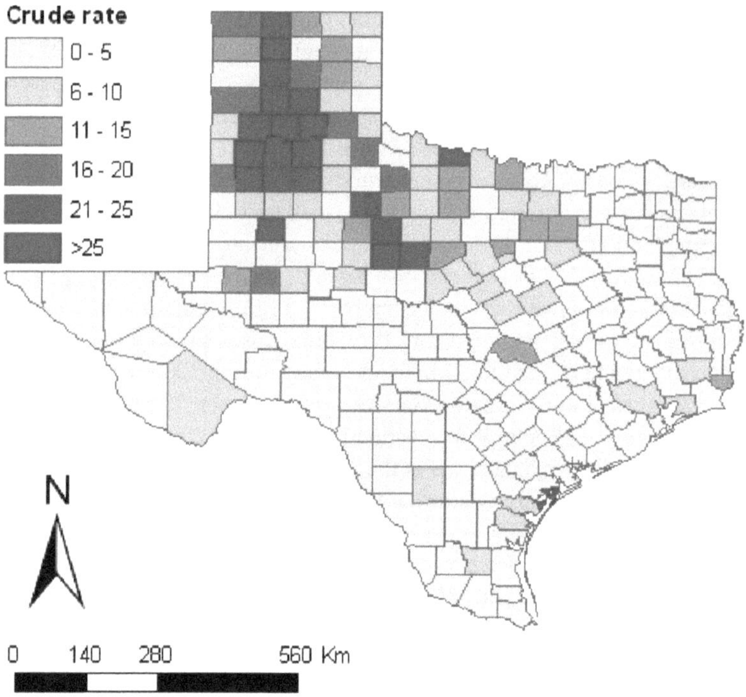

Fig. 13.2 Equine West Nile virus disease reported in Texas during 2002, presented as a choropleth map (cases per 1,000 equine at-risk)

(such as counties and other administrative units) may vary greatly in terms of land area and population size, disease rates and proportions presented in a choropleth map can hide vastly different levels of confidence implicit in these metrics. One solution to this problem is to use empirical Bayes smoothing. This procedure adjusts estimates for individual area units based on the overall (global) disease rate estimated for the entire study region (the 'prior' distribution). Those units with small populations at-risk are adjusted more than those with large populations, reflecting the statistical reliability of the estimates. As a result, disease rates and proportions are made more stable and less variable.

The geometric centroid (latitude, longitude) of each Texas county was identified.[2] County-specific WNV attack rates were smoothed, using an empirical Bayes method.[3] This approach is being increasingly used in spatial epidemiology when rates for areas are derived using varying denominator information (population at-risk), to avoid issues of varying confidence in estimated disease rates and proportions. The rate in

[2] Spatial Statistics: ArcGIS™ 9.0. ESRI, Redlands CA.

[3] Space-time Intelligence System version 1.0.6. TerraSeer, Ann Arbor MI.

each county was smoothed using data from its nearest ten neighboring counties (estimated by inverse squared Euclidean distance). Using this smoothing algorithm, the range of estimated attack rates was reduced from 65 cases per 1,000 horses at-risk to 27 cases per 1,000 horses at-risk, and the standard deviation of the mean attack rate (eight cases per 1,000 horses at-risk) was reduced from 10.04 to 6.59. This demonstrates a desirable characteristic of the algorithm—to reduce the variation in disease rate or proportion data that might result from including areas within differing populations at-risk. This provides more robust data for use in interpolation methods.

A semivariogram of smoothed county attack rates was created.[4] Semivariance is a measure of average dissimilarity between observations as a function of their separation in distance and direction. A semivariogram is a plot of the semivariance of all pairs of locations at a series of defined distances (lags). For locations close to each other, values (for example, sample size) are expected to be similar and the semivariance will be low (values will be highly correlated). As locations get farther apart, values are expected to become more dissimilar and thus the semivariance increases. The rate of increase in semivariance as distance increases, and the distance at which locations can essentially be considered independent, characterize the spatial pattern of the event-of-interest. Estimating the parameters of the line of best fit of an empirical semivariogram allows the distribution to be modeled and interpolated with geostatistical techniques such as kriging. The semivariance of smoothed county attack rates (Fig. 13.3) were interpolated by

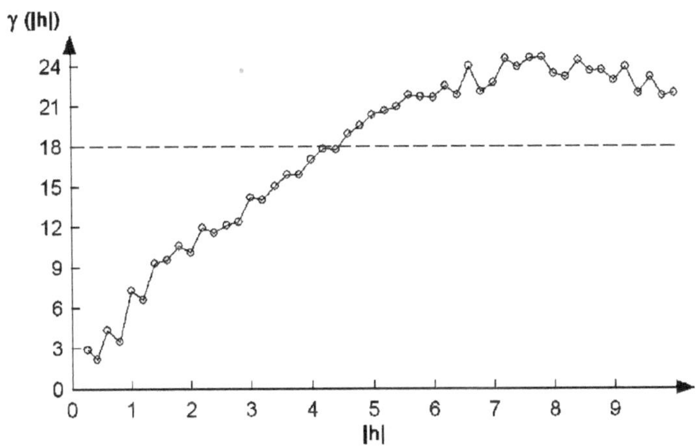

Fig. 13.3 Semivariogram of equine West Nile virus disease rates (cases per 1,000 equine at-risk) reported from Texas counties during 2002. Rates were smoothed using an empirical Bayes procedure

[4] VarioWin 2.2 version. Yvan Pannatier. University of Lausanne, 1995.

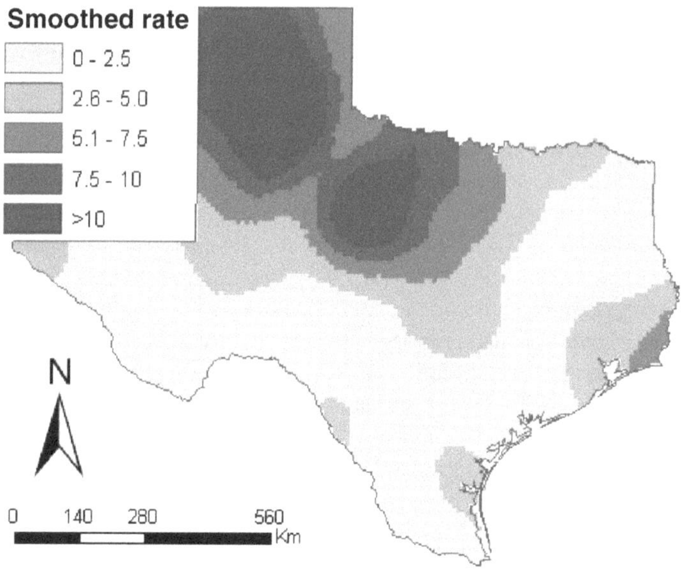

Fig. 13.4 County equine West Nile virus disease attack rates (cases per 1,000 equine at-risk), Texas, 2002, smoothed using a empirical Bayesian approach and interpolated using kriging

fitting a Gaussian model. Smoothed county attack rates were then interpolated by kriging (Fig. 13.4).

By comparing the choropleth map of crude WNV encephalomyelitis attack rates (Fig. 13.2) with the isopleth map (Fig. 13.4) of smoothed, interpolated attack rates, the remarkable increase in visual information provided can be appreciated. This isopleth map is an example of a disease risk map: it concisely represents the areas of Texas in which the risk of equine WNV encephalomyelitis was greatest during 2002. Such information may then lead the researcher or public health practitioner to develop hypotheses of causation, and to develop disease surveillance systems and preventive program. The more routine analysis of spatial disease data and creation of risk maps is one of the recent, major advances in spatial epidemiology that has occurred within the last two decades.

13.3 Endemic Canine Leptospirosis

Leptospires are spirochete bacteria that infect many animal species and humans. *Leptospira* organisms are shed in the urine of infected animals, and may survive for long periods in surface water. Susceptible animals and humans are infected through contact with contaminated water, and localized outbreaks of leptospirosis can occur when susceptible animals and humans are infected via contaminated environments

(Bolin 1996). Since 1983, the prevalence of leptospirosis in dogs has increased (Ward et al. 2002). The reason for this increasing prevalence is unknown, but may include urbanization of rural areas (increasing during the 1980s and 1990s) that provides more opportunity for contact between dogs and wildlife, such as skunks, raccoons, and opossums (Birnbaum et al. 1998; Hanson 1982; Ward 2002a, b). Through the use of geographic information system (GIS) technology, landscape features that are associated with leptospirosis can be identified and this information can be used to implement control programs to reduce disease.

A study in the State of Indiana was undertaken to identify environmental risk factors for canine leptospirosis. Confirmed cases of leptospirosis were identified from patients admitted to the Purdue University veterinary teaching hospital between 1997 and 2002 (Ward et al. 2004b, c). All dogs admitted to the same hospital during the same period and tested seronegative were used as controls. Age, sex, breed, and owner address information was extracted from recorded information for all dogs included in the case-control study. From recorded address information, locations were geocoded[5] by use of a five-digit zipcode, street name, and street number.[6] Locations were mapped,[7] by use of a Universal Transverse Mercator grid system (zone 16) shapefile (GIS-based map) of Indiana counties.[6] A range of environmental factors at study locations was identified by overlaying location data on polygon, line, or point layers. Primary data sources included State Soil Geographic data base for Indiana,[8] metropolitan areas,[6] National Wetlands Inventory,[9] Indiana recreation facilities inventory,[10] and 1961 to 1990 average annual precipitation.[11] Several data sets, including urbanized areas, 1990 to 2000 (areas that were rural according to the 1990 national census, but that were metropolitan according to the 2000 census), county hog density (county hog population divided by county land area), county cattle density (county cattle population divided by county land area), county beef cattle density (county beef cattle population divided by county land area), county dairy cattle density (county dairy cattle population divided by county

---

5 ArcView, version 3.3, Environmental Systems Research Institute Inc, Redlands Calif.

6 Center for Advanced Applications in Geographic Information Systems, Purdue University, West Lafayette Ind. Available at: http://danpatch.ecn.purdue.edu/~caagis/ftp/gisdata/data.html. Accessed on 20 February 2003.

7 ArcView, version 3.3, Environmental Systems Research Institute Inc, Redlands Calif.

8 US Department of Agriculture, Soil Conservation Service, State Soil Geographic (STATSGO) database for Indiana, Fort Worth Tex. Available at: http://danpatch.ecn.purdue.edu/~caagis/ftp/gisdata/data.html. Accessed on 3 March 2003.

9 US Fish & Wildlife Service, National Wetlands Inventory, St.Petersburg Fla. Available at: http://danpatch.ecn.purdue.edu/~caagis/ftp/gisdata/data.html. Accessed on 3 March 2003.

10 Indiana State Department of Natural Resources, Indiana recreation facilities inventory, Indianapolis Ind. Available at: http://danpatch.ecn.purdue.edu/~caagis/ftp/gisdata/data.html. Accessed on 3 March 2003.

11 US Department of Agriculture, National Resources Conservation Service, National Cartography and Geospatial Center, Fort Worth Tex. Available at: http://danpatch.ecn.purdue.edu/~caagis/ftp/gisdata/data.html. Accessed on 3 March 2003.

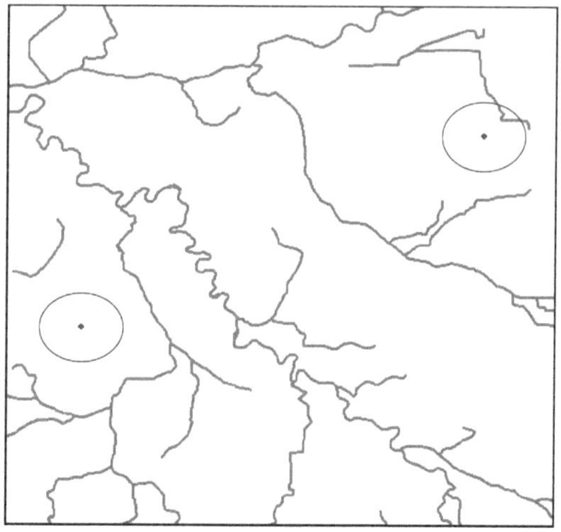

Fig. 13.5 An exampling of GIS buffering. The case in the upper right quadrant lies within 1,000 m of a river or stream and would be considered exposed in an epidemiologic study; the other case is more than 1,000 m from the nearest river or stream and would be considered unexposed

land area), and county forest density (number of trees divided by county land area) were derived from the above data sources because primary data were unavailable. Because of limitations on location accuracy and to account for a possible home range of dogs included in the study, a buffer of 1,000 m was used. For example, for the GIS variable 'rivers and streams,' exposed locations were within 1,000 m of a river or stream; unexposed locations were those > 1,000 m from such geographic features (Fig. 13.5). Logistic regression[12] was used to estimate the risk (odds) of being an exposed case versus an unexposed case (odds ratio [OR]).

A total of 36 case dogs and 138 control dogs were identified. The risk of being a case was most strongly associated with having a residential address in an area that was classified rural according to the 1990 census, but urban according to the 2000 census: seven of the 36 (19 percent) case dogs were located in these newly urbanized areas, compared to only four of 131 (3 percent) control dogs (OR 8.09; 95 percent confidence interval, 2.22–29.44; P = 0.0015). An example of this exposure is shown in Fig. 13.6.

After adjusting for the effect of the age and gender of dogs selected for inclusion in this study, the association between risk of being diagnosed with leptospirosis and having a residential address within a newly urbanized area was found to be even stronger (OR 26, 95 percent CI, 4–160).

---

[12] SPSS for Windows, version 10.1.0, SPSS Inc, Chicago IL.

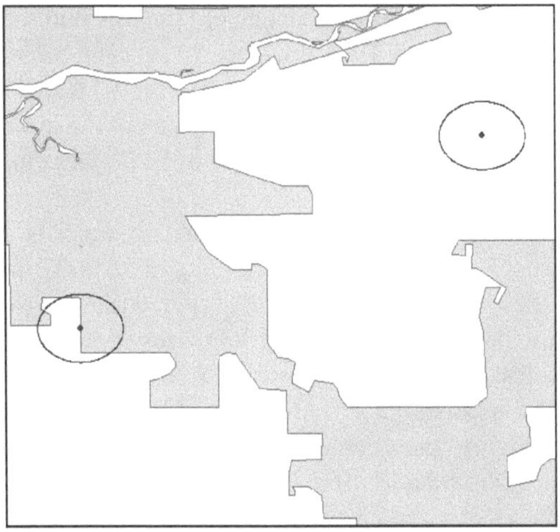

Fig. 13.6 An exampling of GIS buffering. Cases of canine leptospirosis identified in Indiana during the period 1997–2002 were eight times more likely to lie within areas (or within 1,000 m of areas) which were rural in 1990 and urbanized in 2000 (shown as the shaded polygon) than were controls (the point shown in the upper right corner of the area)

This example illustrates our ability to examine and detect environmental factors that increase the risk of disease occurrence. With this knowledge, disease prevention programs may be developed. For example, public awareness campaigns can be initiated to warn dog owners in recently urbanized areas of the risk of leptospirosis, diagnostic testing procedures can be modified based on a patient's history, and targeted surveillance can be undertaken. Because leptospirosis is a zoonotic disease, surveillance of the dog population as a sentinel may also protect public health.

13.4 Spatial Simulation Models of Foot-and-Mouth Disease

Foot-and-mouth disease (FMD) is a highly contagious disease of cloven-hoof animals. Outbreaks can be economically devastating in countries free of infection (Ferguson et al. 2001), mostly due to the economic impact resulting from trade restrictions imposed for up to 12 months following eradication of an incursion. A major complication of FMD control is the existence of non-domesticated animal reservoirs of disease (Pinto 2004; Sutmoller et al. 2000). During the 19th and early 20th centuries, FMD virus was eradicated from the United States six times, the last outbreak occurring in 1925. During this phase of surveillance and eradication, wildlife and feral animal disease reservoirs presented a barrier to successful disease control. For example, during the 1924 outbreak of FMD in California, mule deer in the central portion of the state were infected. It took two years to eradicate FMD virus from the local deer population in one national park, and 22,000 deer were slaughtered. About ten percent of deer slaughtered in this outbreak showed lesions typical of FMD (McVicar et al. 1974; Fletcher 2004). The presence of a feral pig population complicated FMD control measures during a 1924–1925 outbreak on a ranch near Houston (Fletcher 2004). FMD virus is one of the most important potential agroterrorism agents. The deliberate release of FMD virus in the United States could be devastating.

Although spatial models have been developed to describe the farm-to-farm spread of FMD virus (Gerbier et al. 2002; Keeling et al. 2001; Morris et al. 2001; Bates et al. 2003), these models have generally ignored the involvement of feral and wild animal species. Published models of FMD virus spread that do involve these potential host species (Dexter 2003; Pech and Hone 1988) have generally not included the spatial component of spread. Artificial life models such as geographic-automata—that explicitly incorporate spatial relationships—are an alternative modeling approach (Doran and Laffan 2005). These models of physical systems treat space and time as discrete units and interactions occur between local neighbors (Torrens and Benenson 2005). Geographic-automata are generalizations of cellular-automata: they are not restricted to a regular lattice of cells (geographic locations). Each population interacts with neighboring populations based on a set of rules and states at earlier time steps. The repetitive application of transmission rules within this local neighborhood allows the replication of complex spatial behavior such as occurs in disease outbreaks. Geographic-automata models can deal with complex initial conditions and geographical boundaries, are conceptually simple, and can model complex spatial interactions.

A susceptible-latent-infected-recovered model of the spread of FMD virus within and between populations of feral pigs and wild deer, and herds of domesticated cattle has been developed, using a geographic-automata framework (Ward et al. 2007). The study site is southern Texas (a nine-county area referred to as the 'Texas Pilot Study Area' Fig. 13.7). Wild-deer densities are high, most deer exist on private land, and deer have formed metapopulations (the fusion of separate populations to form essentially a large, linked population) because of extensive land-use changes. In addition, the high density of feral pigs in southern Texas (> 50 per km^2 in some areas) would provide an ideal mechanism for spreading the infection. The potential for transmission of FMD virus from wildlife to domesticated livestock is heightened in areas such as Texas where wildlife receive supplemental feed and are hunted for sport. The study area (Fig. 13.7) is approximately 24,000 km^2, and contains an estimated 85,000 domestic cattle (3.5/km^2), 134,000 feral pigs (5.6/km^2) and 395,000 deer (16.4/km^2).

Several factors—including animal population density and distribution, habitat requirements, social organization, age structure, home range, and barriers to dispersal—determine whether an infection will be maintained within feral or wild animal populations, and therefore whether these populations might act as reservoirs of infection for domesticated species. These factors must be incorporated into models, either implicitly or explicitly. A series of models were run to gain an understanding of FMD-virus spread through space and time (Ward et al. 2007). A purpose-built geographic-automata model, implemented using the Perl programming language,

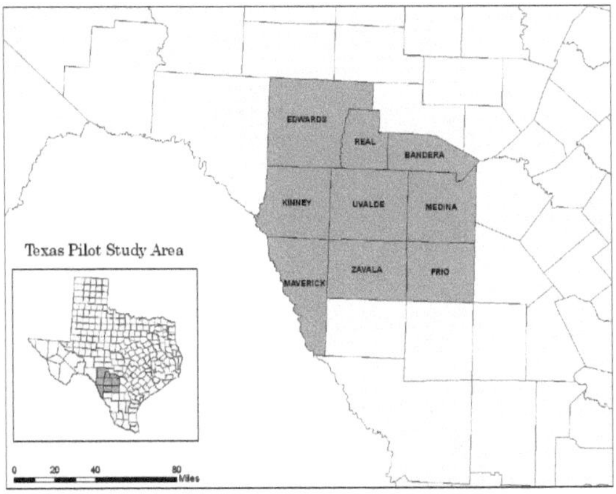

Fig. 13.7 The study area used to simulate the spread of foot-and-mouth disease virus in wild deer, feral pigs and domesticated cattle, using a geographic automata spatial simulation model

was used. Animal distributions were represented as number of animals per unit area (km^2), displayed as raster surfaces (cell size approximately $1,000 \times 1,000$ m, derived from land cover information and expert opinion) developed within a geographic information system (Ward et al. 2007). Animals (feral pigs, wild deer, or cattle) within each unit area represented a herd or social group. In the model, herds can pass, sequentially, through four model states: from susceptible to latent, to infectious, to immune and then to susceptible (recovered) again. Mortality was also included in subsequent model extension. The first transition depends on contact rates between susceptible and infected herds in the previous time step. The probability of transmission from one herd to another is the product of the relative animal densities of the two herds, modified by the distance (km) by which they are separated. Transmission probabilities are reduced as neighboring herds are located further away from each other. Within an a priori specified maximum neighborhood distance and up to a maximum number of neighbors, interactions between each infectious herd and its neighbors were evaluated. For baseline simulations, no interactions took place beyond a distance of two km from each infected herd, with interactions limited to these nearest eight herds surrounding each herd-of-interest. To incorporate chance into the model, interactions between an infectious herd and a susceptible neighbor occurred when a random value from a pseudo-random number generator (PRNG) was below their joint probability threshold. The densities of feral pigs and wild deer were derived using land-use data and estimated ecological site carrying capacity. Cattle densities (Fig. 13.8) were based on county estimates of cattle numbers, disaggregated based on land-use (open grassland, pastureland, shrubland, and forestland) information and likely stocking rates. An example of the predicted spread of FMD virus in cattle, based on the introduction via wild deer at one site in the central part of the study area, is shown in Fig. 13.8.

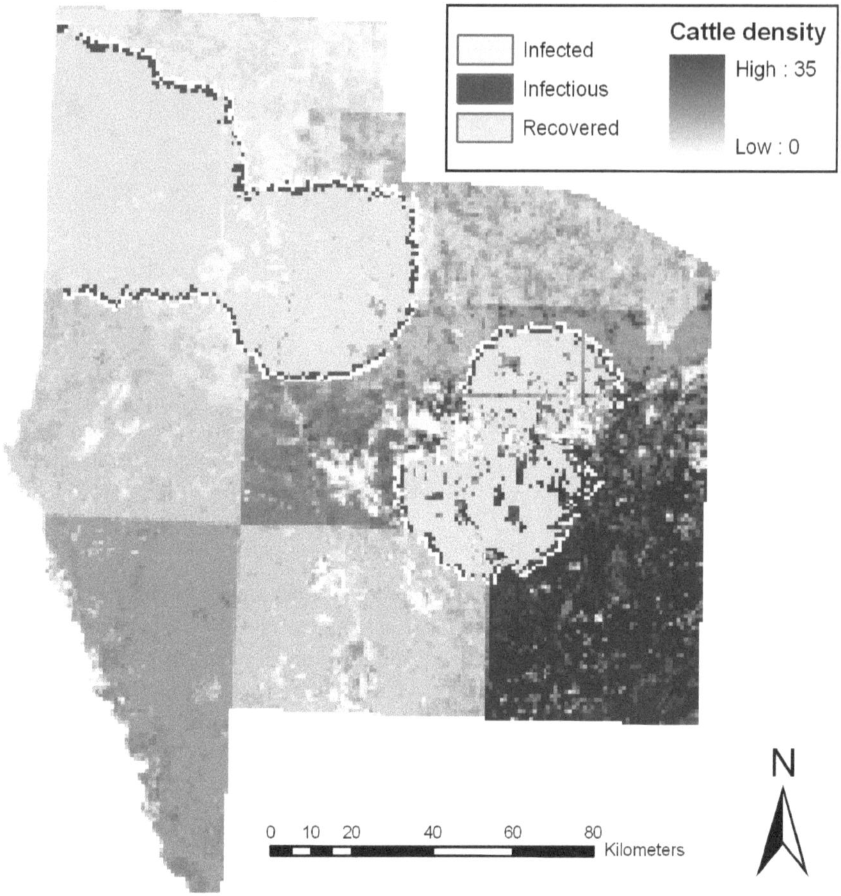

Fig. 13.8 Spatial distribution of infected cattle populations in southern Texas, following simulated incursions of foot-and-mouth disease virus in wild deer, using a two-species susceptible-latent-infected-recovered geographic automata model

An uncontrolled FMD outbreak initiated in feral pigs and in wild deer might infect up to 698 (95 percent prediction interval, 181–1387) and 1,557 (823–2118) cattle and affect an area of 166 (53–306) and 455 (301–588) km^2, respectively. The predicted spread of FMD virus infection is influenced by assumptions made regarding the number of incursion sites and the number of neighborhood interactions between herds (Ward et al. 2007). In a substantial number of model runs, an outbreak of FMD failed to develop in either wild deer or feral pigs, and therefore an outbreak of FMD in cattle did not occur. This was particularly noticeable when the incursion occurred at four or fewer feral pig herd locations.

This spatial modeling approach explicitly incorporates the spatial relationships between domesticated and non-domesticated animal populations, providing a new framework to explore the impacts, costs, and strategies for the control of foreign animal diseases with a potential wildlife reservoir. Spatial disease models help guide decision-makers on disease planning and response. A range of 'what–if' scenarios, only limited by the imagination of the decision-maker, can be explored. Spatial disease models can also be used in 'tabletop' exercises to train emergency responders. For example, responders could be presented with several disease outbreak scenarios and be asked to plan appropriate response strategies. The model could then be simulated to assess how these responses might affect the spread of the disease. Capacity planning to deal with such outbreaks could also be undertaken. For example, the ability to depose of the predicted number of infected animals (via burial, for example), could be assessed. Another area of interest is the ability to have sufficient vaccine stock available and the ability to deliver this vaccine, should such a decision be taken, in a timely and cost-effective manner. This is a relatively new frontier in spatial epidemiology that has great potential to contribute to the discipline.

13.5 Epidemic Avian Influenza H5N1

Avian influenza has recently become an emerging issue for world health: the pathogenic H5N1 influenza strain circulating in Asia, Africa, the Middle East, and Europe has caused numerous disease outbreaks in domestic poultry and wild bird populations, and threatens human health. As of 12 June 2007, a total of 190 (61 percent) of 312 humans known to have been infected with H5N1 since 2003 and reported to The World Health Organization have died in ten countries in southeast Asia, China, and the Middle East.[13] There is a fear that H5N1 could become the next pandemic influenza strain. Avian influenza virus infection is endemic in a range of free-living bird species world-wide (Alexander 2000 and 2001; Rosenberger et al. 1974), particularly species associated with water (Stallnecht and Shane 1988). Waterfowl and shorebirds can be infected by all subtypes of type A influenza viruses with few or no symptoms (Woebser 1997). These species are probably responsible for the spread of viruses between regions (Krauss et al. 2004). Research suggests that waterfowl and shorebirds maintain a separate reservoir of viral gene pools from which new virus subtypes emerge (Webster et al. 1992). In the northern hemisphere, influenza virus infection rates are highest during spring migration for shorebirds, whereas waterfowl infections peak in late summer and early fall (Krauss et al. 2004). Avian influenza outbreaks—both high and low pathogenic—in poultry are often assumed to occur from exposure to wild avian species.

---

[13] World Health Organization (WHO) 2007. Cumulative number of confirmed human cases of avian. influenza A/(H5N1) reported to WHO, 12 June 2007. WHO, Geneva (www.who.int/csr/disease/avian_influenza/country/cases_table_2007_06_12/en/index.html accessed on 16 June 2007).

Two areas in which spatial epidemiology can potentially have an impact on the threat that avian influenza H5N1 poses to human populations and animal production are identifying environmental risk factors for disease outbreaks, and the design of more sensitive and cost-effective surveillance systems. By knowing where and when the disease is more likely to occur, preventive measures (including the increased biosecurity provided by keeping poultry indoors and limiting live bird markets and the transportation of domestic poultry species, and consideration of the potential advantages of using prophylactic vaccination) can be implemented to reduce the risk of an outbreak. Spatial analysis techniques can be used to target surveillance activities, so those areas of the landscape in which incursions of H5N1 avian influenza are more likely can be preferentially sampled.

Highly pathogenic avian influenza virus subtype H5N1 was first detected in Romania on 7 October 2005, in the eastern county of Tulcea. Subsequently, a total of 165 village poultry outbreaks were reported in Romania. The epidemic ended in June 2006. Tulcea county is bordered to the east by the Black Sea and to the north and west by the Danube River. The eastern part of the county consists of an extensive wetland system, part of the Danube River delta. It is a major breeding area and point of congregation for migratory birds on the Black Sea–Mediterranean flyway, which extends from west Africa to central Asia. Outbreaks of H5N1 were controlled by depopulation of poultry in affected villages, disinfection, and surveillance of sentinel chickens in de-populated villages and serological surveillance in selected areas of the county.

Although GIS is a powerful tool for designing, assessing and implementing surveillance systems, many issues need to be considered prior to implementing a GIS-based surveillance system. The data that are collected in surveillance systems needs to be spatially accurate and timely. We need to consider the type of surveillance system used to collect the data (passive or active), location accuracy, the spatial level of aggregation, the use of administrative units versus the aim of the system, edge effects, and the modifiable unit area problem (MAUP). The MAUP is particularly of concern in spatial epidemiologic studies in which data is analyzed as choropleth maps and disease causation is a focus. In this case, the results of statistical analysis will depend on the areal units used (O'Sullivan and Unwin 2003). In general, as data is aggregated, it is more likely that an association will be found. Analytical issues include the detection of clusters (groups of cases of disease occurring in time and space at a rate higher than expected and with a probable common cause) versus trends and patterns (a continuous or steady increase or decrease in the occurrence of disease over time and space), spatial versus temporal and spatio-temporal clusters, statistical power, and multiple statistical testing of all the potential combinations of disease patterns. Some of the issues to consider regarding GIS-based surveillance output include consistency, cartography, and the choice of interpolation techniques.

As part of a surveillance system for H5N1 avian influenza in operation between January and August 2006, sera were collected from a total of 12,172 domestic poultry species in Tulcea county. Variography was used to evaluate serological surveillance for avian influenza antibodies in Tulcea county during

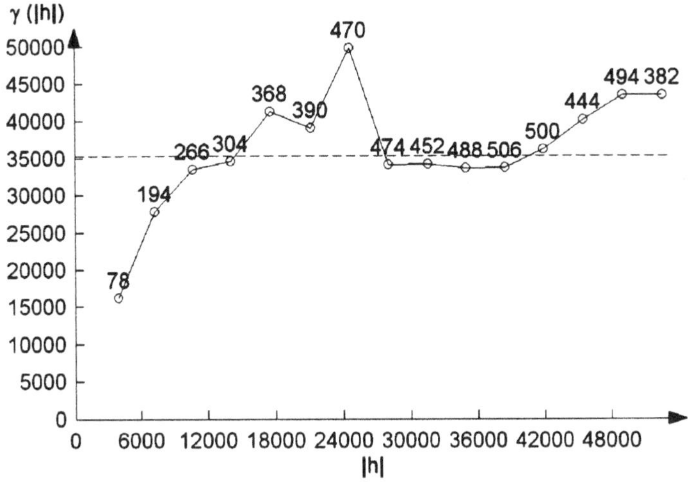

Fig. 13.9 Semivariogram of the number of samples collected for avian influenza virus subtype H5N1 surveillance from villages in Tulcea county, Romania, January to August 2006

this period. The location of all villages (n = 141) in Tulcea county were identified by longitude and latitude coordinates. The total number of domestic poultry from which sera were collected during the study period was calculated. All possible unique pairs of village locations (n = 9,870) were formed, and a semivariogram was constructed (Fig. 13.9).

A range of lag numbers and lag spacings (m) were chosen to produce a semivariogram which could be described by one of a number of a priori models. Using a line-of-best-fit approach, the parameters of the selected model (exponential, spherical, Gaussian) were estimated.[14] These parameters (nugget, range, sill) were used to produce an interpolated map of serological sample size in Tulcea county,[15] and an error (sample variance) map for this interpolated surface (Fig. 13.10). These maps were overlaid on the location of villages to identify localities where surveillance sampling appeared suboptimal to identify avian influenza virus antibodies.

Investigations of spatial disease distribution generally focus on whether disease cases are likely to have occurred at random. If this null hypothesis is rejected, disease cases may be overdispersed (for example, a uniform distribution) or clustered. Our interest is usually in clustering, since this implies that common population characteristics, a source of exposure, or common environmental characteristics have lead to foci of disease. Identifying these foci is the first step in elucidating etiology and thus designing control, prevention, and surveillance programs. In the clustered distribution, there is a definite, discernible aggregation of points above

[14] VarioWin 2.2 version. Yvan Pannatier, University of Lausanne, 1995.

[15] Spatial Statistics: ArcGIS™ 9.0. ESRI, Redlands CA.

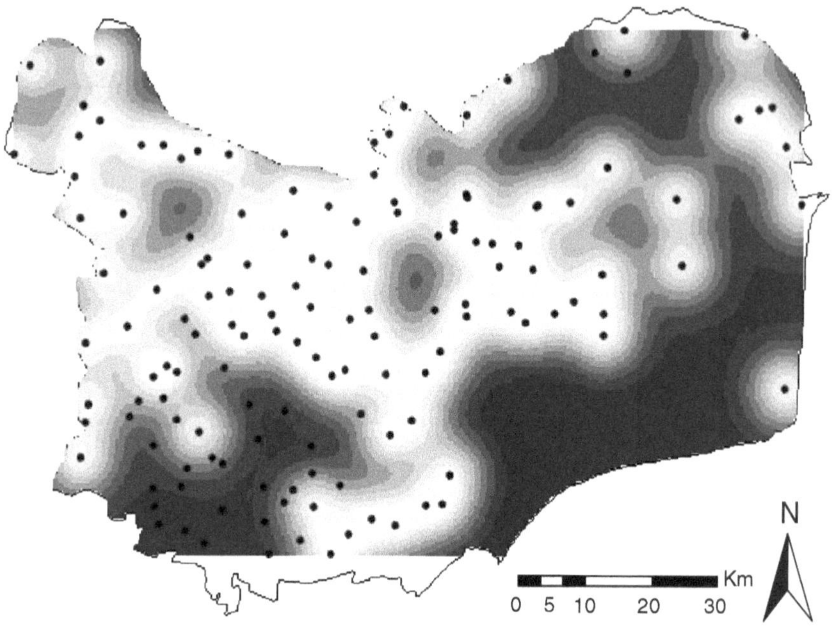

Fig. 13.10 Interpolated variance of sampling for avian influenza virus subtype H5N1 surveillance in Tulcea county, Romania, January to August 2006. Kriging was used as the interpolation method. Village locations are indicated by points

that which would be expected given an underlying population at risk. One of the most common methods of describing the distribution of a set of disease cases, whether measured as points or polygons, is Moran's autocorrelation statistic. This is an example of a global spatial test, and is similar to the traditional Pearson correlation coefficient, except that the correlation of values of the same variable at different spatial locations is examined, with a weight matrix being included to define the spatial relationships between polygons. This weight matrix is often based on Euclidean distance, but may be modified to take into account neighborhood relationships (for example, one (1) for adjacent pairs and zero (0) otherwise). A positive autocorrelation implies clustering. Moran's autocorrelation is sensitive to the spatial distribution of the underlying population at-risk of disease. If the population at-risk is clustered (for example, people living in cities or villages surrounding a lake), then disease cases arising from that population are also expected to be clustered, even if the disease occurrence is not clustered per se. For this reason, Moran's I is generally computed for disease rates or incidence proportions.

Cuzick and Edwards' test for inhomogenous populations accounts for uneven population at-risk distributions. This test compares the locations of case and control locations. Controls are drawn from the same underlying population as cases, thereby accounting for clustering that may occur in the population regardless of the clustering of cases. The test statistic is the number—summed over all cases and controls—of cases that are nearest neighbors to each individual case.

The spatial distribution of surveillance sampling in Tulcea county for H5N1 avian influenza was assessed with both Moran's autocorrelation and the Cuzick and Edward's test.[16] The number of samples collected per village did not show strong evidence of clustering (Moran's autocorrelation statistic 0.026, P = 0.012), but villages from which samples were collected were clustered (Cuzick and Edwards' test Bonferroni P = 0.010), compared to those villages where sampling was not conducted during this period (Fig. 13.11).

A technique that allows detection of both global clustering and the identification of the location of specific clusters, and clustering in time, space and in time and space, is the spatial scan statistic. The spatial scan statistic (Kulldorff and Nagarwalla 1995) uses a theoretical circular window placed on a map of all locations included in a study. This scanning window is sequentially centered around one of many possible centroids in the study area. For each centroid, the window radius may vary continuously from zero to some upper limit. Thus, the procedure creates—in theory—an infinite number of distinct geographical circles, containing within them different sets of neighboring locations. Each set of locations is a possible candidate for a cluster. However, since discrete locations (longitude, latitude)

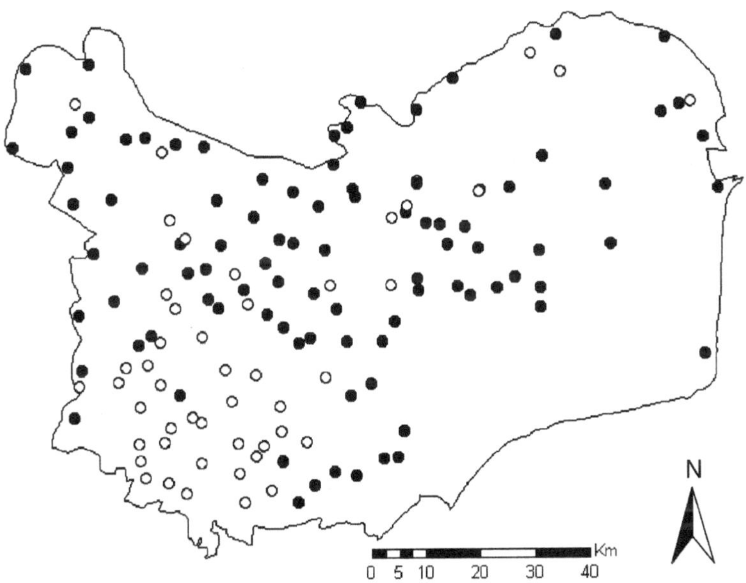

Fig. 13.11 Villages in Tulcea county, Romania from which samples were (●) or were not (○) collected between January and August, 2006 as part of a surveillance program for avian influenza virus subtype H5N1

[16] ClusterSeer 2.0. TerraSeer, Inc., Ann Arbor MI.

or the centroid of areas within a study are used in spatial analysis, the number of candidate circles that must be assessed is finite (Kulldorff 2006).

The scan procedure is flexible: data can be analyzed using two different probabilistic models, based on the Bernoulli or Poisson distributions. For the Bernoulli model, the data has the form of cases and non-cases (controls). Cases and non-cases may be selected from the study population, or may represent the entire study population. For the Poisson model, the number of cases at each location or within each area is assumed to be Poisson distributed. Under the null hypothesis, the expected number of cases at each location is proportional to the population size or population-time at-risk at that location (Kulldorff 2006).

The spatial distribution of human and animal populations is almost always heterogenous. This has implications for selecting a spatial cluster statistic. The question of interest is usually 'Does spatial clustering occur above and beyond the spatial clustering of cases that arises due to spatial variation in population density?' Regardless of the model—Bernoulli or Poisson—used in the spatial scan procedure, adjustment for lack of population homogeneity is achieved by conditioning on the total number of cases observed to calculate the expected number of cases for each location. The spatial scan statistic methodology has been extended to analyze data for clusters in time and space (Kulldorff et al. 1998).

Data describing Tulcea village poultry outbreaks of H5N1 avian influenza were scanned for time-space clusters. A Bernoulli (case-control) model was used, and data were scanned using both circular and ellipsoid windows up to 50 percent of the study area and a time period of up to 30 days. Results are shown in Fig. 13.12. Identification of disease clusters, particular those cases of disease that appear unusually close in both time and space, allow public health practitioners to better identify the underlying causes of disease and therefore take action to control current outbreaks and prevent the occurrence of future outbreaks. This information may also assist in the more efficient provision of health care services.

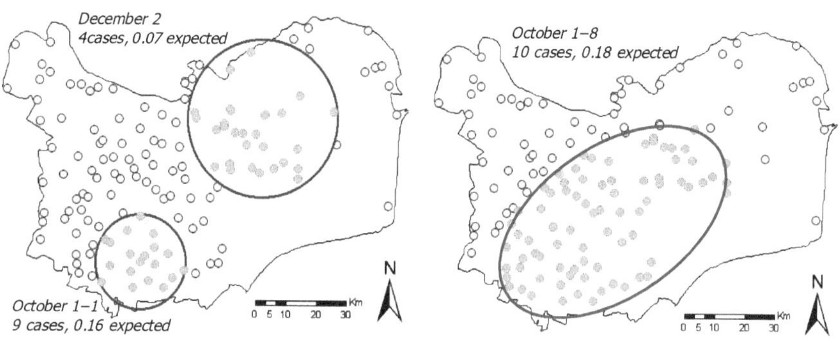

Fig. 13.12 Spatio-temporal clusters of outbreaks of avian influenza virus subtype H5N1 in village poultry in Tulcea county, Romania, 2005–2006, detected using the time–space scan statistic with scanning windows up to 50% of the study area (left: circular; right: ellipsoidal) and 30 days duration

13.6 Conclusions

Modern epidemiology is founded on a tradition that recognizes the central role of geography and the environment in disease causation. The spatial epidemiology tradition in medical geography has developed into a field within its own right, and during the past two decades advances in geographic information systems technology and statistical methods for analyzing spatially correlated health data have allowed epidemiologists to routinely perform spatial analyses. Some specific major advances that have impacted this field include data visualization, detection of disease clusters, identification of spatial risk factors, application of predictive models, and the routine incorporation of GIS into disease surveillance programs. Despite these advances, the availability and quality of spatial disease data and information on the distributions of the populations at-risk remains a major obstacle to the consolidation of spatial analysis as a foundation of modern epidemiology. The integration of spatial statistical and geostatistical methods seamlessly into epidemiologic software packages is another obstacle that remains. Finally, how techniques (that are often developed in the geosciences) should be validly applied, when disease distribution data may be either non-continuous or non-normally distributed, to help solve disease issues is another source of confusion and frustration.

Use of GIS in disease surveillance programs in recent years highlights advances that have been made. An example of this application is the use of GIS to facilitate spatial risk assessments for surveillance sampling to detect potential incursions of highly pathogenic avian influenza H5N1 (East et al. 2007; Miller et al. 2007; Ward and Rollo 2007). The ability to collate spatial data describing risk factors for H5N1 introduction and spread (for example, poultry production, migratory waterfowl, poultry distributors, presence of wetlands) can greatly aid prioritization to improve the sensitivity of surveillance systems (via targeted surveillance) and to minimize costs. Information generated by such applications of GIS in surveillance systems is being used by the US Department of Agriculture and the Australian Department of Agriculture, Fisheries and Forestry as examples, to guide policy with respect to avian influenza H5N1 preparedness. Another example relevant to homeland security is the use of spatial simulation models of foot-and-mouth disease spread to inform response policies being developed in the United States (Ward et al. 2007) and Australia (Garner and Beckett 2005).

Despite the rapid increase in the application of geographical information system technology, examples of the use of statistical tests to investigate clustering of disease and modern methods of data visualization remain relatively uncommon in public health. Analysis of data in GIS does not routinely employ statistical tests of spatial clustering. Rather, GIS have generally been used to analyze (through visual interpretation) the relationships between potential risk factors and the occurrence of disease (incidence or prevalence). The lack of availability of and user familiarity with statistical software has, up until recently, restricted the spatial and temporal analyses of data sets for disease clustering. An impediment to the use of spatial statistical software might also be the perception among epidemiologists that spatial statistical techniques have limited utility. One of the barriers to the use of advanced spatial analytical techniques has been the lack of compatibility between GIS and specialist statistical software. However, this is changing. For example, ArcGIS versions 9.0, 9.1 and 9.2 have many spatial statistical tools included, and GIS file formats can be used in other specialist spatial statistical packages (for example, ClusterSeer and GeoDa) Nevertheless, the statistical analysis of spatial distributions remains a weak point in the application of GIS technology. If GIS technology is to fulfill its potential as a general purpose tool for handling spatial data it needs stronger analytical capabilities (Paterson 1995). Development of statistical software to investigate disease clustering, and integration of these routines into GIS, has improved the ability of epidemiologists to identify and describe determinants of disease.

An important tool that GIS offers is data visualization. However, sampling for disease occurrence is often spatially haphazard. As a result, data are often aggregated and choropleth maps are used to present and facilitate interpretation of disease data, often preventing adequate visualization and hypothesis generation and testing, leading to inefficient resource allocation. Smoothing and interpolation methods reduce artificial effects of boundaries and facilitate identification of patterns by estimating disease occurrence at a given location using data from surrounding locations. A recent development in the assessment of the ability of surveillance programs to detect disease is risk mapping—the production of isopleths maps that show disease risk (prevalence or incidence) surfaces. The need to represent disease distributions is driven by a desire to identify disease foci, visualize disease spread, and identify risk factors. For resource allocation, there is a need for surveillance planning, disease control and prevention. Isopleth maps aid visualization of latent risk throughout the area of interest, without being restricted by artificial administrative boundaries, and address other problems associated with choropleth mapping, specifically visual bias resulting from the uneven shape and size of administrative units in most countries and regions, whereby large areas may visually dominate maps. Although interpolation techniques have been used for some time to produce smoothed maps of disease risk (incidence or prevalence), these techniques have rarely been used to interpolate sampling schemes and assess the spatial sampling coverages within disease surveillance systems. In addition, some of the fundamental assumptions implicit in risk mapping, such as isotrophy (spatial directionality in disease occurrence), are difficult to meet in many disease analysis situations. Considerable work remains to be done to improve and to integrate interpolation techniques into a disease risk mapping analytic framework.

More generally, the spatial distributions of disease occurrence often are complex, particularly in the case of disease outbreaks. The distribution might not be continuous, disease may spread at different rates in different directions, and complex contact structures (caused by social interactions in human populations and specific economic conditions in livestock populations) within the population at-risk can result in unusual spatial structures in which more distant members are more alike than are closer members. Thus, it is particularly important to initially explore and describe disease data prior to using spatial statistics and geostatistical methods.

In some cases, it may be found that the use of these techniques is not appropriate, or the data may need to be modified or transformed before such analyses. More techniques designed specifically for disease data are needed. Some recent examples include Cuzick and Edwards' test for inhomogeneous populations (Cuzick and Edwards 1990) and the spatio-temporal scan statistic (Kulldorff et al. 1998).

Another continuing barrier to the application of spatial epidemiologic techniques is lack of accurate and detailed information on the populations at-risk of disease. For example, identification of risk factors for highly pathogenic avian influenza H5N1 outbreaks has been hampered by a lack of information on the poultry populations at-risk in many countries. In most developing countries, poultry production is based within villages, and the size and specific location of the population at any given time is not well characterized or available in electronic format. Even in developed countries such as the United States, the number of poultry raised in non-commercial, 'backyard' environments is unknown. In addition, the movement of members of these populations via live bird markets and other contact points (for example, illegal cock fighting venues) is unknown. Without good demographic information, spatial epidemiologic cannot be effectively applied to identify spatial risk factors.

Investigations of spatial disease patterns can be greatly enhanced through the use of a variety of analytical techniques. These techniques add considerable information to disease investigations and medical research, providing the epidemiologist and public health practitioner with a firm foundation on which to build causal hypotheses and implement control strategies. For example, if specific environments increase the risk of disease occurrence, then these environments might be modified or avoided to reduce disease risk. Another application is forecasting when such environmental risk might occur, so that preventive measures (for example, vaccination, prophylaxis, vector control) can be implemented. This is illustrated by John Snow and the Broad Street pump: removal of the pump handle apparently controlled an outbreak of cholera. A challenge is to improve the quality of spatial disease data that is routinely collected and the analytic techniques available, so that the promise of spatial epidemiology, formally recognized at least 150 years ago, can be realized.

References

Alexander, D.J. (2000). A review of avian influenza in different bird species. *Veterinary Microbiology*, 74, 3–13

Alexander, D.J. (2001). Orthomyxoviridae–Avian influenza. (In F. Jordan, M. Pattison, D. Alexander, & T. Flanagan (Eds.), *Poultry diseases* (5th ed.) London: WB Saunders)

Anselin, L. (1995). Local indicators of spatial association-LISA. *Geographical Analysis*, 27, 93–115

Bates, T.W., Thurmond, M.C. & Carpenter, T.E. (2003). Description of an epidemic simulation model for use in evaluating strategies to control an outbreak of foot-and-mouth disease. *American Journal of Veterinary Research*, 64, 195–204

Birnbaum, N., Barr, S.C., Center, S.A., Schermerhorn, T., Randolph, J.F. & Simpson, K.W. (1998). Naturally acquired leptospirosis in 36 dogs: Serological and clinicopathological features. *Journal of Small Animal Practice*, 39, 231–236

Bithell, J.F. (1995). The choice of test for detecting raised disease risk near a point source. *Statistics in Medicine*, 14, 2309–2322

Bithell, J.F., Dutton, S.J., Draper, G.J. & Neary, N.M. (1994). The distribution of childhood leukemias and non-Hodgkin's lymphomas near nuclear installations in England and Wales. *British Medical Journal*, 309, 501–505

Bolin, C.A. (1996). Diagnosis of leptospirosis: A reemerging disease of companion animals. *Seminars in Veterinary Medicine and Surgery (Small Animal)*, 11, 166–171

Castillo-Olivares, J. & Wood, J. (2004). West Nile virus infection of horses. *Veterinary Research*, 35, 467–483

Clark, P.J. & Evans, F.C. (1954). Distance to nearest neighbor as a measure of spatial relationships in populations. *Ecology*, 35, 445–453

Cuzick, J. & Edwards, R. (1990). Spatial clustering for inhomogenous populations. *Journal of the Royal Statistical Society*, Series B 52, 73–104

Dexter, N. (2003). Stochastic models of foot and mouth disease in feral pigs in the Australian semi-arid rangelands. *Journal of Applied Ecology*, 40, 293–306

Diuk-Wasser, M.A., Gatewood, A.G., Cortinas, M.R., Yaremych-Hamer, S., Tsao, J., Kitron, U., Hickling, G., Brownstein, J.S., Walker, E., Piesman, J. & Fish., D. (2006). Spatiotemporal patterns of host-seeking Ixodes scapularis nymphs (Acari: Ixodidae) in the United States. *Journal of Medical Entomology*, 43, 166–176

Doran, R.J. & Laffan, S.W. (2005). Simulating the spatial dynamics of foot and mouth disease outbreaks in feral pigs and livestock in Queensland, Australia, using a Susceptible-Infected-Recovered Cellular Automata model. *Preventive Veterinary Medicine*, 70, 133–152

East, I.J., Hamilton, S.A. & Garner M.G. (2007). Identification of Australian poultry rearing areas at high risk of exposure to avian influenza. *Proceedings of the GisVet'07 Conference*, Copenhagen, 22–24 August

Ederer, F., Myers, M.H. & Mantel, N. (1964). A statistical problem in space and time: Do leukemia cases come in clusters? *Biometrics*, 20, 626–638

Elliot, P., Wakefield, J.C., Best, N.G. & Briggs, D.J. (2000). Spatial epidemiology: Methods and applications. (In P. Elliot, J.C. Wakefield, N.G. Best & D.J. Briggs (Eds.), *Spatial epidemiology: Methods and applications*. New York: Oxford University Press)

English, D. (1992). Geographical epidemiology and ecological studies. (In P. Elliott, J. Cuzick, D. English & R. Stern (Eds.), *Geographical and environmental epidemiology: Methods for small-area studies*. New York: Oxford University Press)

Fletcher, J. (2004). Foot and mouth disease in deer. (In *Proceedings of the Deer Branch of the New Zealand Veterinary Association*. World Deer Veterinary Congress)

Ferguson, N.M., Donnelly, C.A. & Anderson, R.M. (2001). The foot and mouth epidemic in Great Britain: Pattern of spread and impact of interventions. *Science*, 292, 1155–1160

Garner, M.G. & Beckett, S.D. (2005) Modelling the spread of foot-and-mouth disease in Australia. *Australian Veterinary Journal*, 83, 758–766

Geary, RC (1954). The contiguity ratio and statistical mapping. *The Incorporated Statistician*, 5, 115–145

Gerbier, G., Bacro, J.N., Pouillot, R., Durand, B., Moutou, F. & Chadoeuf, J. (2002). A point pattern model of the spread of foot-and-mouth disease. *Preventive Veterinary Medicine*, 56, 33–49

Getis, A. & Ord, J.K. (1992). The analysis of spatial association by use of distance statistics. *Geographical Analysis*, 24, 189–206

Hanson, L.E. (1982). Leptospirosis in domestic animals: The public health. *Journal of American Veterinary Medicine Association*, 181, 1505–1509

Keeling, M.J., Woolhouse, M.E.J., Shaw, D.J., Matthews, L., Chase-Topping, M., Haydon, D.T., Cornel, S.J., Kappey, J., Wilesmith, J. & Grenfell, B.T. (2001). Dynamics of the 2001 UK foot and mouth epidemic: Stochastic dispersal in a heterogeneous landscape. *Science*, 294, 813–817

Knox, G. (1964). The detection of space-time interactions. *Applied Statistics*, 13, 25–29

Krauss, S., Walker, D., Pryor, S.P., Niles, L., Chenghong, L., Hinshaw, V.S. & Webser, R.G. (2004). Influenza A viruses of migrating wild aquatic birds in North America. *Vector-borne and Zoonotic Diseases*, 4, 177–189

Kulldorff, M. (2006). *SaTScan™ version 6.1.2. Software for the spatial and space-time scan statistics* [Electronic version]. Retrieved from http://www.satscan.org/

Kulldorff, M. & Nagarwalla, N. (1995). Spatial disease clusters: Detection and inference. *Statistics in Medicine*, 14, 799–810

Kulldorff, M., Athas, W.F., Feuer, E.J., Miller, B.A. & Key. C.R. (1998). Evaluating cluster alarms: A space-time scan statistic and brain cancer in Los Alamos. *American Journal of Public Health*, 88, 1377–1380

Last, J.M. (2001). *A dictionary of epidemiology* (4th ed.). (Oxford: Oxford University Press)

McVicar, J.W., Sutmoller, P., Ferris, D.H. & Campbell, C.H. (1974). Foot and mouth disease in white-tailed deer: Clinical signs and transmission in the laboratory. *Proceedings of the 78th Annual Meeting of the US Animal Health Association*, 169–180

Miller, R., Farnsworth, M., Kendall, W., Doherty, P., Nichols, J., White, G., Burnham, K., Franklin, A. & Freier, J. (2007). Risk-based targeted surveillance: Identifying areas and populations of importance for surveillance of High Path Avian Influenza in the United States. *Proceedings of the GisVet'07 Conference*, Copenhagen, 22–24 August

Moran, P.A.P. (1950). Notes on continuous stochastic phenomena. *Biometrika*, 37, 17–23

Morris, R.S., Wilesmith, J.W., Stern, M.W., Sanson, R.L. & Stevenson, M.A. (2001). Predictive spatial modelling of alternative control strategies for the foot-and-mouth disease epidemic in Great Britain. *Veterinary Record*, 149, 137–144

Naus, J.I. (1965). The distribution of the size of the maximum cluster of points on a line. *Journal of the American Statistical Association*, 60, 532–538

Naus, J.I. (1966). A power comparison of two tests of non-random clustering. *Technometric*, 8, 493–517

Oden, N. (1995). Adjusting Moran's I for population density. *Statistics in Medicine*, 14, 17–26

Office International des Epizooties. (1999). (In B.Toma, J-P. Vaillancourt, B. Dufour, M. Eloit, F. Moutou, W. Marsh, J-J. Bénet, M. Sanaa & P. Michel (Eds.), *Dictionary of veterinary epidemiology*. Ames, IA: Iowa State University Press)

O'Sullivan, D. & Unwin, D.J. (2003). *The pitfalls and potential of spatial data. Geographic information analysis*. (Hoboken, NJ: Wiley)

Paterson, A.D. (1995). Problems encountered in the practical implementation of geographical information systems (GIS) in veterinary epidemiology. (In *Proceedings of the Society of Epidemiology and Preventive Medicine Meeting*. Reading, United Kingdom, p. 162)

Pech, R. & Hone, J. (1988). A model of the velocity of advance of foot and mouth disease in feral pigs. *Journal of Applied Ecology*, 25, 63–77

Pinto, A.A. (2004). Foot and mouth disease in tropical wildlife. *Annals of the New York Academy of Science*, 1026, 65–72

Porter, M.B., Long, M.T., Getman, L.M., Giguere, S., MacKay, R.J., Lester, G.D., Alleman, A.R., Wamsley, H.L., Franklin, R.P., Jacks, S., Buergelt, C.D. & Detrisac, C.J. (2003). West Nile Virus encephalomyelitis in horses: 46 cases (2001). *Journal of the American Veterinary Medicine Association*, 222, 1241–1247

Rosenberger, J.K., Krauss, W.C. & Slemmons, R.D. (1974). Isolation of newcastle disease and type-A influenza viruses from migratory waterfowl in the Atlantic flyway. *Avian Diseases*, 18, 610–613

Salazar, P., Traub-Dargatz, J.L., Morley, P.S., Wilmot, D.D., Steffen, D.J., Cunningham, W.E. & Salman, M.D. (2004). Outcome of equids with clinical signs of West Nile virus infection and factors associated with death. *Journal of the American Veterinary Medicine Association*, 225, 267–274

Schwabe, C.W. (1984). *Veterinary medicine and human health* (3rd ed.) (Baltimore, MD: Williams & Wilkins)

Stallnecht, D.E. & Shane, S.M. (1988). Host range of avian influenza virus in free-living birds. *Veterinary Research Communications*, 12, 125–141

Sutmoller, P., Thomson, G., Hargreaves, S., Foggin, C.M. & Anderson, E.C. (2000). The foot and mouth disease risk posed by African buffalo within wildlife conservancies to the cattle industry in Zimbabwe. *Preventive Veterinary Medicine*, 44, 43–60

Thomson, M.C., Connor, S.J., D'Alessandro, U., Rowlingson, B., Diggle, P., Cresswell, M. & Greenwood, B. (1999). Predicting malaria infection in Gambian children from satellite data and bed net use surveys: The importance of spatial correlation in the interpretation of results. *American Journal of Tropical Medicine and Hygiene*, 61, 2–8

Torrens, P.M. & Benenson, I. (2005). Geographic automata systems. *International Journal of Geographic Information Science*, 19, 385–412

Vazquez-Prokopec, G.M., Cecere, M.C., Canale, D.M., Gurtler, R.E. & Kitron, U. (2005). Spatiotemporal patterns of reinfestation by Triatoma guasayana (Hemiptera: Reduviidae) in a rural community of northwestern Argentina. *Journal of Medical Entomology*, 42, 571–581

Ward, M.P. (2002a). Seasonality of canine leptospirosis in the United States and Canada and its association with rainfall. *Preventive Veterinary Medicine*, 56, 203–213

Ward, M.P. (2002b). Clustering of leptospirosis among dogs in the United States and Canada. *Preventive Veterinary Medicine*, 56, 215–226

Ward, M.P. & Carpenter, T.E. (2000). Techniques for analysis of disease clustering in space and in time in veterinary epidemiology. *Preventive Veterinary Medicine*, 45, 257–284

Ward, M.P. & Rollo, S. (2007). County-level risk assessment of avian influenza introduction and spread. *Proceedings of the 62nd International Conference on Diseases in Nature Communicable to Man*, Madison WI, 12–14 August. Abstract #14

Ward, M.P., Glickman, L.T.G. & Guptill, L. (2002). Prevalence of and risk factors for leptospirosis among dogs in the United States and Canada: 677 cases (1970–1998). *Journal of the American Veterinary Medicine Association*, 220, 53–58

Ward, M.P., Levy, M., Thacker, H.L., Ash, M., Norman, S.K.L., Moore, G.E. & Webb, P.W. (2004a). An outbreak of West Nile Virus encephalomyelitis in a population of Indiana horses: 136 cases. *Journal of the American Veterinary Medicine Association*, 225, 84–89

Ward, M.P., Guptill, L.F., Prahl, A. & Wu, C.C. (2004b). Serovar-specific prevalence and risk factors for leptospirosis among dogs. *Journal of the American Veterinary Medicine Association*, 224, 1958–1963

Ward, M.P., Guptill, L.F. & Wu, C.C. (2004c). Geographical risk factors for leptospirosis among Indiana dogs. *Journal of the American Veterinary Medical Association*, 225, 72–77

Ward, M.P., Schuermann, J.A., Highfield, L. & Murray, K.O. (2006). An outbreak of West Nile virus encephalomyelitis in Texas equids: 1,698 cases. *Veterinary Microbiology*, 118, 255–259

Ward, M.P., Laffan, S.W. & Highfield, L.D. (2007). The potential role of wild- and feral-animals as reservoirs of foot-and-mouth disease. *Preventive Veterinary Medicine*, 80, 9–23

Webster, R.G., Bean, W.J., Gorman, O.T., Chambers, T.M. & Kawaoka, Y (1992). Evolution and ecology of influenza A viruses. *Microbiological Reviews*, 56, 152–179

World Health Organization. (2007). Update: WHO-confirmed human cases of avian influenza A (H5N1) infection, 25 November 2003 – 24 November 2006. *Weekly Epidemiological Record*, 82, 41–48

Woebser, G.A. (1997). *Avian influenza, Newcastle disease, and other paramyxoviruses in diseases of wild waterfowl*. (New York: Plenum)

Chapter 14
The Role of Geosurveillance and Security in the Politics of Fear

Jeremy W. Crampton

Abstract This chapter examines the role of geographic information technologies (GIT) in the production of the politics of fear. While technologies such as mapping and GIS appear to offer a fix or solution to problems of terrorism, crime, or disaster, they can contribute to the use of fear for political exploitation. What sustains this politics of fear? This chapter suggests that if GIT continue to produce knowledge of populations in terms of risk, then a politics of fear can be exploited to justify mass geosurveillance. In this light, two case studies are examined; nineteenth century mapping and contemporary crime mapping.

Keywords Biopower, cartography, critical cartography, critical GIS, Foucault, geosurveillance, GIS, governmentality, normalization, politics of fear, rationalities of technology, risk, security

14.1 The Politics of Fear

This chapter examines the role of geographic information technologies (GIT) in the production of the politics of fear. While technologies such as mapping and GIS appear to offer a fix or solution to problems of terrorism, crime, or disaster, they can contribute to the use of fear for political exploitation. In examining what it is that sustains this politics of fear, I highlight the political rationalities of technology. My argument is that GIT can be used to produce knowledge of human populations in terms of risk, and that in doing so, fear of these risks can be exploited to justify deployment of mass geosurveillance and data mining. I discuss the weaknesses of using risk analysis in GIT in two case studies: nineteenth century mapping and contemporary crime mapping. It is through an examination of the 'grounds' of the production of fear that we may hope to mitigate it (Sparke 2007).

Georgia State University

D.Z. Sui (ed.) *Geospatial Technologies and Homeland Security*,

One of the dominant narratives that followed September 11, 2001 was that fear had now breached the sanctity of the American geographical homeland. The Fire Chief of the Livermore/Pleasanton Fire Department expressed it this way on the 'Homeland Security' CD–ROM distributed by ESRI shortly after 9/11:

> I think now that everyone's reminded that anytime, anywhere, a significant catastrophic event can occur. An industrial accident, internal sabotage, external terrorism, a bad weather that has not come in a hundred years, and that our citizens expect everybody to be prepared for that (ESRI 2002).

This narrative, in other words, portrays pre-9/11 America as complacent, perhaps born of the peace dividend and the cold war victory over Communism. 9/11 served as a fearful wake-up call when 'everything changed.' The geopolitical argument made by the political elites following 9/11 was framed around the need to return to a binary viewpoint constructed from 'friends' and 'enemies' (compare, for example, the geopolitical rhetoric of the terms 'axis of evil' used by President Bush in his 2002 State of the Union address, and 'outposts of tyranny' used by Secretary of State Condoleezza Rice in 2005).

Yet fear is a political idea with a considerable history (Robin 2004). As Robin argues, the narrative of fear can be traced from Michel de Montaigne (who declared 'the thing I fear most is fear') through the work of Hannah Arendt and the McCarthy era (Robin 2004: 3). The attacks of 9/11 were used to renew the political narrative of fear. As Agamben (2005) shows in his history of the suspension of law (what he calls the 'state of exception') 9/11 can be seen as only one among many such suspensions dating back to the French revolution. The passage of the USA PATRIOT Act in October 2001 was motivated by a desire to restore many of the powers of a sovereign who operates under the state of exception:

> [President] Bush is attempting to produce a situation in which the emergency becomes the rule, and the very distinction between peace and war (and between foreign and civil war) becomes impossible (Agamben 2005: 22).

Moreover, fears can be fed or killed (Lawson 2007). Political fears can be used to motivate acceptance of a series of responses (including the state of exception) that would seemingly quench fears but which actually feed and enlarge them (Siegel 2005). One way fear can be politically exploited in this manner is to frame a choice between more security or being at risk. Most rational people will opt for the perceived security (and the forms of surveillance it necessitates) rather than the risks. Since the security is onerous, however, and therefore potentially subject to rejection by the population, more fear must be generated in order to justify it. Yet while fear becomes more pervasive throughout this country, statistically we have never been safer. We live longer, healthier lives (some 60 percent longer in 2000 than 1900), have better access to clean water and food, and enjoy safer workplaces (Siegel 2005). There is an increasing mismatch between perceived danger and actual risk. Siegel cites the fact that between 1990 and 1998 'the murder rates decreased by 20 percent, while murder stories on media newscasts increased by 600 percent (not even counting O.J. Simpson)' (Siegel 2005: 56–57).

That our fears are constructed can be easily shown by the fact that we have a poor idea of the difference between real and potential risk. We worry about avian flu (AH5N1) which according to the World Health Organization (WHO) killed 80 people in 2006—most of them elderly and in countries with overburdened health care systems—and ignore human influenza, which in the US alone kills 35,000–40,000 people every year. Researchers call this the 'dread risk' effect, where we over-respond to a high profile but low threat risk (Gigerenzer 2004). Following 9/11 for example, many people avoided flying and drove instead. Given that driving is much less safe than flying, this resulted in an estimated 1,500 additional deaths in the year following 9/11 (Gigerenzer 2006). More generally, as the well-known work of Tversky and Kahneman has shown, human decision-making is impaired by anchoring (unshakeable focus on marginal data), the base rate fallacy (ignoring the fact that many events are improbable), and framing (over-attention to a framing narrative) during judgments under uncertainty (Tversky and Kahneman 1974). Politically this means that people respond to fear by accepting the response of security (as if those are the only two options), and perhaps even 'hawkish' over 'dovish' behavior (Kahneman and Renshon 2007). In other words, a risk-based approach is unlikely to assist in distinguishing between realistic and unrealistic fears.

In their examination of responses following 9/11, Gregory and Pred (2007) identify two that especially serve to give fear its performative opening: on the one hand 'those who enlisted the rhetoric of the 'war on terror' as a means of legitimizing and intensifying their own apparatus of repression' and on the other:

> those who proposed a purely technical or instrumental response to 9/11, drawing on political technologies (that were also geographical technologies) to profile, predict, and manage the threat of terrorism as an enduring mode of late-modern government (Gregory and Pred 2007: 1).

In this chapter I examine these geographical information technologies—mapping and GIS—in more detail, and discuss their assumptions and consequences.

14.2 The Geographical Imagination and 9/11

Discussions of al-Qaeda, Abu Ghraib, warrantless wiretaps, torture memos, Baghdad and terrorism all have something in common: they are based on knowledge. Much of this knowledge is geographic, not just in the traditional sense of where things are and where things go, but in the sense of how identity is formed. This emphasis on knowledge, which for many is associated with writers such as Foucault and his formulation of 'power-knowledge,' is nevertheless a standard trope of science. In their recent book on geographical methodologies for example, Montello and Sutton define the scientific method as 'the creation and evaluation of knowledge' (Montello and Sutton 2006: 3). The particular categories of knowledge that are created through and by GIS and mapping are the subject matter of 'critical GIS' and 'critical cartography' (for recent progress reports see O'Sullivan 2006;

Perkins 2004). These modes of inquiry have come rather late to the geographical enterprise and seek to negotiate the tricky terrain between critical geography and geospatial technologies, and perhaps because they do so remain a 'distinctly minority pursuit' in the words of O'Sullivan (2006: 783). Yet the events of 9/11 were replete with the production of GIS and cartographic knowledge and the subsequent geographical imaginary, that then played an important role in crafting the 'response' so adamantly opposed by Gregory and Pred.

Following 9/11, one GIS company (ESRI) offered a series of seminars around the country on how GIS could assist in emergency prevention and response, published white papers, produced a CD–ROM on security, and established a website for GIS and security. The company also awarded $2.3 million in 'Homeland Security Grants' to cities and agencies across the USA.[1] The Association of American Geographers (AAG) meanwhile launched a workshop funded by the National Sciences Foundation (NSF) on 'geographical dimensions to terrorism' and established a list of priority action and research items. Mapping and GIS are key components of this effort. The first priority action item listed is to '[e]stablish a distributed national geospatial infrastructure as a foundation for homeland security' (Cutter et al. 2002: 2), which would include geospatial databases and GIS analysis. This effort is based on the DHS 'Information Analysis and Infrastructure Protection' (IAIP) Directorate's budget request for 'development and maintenance of a *complete and accurate mapping* of the Nation's critical infrastructure and key assets' (US Government Office of the President 2003: 472 emphasis added). The IAIP Directorate is charged with:

> Analyzing law enforcement, intelligence, and other information to evaluate terrorist threats to the homeland; Assessing the vulnerabilities of key US resources and critical infrastructures; Mapping threat information against our current vulnerabilities; and, Working with federal, state, local, and private stakeholders to issue timely warnings and take or effect appropriate preventive and protective action (US Government Office of the President 2003: 471).

The Association of American Geographers also moved quickly to show the relevance of geography in combating terrorism, but unfortunately the scope and assumptions of the outcome, a NSF-funded book (Cutter et al. 2003), lead many to regret it, rather than celebrate it. Writing in the American Geographical Society's *Geographical Review*, one Middle East expert concluded that:

> They failed to remember that conflict and terrorism are the result of human agency and not conducive to modeling the way natural hazards such as El Nino may be. The volume failed to offer any understanding of the societal context that has produced many of today's Middle-East terrorist groups; the terms 'Islam' and 'Middle East' do not even appear in the volume's index (Stewart 2005: iv).

The book also received a lukewarm response in the AAG's own flagship journal (see e.g. de Blij 2004; Johnston 2004). These reactions were born out of a frustration with the 'geographical imagination' failing to engage the full context of terrorism

---

[1] See http://www.esri.com/industries/homelandsecurity

as a political problem, but approached it as one of hazards and risks, and because it situated technology as 'non-political' rather than treating geospatial technology as part of the political decision-making process. By contrast, disciplinary responses from organizations such as the American Sociological Association (ASA) emphasized 'religious and cultural perspectives' on terrorism, passed a resolution calling for open access to data sets that were being removed (such as GIS data) in 2002, and in spring 2003 voted on a resolution concerning the US invasion of Iraq (Rosich 2005). During 2006 when the Bush administration threatened to defund part of the Census Bureau's collection of income and poverty data, dozens of national organizations organized to resist the cut, but not the AAG (Center for Economic and Policy Research 2006). This narrow focus on technology coupled with a consistent political hands-off approach only serves to make the geographical imagination less powerful in the 21st century.

14.3 The 'Risk' of Risk

What happens when a narrative of risk is established? I argue that it gives rise to negative unintended consequences. These consequences include profiling (which in turn depends on racism, stereotyping, and normalization), geosurveillance, and the use of fear as a tactic of governance. These consequences apply as much to the putative 'enemy' or threat-source as they do to those in the homeland (for example, through the constant and increasing penetration of the lives of millions of Americans by surveillant technologies such as warrantless wiretaps). Yet risk assessment is a key component of GIS-supported efforts to improve homeland security.

There are several components of risk that are worth noting in this context. Some are statistical and involve the problem of false positives and base rates. Even seemingly very accurate tests can yield far more false positives than true hits, especially if the base rate is low. This is especially a problem if data-mining surveillance is pursued (such as the wide-scale warrantless wiretaps carried out by the USA). The mathematician John Allen Paulos describes the weakness of this surveillance with an example of a profiling test that is 99 percent accurate in the following sense: the profile will correctly detect terrorists 99 percent of the time, and correctly detect non–terrorists 99 percent of the time. Assume a base rate of 1 in a million people in America is a terrorist (about 300 people). The profile will find 297 of the terrorists (99 percent). But it will also find that 1 percent of the rest 'fit the profile,' or in other words *some 3 million false positives* (Paulos 1996, 2006a). Since it will not be known which of the positive hits are false and which true, all the positive hits will have to be investigated and surveilled. That is the logic that supports mass surveillance. For the same reason many doctors do not advocate inappropriate screening (for example, breast cancer screening): it will indicate many false positives and cause unnecessary worry and distress.

A related issue concerns the work on human perception of risk discussed above. How well do we correctly assess risk? In the aftermath of 9/11, Vice-President Dick

Cheney developed his 'one percent doctrine' namely that if there is 'just a one percent chance of the unimaginable coming due, act as if it is a certainty' (Suskind 2006: 62). This is exactly the low-probability, high-impact event known as the dread risk; a risk we tend to wildly exaggerate. If the one percent doctrine were shared by doctors, gamblers, or scientists it could have disastrous consequences. In international relations it would lead to a huge number of threats being misconstrued (Paulos 2006b). In scientific work it is typical to demand probabilities not of one percent, but rather of at least 95 percent. Even the latter means we will be wrong in one out of 20 cases.

A third issue arises surrounding normalization. To assess risk it is necessary to know what comprises a normal state of affairs and when that state of affairs has been deviated from. This might seem unobjectionable, but as more than one writer has discussed, the establishment of norms can have debilitating effects on those who are outside of those norms. As the history of racism, homosexuality, immigrant groups, and the 'feeble-minded' demonstrates, the 'abnormals' are subjected to exclusion, mistreatment, peer pressure, and medical experimentation. During the nineteenth century a whole array of techniques were formalized to assist in the establishment of norms, including probability theory and the normal distribution curve. Many forms of mapping were also invented to establish what was normal across the geography of the nation. These new techniques were used to create profiles of groups. Then, if you belonged to the group, it was inferred that you fit the profile. Analysis was at the level of the group. This was easier than tracking people individually (thematic maps, for example, tend to show distributions of populations not individuals).

So approaching terrorism through a framework of risk and threat has the following negative unintended consequences: it centers risk (which we misperceive and exaggerate); it produces massive numbers of false positives; it normalizes (and abnormalizes) through profiles; and it requires ubiquitous surveillance to collect data on the normal and abnormal. When you live by fear, everything is a risk.

We need to understand how mapping and other sources of geographical knowledge act to produce this politics of fear. The answer is not to cease using GIS and mapping technologies (or only to use the 'good ones'), but rather to be careful and critical about the knowledge that isconstructed with them and the subsequent political rationalities that are supported by them. This claim might seem unobjectionable, but in fact, it is often ignored. For example, in a major report in 2006 the National Research Council of the National Academies investigated the implications of new technologies in GISci, and wrote that 'as is true of any technology' GIS is 'neutral in and of itself' (Committee on Beyond Mapping 2006: 47). Such a viewpoint traduces two decades of work in critical GIS and cartography. Surely, it is not the neutrality of technology but the very filigrees of interrelationships between technology, power-knowledge, and society—their geographically-situated and inherently political nature—that makes them so interesting and vital (Livingstone 2003). We cannot understand how technologies work nor assess the rationalities they operate under if the context of their political deployment is not examined. In this light the recent paper by Klinkenberg (2007) is crucial, for he acknowledges that GIT are always caught in the interplay of political applications, and it is by gaining insight into these applications that we may promote hope rather than fear. In pursuit of this goal, he predicts that 'in the future…[f]orming an integral part of a multiple-methods approach to research, [GIT] will be situated within a broader, socially-aware context' (Klinkenberg 2007: 356). I extend Klinkenberg's analysis here by identifying risk analysis as being especially susceptible to the political production of fear.

14.4 Maps as Government: Biopower

Today's renewed emphasis on security and surveillance is part of a longstanding series of historical linkages between government, knowledge, and technologies of power. These historical linkages were forged during the rise of modern industrial societies in the eighteenth century. Politics depends on the sorts of geographic knowledge that are deployed, yet at the same time provides a crucial context for some knowledge to prevail over others. Maps are a form of government.

The study of how people govern themselves and others is known as 'governmentality' (Foucault 1991) and it has proven to be a fruitful area of study in a range of disciplines including geography (Elden 2007). More specifically it is the study of the relations between power and knowledge and the rationalities (ways of thinking) that permeate them. Foucault's analysis of government was concerned with how individuals and populations were divided and grouped according to norms. When this occurred with groups or populations, he called it 'biopower' (Foucault 1978). There are specific data gathering exercises to produce knowledge of populations—the census, thematic mapping, statistics to measure and record birth and death rates, crime, disease, and so on. The target of biopower is the distribution of the population over its territory. Although Foucault looked at particular practices in their time and place, he understood them as constitutive of larger ways of thinking, or rationalities. These rationalities come into being and reach dominance at certain moments. But they can change—or be changed. The more we know about them the more we can resist them.

In order to understand how governmentality arose we can examine discipline and biopower in the context of historical changes in juridicality and criminality. Prior to the legal reforms of the 18th and early 19th centuries, as Foucault argued, the law focused on the nature of the crime committed, the evidence of guilt or innocence, and the system of penalties to be applied. In other words: crime and punishment. The person of the criminal was important and would be scrutinized only insofar as he or she was the individual to which the crime would be attributed. With the reforms, this hierarchy was reversed, the crime was merely an indicator of something more significant—the 'dangerous individual' (Foucault 1977: 252). The law was now interested in the potential danger of the individual: 'The idea of *dangerousness* meant that the individual must be considered by society at the level of his potentialities, and not at the level of his actions; not at the level of the actual violations of an actual law, but *at the level of the behavioral potentialities they represented*' (Foucault 2000b: 57 original emphasis). In terms of surveillance then, the switch that has been effected is one that shifts scrutiny from the accused individual to the potentially dangerous or risky population group (which has, nevertheless, not committed any criminal acts). Surveillance moves from the actual suspect to a sort of mass 'pre-criminal'. Punitive responses thus had to be appropriately tailored to *perceived* threat measured in terms of risk (e.g. a 'risk surface' in GIS). While risk analysis is typical in environmental or natural systems using techniques such as kriging and kernel density estimation modeling (Schröder 2006), its application to society is much more recent.

We can use Foucault's historical method to study how mapping and GIS are used in contemporary surveillance and security. In particular, a parallel from early 19th century cartography is informative because it casts light on the 21st century politics of fear. How so? First, security and risk were used to think of space and people as resources that required management and protection. Second, space and individuals were understood through a normalizing surveillance. Surveillance (including 'geosurveillance' specifically concerned with locations and distributions across spatial territories) was therefore an important technology of government tied to discourses of resource management and normalization. We can conclude from this historical comparison that it is not technologies of surveillance—mapping or GIS per se—that are problematic, but rather the *underlying political rationality of normalization* which constituted people and the environment as threatened resources under risk of hazard. This political rationality is the context in which we can understand technologies of surveillance.

14.5 Maps as Government: Moral Statistics in Early 19th Century Europe

In early 19th century Europe a completely new form of mapping was devised—thematic or statistical mapping. Thematic maps were invented precisely when population management and counting became problematic; and they are critical to censuses, census mapping, and distributions of populations across territories.

In 1829, fear over the threat of crime had reached such heights that when a map was published in France that showed no relationship between crime rates and education levels there was a huge outcry. Education was commonly thought to be an effective preventative measure against crime. Areas with higher educational levels would have lower crime rates. Crime was an activity of the uneducated lower classes; they had a 'penchant au crime' (Robinson 1982: 161). However, the 1829 maps, which employed the latest techniques of 'comparative statistics,' showed the precise opposite—areas with high education levels had high crime levels. As one commentator described it:

> Such a conclusion was sensational. Paris saw itself as being in the grip of a terrible crime wave. Ask a New Yorker of today [i.e. 1990] about muggings, then double the fear: that was how Parisians felt. The [illustrated] police gazettes, rich in reports of crimes, were

> taken in weekly … naturally one supposed that the degeneracy and ignorance of the working classes was the source of their criminal propensity (Hacking 1990: 78).

If education was not the cause of crime, then what was? The startling possibility arose that crime could occur anywhere. These crime maps were published by the Italian and French statisticians Adriano Balbi and André Michel Guerry, who had deep interest in 'moral statistics' or social problems (e.g., crime, education, birth rates, suicide). The maps were remarkable for another reason too; they were one of the first examples of the choropleth technique which had been invented by Charles Dupin just three years earlier (Robinson 1982). Dupin's choropleth maps were exceptionally popular methods for revealing the moral statistics of his day, and they were extensively emulated. After Balbi and Guerry (who was awarded a special prize in 1864 by the Academy of Sciences for his work) came D'Angeville with health and wealth choropleths in 1836, Charles Joseph Minard, who popularized proportional symbol maps in the mid-19th century, and many others. So once social problems could be grasped in their distribution across territories, policies could be implemented to address them.

Policies are needed to govern and regulate (the word shares the same origin as 'politics' and 'police'). Gordon (2000) argued that the 17th century developed 'a program of exhaustive, detailed knowledge and regularization' (p. xxvii) that assessed threat or 'dangerousness' of individuals, and produced technologies that would help maintain social order through surveillance (Foucault 2000a).

Maps have long been associated with this effort because they provide a picture of where things are so that there can be a 'right disposition' of resources and people over the territory (Foucault 1991: 93). This idea of a rightful distribution is important because it requires comparison to some norm. Territorial mapping has occurred for thousands of years to assist in inventories and taxation, and it is perhaps surprising that it was only in the early 19th century that thematic maps were invented. Why were they not deployed previously? In fact, it turns out that thematic or statistical maps were part of a more general effort to govern by means of statistical analysis. It was only with the development of descriptive and probabilistic statistics, and the formulation of society in terms of likelihoods and norms, that thematic maps could emerge. Thematic statistical maps appeared at precisely the same moment that society came to understand itself in statistical terms for purposes of regulation (policing in the larger sense) and management. A few examples will illustrate how this occurred.

In the 1820s the Belgian statistician Adolphe Quetelet derived the new analytics of probability theory and the normal distribution curve. These advances were keyed to societal problems that were thought to be amenable to governmental intervention. Quetelet was concerned about the social upheavals in Europe during the 1830s and centered his analysis of social variation around *l'homme moyen*, or the average man (his needs and typical actions and the nature of error or deviation away from this norm). Total human variation could thus be justifiably reduced to divergence around a norm. If these norms could be properly and reliably determined, then this would be extremely useful in dealing with the 'great masses of registered facts' about populations as Sir John Herschel put it in 1857 (quoted in Atkins and Jarrett 1979).

The positivist conception of science that emerged at the end of the seventeenth century gave epistemological primacy to observable data that was value free, a primacy that is underpinned by statistics. As Atkins and Jarrett show, statistical inference and significance tests on samples also permitted populations to be compared and known (how much they vary around a mean, for example, in their susceptibility to infant mortality). In sum, the newly emerging positive sciences were founded around the governmental concerns of knowledge, statistics, and population.

During the nineteenth century great strides were made in the sciences of statistics, probability, and statistical mapping. These did not occur in isolation from one another, nor more interestingly, from the question of politics—indeed, they were stimulated and put into the service of 'political' problems. Thematic mapping was part and parcel of this political problematic. Godlewska, for example, documents Alexander von Humboldt's recognition in 1811 that 'natural geography, by virtue of its ability to convey natural history's data to number and statistic, could substantially contribute to forming an exact idea of the territorial wealth of a state' (Godlewska 1999: 247). The ability to identify one's resources and thus to exploit them was necessary for the secure governing of the state.

Perhaps the most visible and influential practice of using statistics to help govern the state occurs during the great decennial censuses of many European countries (from 1790 in the United States). Although in Europe these censuses were depicted in maps in the early 19th century, in the USA it was not until the ninth census in 1870 that results were shown cartographically. These maps appeared in 1874 in America's first statistical atlas (Hannah 2000; Walker 1874).

Hannah's excellent analysis of the 1870 census atlas using Foucault's work on governmentality sheds considerable light on the spatial politics of knowledge at this time. The atlas had a tremendous impact on cartographic representations of space in the following decades. In particular, it introduced thematic mapping to the United States in a concerted manner (although several maps from the 1860 census had appeared, see Schwartz and Ehrenberg 2001, plate 177). Maps from the census were first presented at the American Geographical Society (AGS) in 1871, where, according to J.B. Jackson, they received so much attention that the Secretary of the Interior 'was persuaded to authorize a special atlas...Walker was the first American to try to show the spatial dimension of social and economic facts, to relate social problems to their physical setting and thereby throw new light on them' (Jackson 1972: 15). As stated in its Foreword, the atlas was designed to promote political education, and many of its 5,000 copies were sent to schools and colleges (Jackson 1972: 14). The 1874 atlas gave a framework for how to think about space and human occupation and led to the more sophisticated 1883 Scribner's Statistical Atlas of the United States by Fletcher W. Hewes and Henry Gannett (based on the tenth census) as well as Paullin's mighty 1932 *Atlas of the Historical Geography of the United States.*[2]

[2] The influence of the 1870 census atlas is evident in Paullin's population maps. See especially his Plates 67B–70B on the 'Colored Population' and Plates 71–76A on the 'Foreign-Born Population' (Hannah 2000: 152–153) and it is directly acknowledged on p. 48 (Paullin 1932).

The atlas was a profound statement about the relationship between politics and space, and the necessary relationship between the two. It is an exemplary document that illustrates the development of strategies of spatial surveillance for purposes of government. The contemporary role of GIS in geosurveillance and security is situated in the same age-old practice of governmental surveillance established by the first atlases.

14.6 Geosurveillance: A Contemporary Discussion

If we grant that the modern state is predicated on the establishment of norms of dangerousness, these norms need a set of experts to administer them. These experts in turn require tools. In this section, I would like to examine this idea in the context of contemporary examples and draw a parallel between governmental blanket surveillance programs and GIS crime mapping.

The crime map is an important means of constructing knowledge about the city and its inhabitants and for implementing policies to manage a crime situation. Much of this analysis is predictive or preventative in nature and draws on population-level surveillance and data collection. The origin of crime maps are closely tied to the rise of social statistics such as the FBI Uniform Crime Report or UCR (collected since the 1930s), but also local police reports, victim reports, and corporate loss reports. These maps help to construct a discourse of risk that must be surveilled as a potential danger.

One way in which this production of risk assessment works for crime mapping is through 'geoprofiling.' Geoprofiling is a disciplinary technique for determining the typical spatial patterns of an individual with the goal of predicting that person's behavior or targeting them for surveillance. With geoprofiling maps can easily be made of crime hotspots and coldspots. The theory of geoprofiling was developed by Kim Rossmo in 1995 and has since been implemented in a software system called *Rigel* that can make a predictive surface of a criminal's location (Rossmo 2000). Rossmo claims that with five to six incidents traceable to one person, his software can reduce the search area by up to 90 percent.

Crime maps enable geoprofiling to isolate behavior that does not conform to the norm. But profiling can be controversial. After a series of high profile incidents on the New Jersey turnpike in which African American drivers were disproportionately stopped by the highway patrol, it was charged that the police were stopping blacks because of who they were, not because of their actual behavior (Colb 2001). That is, criminality judgments were made on the basis of *potential dangerousness*, rather than actual offenses being committed (i.e. the searches were made without probable cause). In a similar case, the FBI has begun constructing geodemographic profiles of localities that includes a count of the number of mosques in an area (Isikoff 2003). As these examples show, crime is understood as a departure from the normative.

As an example of crime-related geosurveillance technologies, consider offender monitoring. A common technology is an ankle bracelet or tag which emits an RF radio signal that can be detected by a device in the home linked to the phone system. A more advanced approach is to use GPS. It, too, is often based on an anklet worn by the offender which can receive GPS signals and transmit its location (through the cell phone system) to the company's monitoring center. In Iowa, for example, the police have required some offenders to wear a device from a company called iSecureTrac which provides GPS offender monitoring services. This monitoring is geographically flexible: '[e]ach map is tailored for a specific parolee. A map can show, for instance, areas where a paroled pedophile must remain clear of—such as a school—when going to and from an offsite counseling session' (Chabrow 2002). Other devices include home breathalyzers and ignition interlocks for felony DUI offenders, and continuous signaling devices. However, electronic supervision is expensive, and the company reported increasing losses in 2007 (Larson 2007).

Graham (1998) discussed implications of regulating space by what he calls 'surveillant simulation' (Bloomfield 2001) which acts in this disciplinary manner. Graham highlighted four cases of surveillance: as social control especially of criminality; in and around consumption; differential deployment over space (transport informatics); and the utility industry.

Perhaps the most serious question here however is how the narrative of fear and hope has been cast as a choice between surveillance of threatening (i.e. risky) behavior and security. In other words, why it is that the public is generally happy to accept mass surveillant measures. A range of examples illustrate this point more concretely.

On an everyday level there are the no doubt minor irritations with airport security which most people accept. Yet from time to time there are hints that the scope of these measures is wider and deeper than most people realize, as in the case of the US Automated Targeting System (ATS) which ranks every traveler:

> The scores are assigned to people entering and leaving the United States after computers assess their travel records, including where they are from, how they paid for tickets, their motor vehicle records, past one-way travel, seating preference and what kind of meal they ordered (Associated Press 2006).

The report continues that 'travelers are not allowed to see or directly challenge the risk assessments, which the government plans to hold for 40 years.' This surveillance is obviously extensive, but Americans are more than willing to permit this list (and the 'no-fly' list) because of the presumed security benefits it brings.

A CBS investigation in fall 2006 however indicated a number of relevant issues with the no-fly list. First is its burgeoning nature. According to CBS, on September 11, 2001 the list had 16 names on it; by December 2002 it had over 1,000; and by March 2006 it had over 44,000, plus another 75,000 people on a list for additional security screening (CBS 2006). Second is its inaccuracies and potential for false positives. CBS for example discovered that 14 of the 19 deceased terrorists from 9/11 were still on the list in 2006. The media has also reported on a number of other false positives (such as denying entry to the singer Cat Stevens who now goes by the name Yusuf Islam), but since the list is secret its accuracy cannot be reliably assessed. Finally, whatever its error rate, the list is symptomatic of modern surveillance and risk analysis, in that it is based on collecting data about *everyone* and assessing them against a risk profile. Thus an entire country's population—innocent and guilty—is surveilled in order to determine which individuals are risky.

Polls taken after 9/11 showed an almost universal fear of further attacks. A CBS/New York Times poll in October 2001 found that 85 percent of Americans feared a further terrorist attack 'in the next few months'. In September 2007, that number stood at 48 percent, its lowest since 9/11 but still historically high. Additionally, some 90 percent of Americans believe there are members of Al Qaeda in the United States today, according to a poll by Fox News/Opinion Dynamics in September 2007. Polls consistently indicate that the public is willing to submit to surveillant technologies when these are linked to fighting terrorism. For example, a *Newsweek* poll in July 2007 asked if the FBI should wiretap mosques to 'keep an eye out for radical preaching by Muslim clerics'. Over half the respondents (52 percent) agreed. Another poll by the Pew Research Center for the People and the Press in December–January 2006–7 revealed that 40 percent of respondents thought it necessary to give up civil liberties 'in order to curb terrorism.' The reason for these findings is not hard to discern: most Americans think of themselves as law-abiding and therefore these technologies are not likely to affect them personally.[3]

This narrative was highlighted in the debate surrounding another surveillant technology in the United States, that of warrantless wiretapping. In December 2005 the *New York Times* reported that shortly after 9/11, the United States had secretly instituted a practice of monitoring phone calls in the US without a court warrant as required by law (Risen and Lichtblau 2005). When asked, polls showed that Americans were more ambivalent about this program. A *USA Today*/Gallup poll in May 2006 for example found that while 54 percent of respondents thought the program violated the law, and 57 percent thought it violated their personal privacy, 43 percent still approved of this program. (The administration confirmed it was performing the warrantless wiretaps and argued it was a necessary tool. Nevertheless after the 2006 midterm elections it agreed to halt this practice, and the then Attorney General Alberto Gonzales has since resigned. In the summer of 2007, Congress temporarily restored some warrantless surveillance measures, but the future of the program is uncertain.)

Whatever the legality, support, or public policy ramifications of these measures, the point to be emphasized here is that they depend on techniques of mass surveillance and data mining. Operating at the level of the population rather than the accused individual, they take large amounts of surveillance data and sift it for risky behaviors. Therefore citizens are rendered in a 'mappable landscape of expectation' (Hannah 2006) which re-imagines the landscape as a kind of blanket saturated with risk. This 'risk blanket' lies at the heart of many techniques of GIS crime mapping.

The same reasoning applies to the Bush administration's controversial plans for TIPS (Terrorism Information and Prevention System), which was proposed in early

[3] Poll numbers are available at pollingreport.com.

2002 but has since been dropped from Homeland Security. In this plan, citizens and workers who often go into residential neighborhoods (e.g. postal workers, cable TV installers, truck drivers) would be recruited to call a government hotline if they saw suspicious activity. The idea was to benefit from as many as a million sources of surveillance in ten pilot cities (these cities were never specified).

We have thus reached an analogous situation to that faced by the citizens of Paris in 1829 when they were presented with the Balbi and Guerry crime maps: we fear crime and threats to our security from everywhere, and it is no surprise that normative governmental rationality gives rise to widespread geosurveillance in order to manage these threats.

14.7 Conclusion

The purpose of this chapter is to show that a technological response to threat is an insufficient one by itself. Risk and hazards research have a long tradition in geography and are amenable to GIS analyses (a search of the ISI database yields nearly 1,000 articles on 'GIS AND risk' and nearly 400 on 'GIS AND hazard'). Prominent disciplinary responses by the AAG, funded by the NSF (Cutter et al. 2002, 2003, 2004) (co-authored by the AAG's Executive Director), foreground just such a technological approach. Readers may also examine this volume and judge whether this observation still holds true.

But I would like to conclude by re-iterating several points. First, although we have been perhaps led to see technology and politics as alternative approaches, they are not. Technology is part of the political decision-making process, not some neutral activity. In particular, the GIS and mapping industry has long-established relationships with intelligence and military agencies (Cloud 2002). In the USA for example, ESRI is a 'strategic partner' alongside the National Geospatial-Intelligence Agency (NGA) and the CIA of the US Geospatial Intelligence Foundation (USGIF) which sponsors the annual GEOINT Symposium. The CEO of ESRI and some contributors to this volume are Board members of the USGIF. In 2006 the Keynote speaker at GEOINT was John Negroponte, then Director of National Intelligence who underlined the 'value of geospatial intelligence to our national security' (Negroponte 2006: 1).

I have especially highlighted how a rationality of security is constructed in which geosurveillance is deployed as a response to dangerousness, and in which the environment and people are constructed as at-risk resources subject to normalization and management. The question is whether or not we choose to acknowledge and engage with the political implications of our technologies.

Second, and for this reason, an attempt to pick out the 'good' from the 'bad' uses of GIS, as we are encouraged to do by the NRC report (Committee on Beyond Mapping 2006), is to miss the point that GIS produces a distinctive political rationality of government. That rationality is one of biopolitics which is addressed to the problem of 'populations' and their rightful disposition across territories (Legg 2005).

Mass surveillance techniques such as warrantless wiretaps and spatial data-mining based on the creation of norms and profiles are today integral to modern society. Again, the question here is whether we contribute to or resist the production and proliferation of geosurveillance and geo-profiling.

False positives are another danger. Even if data collection methods are almost perfect (99.9 percent accurate), when collecting data with billions of entries, there will be many false 'hits.' These all have to be checked, draining manpower and resources.

I have argued that these factors contribute to a politics of fear. In short, there is a risk of foregrounding a technological analysis of risk as a response to what are complex geopolitical events and processes. The risk lies in the negative unintended consequences of profiling, geosurveillance, and the political use of fear. Fear is a complex human emotion, which once activated is hard to deactivate. In the United States these fears have been exploited to justify a startling range of surveillance programs. Yet geographical expertise, including that of the GIS and mapping community, can provide antidotes to this fear in a number of ways. Researchers could examine whether their work can be used to engage in risk and data-mining research that create profiles. Geographers could also do much more to examine the social and theoretical aspects of mapping technologies. In a review of the literature, Schuurman and Kwan found that less than 4 percent of articles appearing in leading GIScience journals make any such reference (Schuurman and Kwan 2004). In this light, the emerging sub-disciplines of critical GIS and critical cartography and assessments of the production of geographical knowledge (Castree 2006) will play important roles.

In addition to critique it is evident that resistance to the politics of fear will include practice. Here a tentative note of optimism is warranted as we see a re-emergence of the role of public geographies, public debate, and public practices such as community GIS (Murphy 2006) that will help educate people and erase the fearful mystery of the 'other'; the rise of non-traditional 'people-powered' political movements in the netroots and blogs (Armstrong and Zúniga 2006), and even access to open-source data and map mashups (Crampton, forthcoming; Miller 2006). Meanwhile non-traditional military geographies are critiquing the CIA's extraordinary rendition (Paglen and Thompson 2006) and the multiple relations between terror and political violence (Gregory and Pred 2007). And federal judges have twice now struck down provisions of the PATRIOT Act as violating constitutional rights (pertaining to national security letters, and probable cause).

If these developments lead to a sense and reality of more control (and thus less surveillance), and less focus on hyped-up and unlikely 'dread risks,' then it may be possible to start making inroads into today's politics of fear. As Klinkenberg recently argued, geographic information technologies are not just a technique but are 'entire new ways of seeing' that weave together the technology, the political, and the social (Klinkenberg 2007: 357). Our responsibility as users of these technologies is therefore also one that should include paying attention to the political and social deployment of these technologies.

Acknowledgements I would like to thank the members of the Spring 2003 Seminar in Geosurveillance and Security for comments. Parts of this chapter appeared in 2003 as 'Cartographic Rationality and the Politics of Geosurveillance and Security' in *Cartography and GIS*, 30(2), pp. 131–144. My argument in this chapter was clarified with the help of two anonymous referees.

References

Agamben, G. (2005). *State of exception* (K. Attell, Trans.). (Chicago, IL: The University of Chicago Press)

Armstrong, J. & Zúniga, M. M. (2006). *Crashing the gate. Netroots, grassroots and the rise of people-powered politics*. (White River Junction, VT: Chelsea Green Publishing)

Associated Press (2006, December 1). US assigns terror score to international travelers. *The New York Times*, pp. A1, 28

Atkins, L. & Jarrett, D. (1979). The significance of 'significance tests.' (In J. Irvine, I. Miles & J. Evans (Eds.), *Demystifying social statistics* (pp. 87–109). London: Pluto Press)

Bloomfield, B. (2001). In the right place at the right time: Electronic tagging and the problems of social order/disorder. *The Sociological Review*, *49*, 174–201

Castree, N. (2006). Research assessment and the production of geographical knowledge. *Progress in Human Geography*, *30* (6), 747–782

CBS (2006, Updated June 7, 2007). *Unlikely terrorists on no fly list*. Retrieved September 23, 2007, from http://www.cbsnews.com/stories/2006/10/05/60minutes/main2066624.shtml

Center for Economic and Policy Research (2006, Updated April 4, 2006). *Petition to save SIPP*. Retrieved September 20, 2007, from http://www.ceprdata.org/savesipp/orgletter-name.pdf

Chabrow, E. (2002). *Every move you make, every breath you take…* Retrieved September 20, 2007, from http://www.informationweek.com/story/IWK20020830S0027

Cloud, J. (2002). American cartographic transformations during the cold war. *Cartography and Geographic Information Science*, *29* (3), 261–282

Colb, S. F. (2001). *The new face of racial profiling: How terrorism affects the debate*. Retrieved September 20, 2007, from http://writ.findlaw.com/colb/20011010.html

Committee on Beyond Mapping (2006). *Beyond mapping. Meeting national needs through enhanced geographic information science*. (Washington DC: The National Academies Press)

Crampton, J. W. (forthcoming). Will peasants map? Hyperlinks, maps mashups and the future of information. (In J. Turow and L. Tsui (Eds.), *The hyperlinke society. Questioning connections in the digital age*. Ann Arbor, MI: University of Michigan Press)

Cutter, S. L., Richardson, D. B. & Wilbanks, T. J. (Eds.). (2002). *The geographical dimensions of terrorism: Action items and research priorities*. (Washington, DC: Association of American Geographers)

Cutter, S. L., Richardson, D. B. & Wilbanks, T. J. (Eds.). (2003). *The geographical dimensions of terrorism*. (London/New York: Routledge)

Cutter, S. L., Richardson, D. B. & Wilbanks, T. J. (2004). The geographical dimensions of terrorism: Future directions. *Annals of the Association of American Geographers*, *94* (4), 1001–1002

de Blij, H. J. (2004). Explicating geography's dimensions—an opportunity missed. *Annals of the Association of American Geographers*, *94* (4), 994–996

Elden, S. (2007). Rethinking governmentality. *Political Geography*, *26* (1), 29–33

ESRI (2002). GIS solutions for homeland security [CD-ROM]. Redlands, CA: ESRI.

Foucault, M. (1977). *Discipline and punish: The birth of the prison* (1st American ed.). (New York: Pantheon Books)

Foucault, M. (1978). *The history of sexuality* (1st American ed.). (New York: Pantheon Books)

Foucault, M. (1991). Governmentality. (In C. G. G. Burchell & P. Miller (Eds.), *The Foucault effect: Studies in governmentality* (pp. 87–104). Chicago, IL: University of Chicago Press)

Foucault, M. (2000a). 'Omnes et singulatim': Toward a critique of political reason. (In J. Faubion (Ed.), *Power. The essential works of Michel Foucault 1954–1984. Vol. 3* (pp. 298–325). New York: New Press)

Foucault, M. (2000b). Truth and juridical forms. (In J. Faubion (Ed.), *Power. The essential works of Michel Foucault 1954–1984. Vol. 3* (pp. 1–89). New York: New Press)

Gigerenzer, G. (2004). Dread risk, September 11, and fatal traffic accidents. *Psychological Science, 15* (4), 286–287

Gigerenzer, G. (2006). Out of the frying pan into the fire: Behavioral reactions to terrorist attacks. *Risk Analysis, 26* (2), 347–351

Godlewska, A. M. C. (1999). *Geography unbound. French geographic thought from Cassini to Humboldt.* (Chicago, IL: University of Chicago Press)

Gordon, C. (2000). Introduction. (In J. Faubion (Ed.), *Power, essential works of Foucault 1954–1984, Vol. III* (pp. 1–89). New York: New Press)

Graham, S. (1998). Spaces of surveillant simulation: New technologies, digital representations, and material geographies. *Environment and Planning D-Society & Space, 16* (4), 483–504

Gregory, D. & Pred, A. (2007). *Violent geographies.* (New York/London: Routledge)

Hacking, I. (1990). *The taming of chance.* (Cambridge, UK/New York: Cambridge University Press)

Hannah, M. (2000). *Governmentality and the mastery of territory in nineteenth-century America.* (Cambridge: Cambridge University Press)

Hannah, M. (2006). Torture and the ticking bomb: The war on terrorism as a geographical imagination of power/knowledge. *Annals of the Association of American Geographers, 96* (3), 622–640

Isikoff, M. (2003, February 3). The FBI says, count the mosques. *Newsweek*, p. 6

Jackson, J. B. (1972). *American space: The centennial years.* (New York: W.W. Norton)

Johnston, R. (2004). Geography, GIS, and terrorism. *Annals of the Association of American Geographers, 94* (4), 996–998

Kahneman, D. & Renshon, J. (2007). Why hawks win. *Foreign Policy, Jan/Feb* (158), 34–38

Klinkenberg, B. (2007). Geospatial technologies and the geographies of hope and fear. *Annals of the Association of American Geographers, 97* (2), 350–360

Larson, V. (2007, August 10). iSecureTrac loss grows larger. *Omaha World Herald*, p. 2D

Lawson, V. (2007). Introduction: Geographies of fear and hope. *Annals of the Association of American Geographers, 97* (2), 335–337

Legg, S. (2005). Foucault's population geographies: Classifications, biopolitics and governmental spaces. *Population, Space and Place, 11* (3), 137–156

Livingstone, D. N. (2003). *Putting science in its place: Geographies of scientific knowledge.* (Chicago, IL: University of Chicago Press)

Miller, C. C. (2006). A beast in the field: The Google maps mashup as GIS/2. *Cartographica, 41* (3), 187–199

Montello, D. R. & Sutton, P. (2006). *Introduction to scientific research methods in geography.* (Thousand Oaks, CA: Sage)

Murphy, A. B. (2006). Enhancing geography's role in public debate. *Annals of the Association of American Geographers, 96* (1), 1–13

Negroponte, J. D. (2006). *Remarks and Q&A by the director of national intelligence.* Retrieved February 14, 2007, from http://www.dni.gov/speeches/20061116_speech.pdf

O'Sullivan, D. (2006). Geographical information systems: Critical GIS. *Progress in Human Geography, 30* (6), 783–791

Paglen, T. & Thompson, A. C. (2006). *Torture taxi. On the trail of the CIA's rendition flights.* (Hoboken, NJ: Melville House Publishing)

Paullin, C. O. (1932). *Atlas of the historical geography of the United States.* (Washington, DC / New York City: American Geographical Society/Carnegie Institution)

Paulos, J. A. (1996). *A mathematician reads the newspapers*. (New York: Anchor Books)
Paulos, J. A. (2006a). *Of wiretaps, Google searches and handguns*. Retrieved February 14, 2007, from http://abcnews.go.com/Technology/WhosCounting/story?id=1560771
Paulos, J. A. (2006b). *Cheney's one percent doctrine*. Retrieved February 14, 2007, from http://abcnews.go.com/Technology/story?id=2120605&page=1
Perkins, C. (2004). Cartography—cultures of mapping: Power in practice. *Progress in Human Geography, 28* (3), 381–391
Risen, J. & Lichtblau, E. (2005, December 16). Bush lets US spy on callers without courts. *The New York Times*, p. A1
Robin, C. (2004). *Fear. The history of a political idea*. (Oxford: Oxford University Press)
Robinson, A. H. (1982). *Early thematic mapping in the history of cartography*. (Chicago, IL: University of Chicago Press)
Rosich, K. J. (2005). *A history of the American Sociological Association, 1981–2004*. (Washington, DC: American Sociological Association)
Rossmo, D. K. (2000). *Geographic profiling*. (Boca Raton, FL: CRC)
Schröder, W. (2006). GIS, geostatistics, metadata banking, and tree-based models for data analysis and mapping in environmental monitoring and epidemiology. *International Journal of Medical Microbiology, 296* (Suppl. 1), 23–36
Schuurman, N. & Kwan, M.-P. (2004). Guest editorial: Taking a walk on the social side of GIS. *Cartographica, 39* (1), 1–3
Schwartz, S. I. & Ehrenberg, R. E. (2001). *The mapping of America* (2nd. ed.). (Edison, NJ: Wellfleet)
Siegel, M. (2005). *False alarm. The truth about the epidemic of fear*. (Hoboken, NJ: Wiley)
Sparke, M. (2007). Geopolitical fears, geoeconomic hopes, and the responsibilities of geography. *Annals of the Association of American Geographers, 97* (2), 338–349
Stewart, D. J. (2005). Geography and the Middle East. *Geographical Review, 95* (3), iii–vi
Suskind, R. (2006). *The one percent doctrine*. (New York: Simon & Schuster)
Tversky, A. & Kahneman, D. (1974). Judgment under uncertainty: Heuristics and biases. *Science, 185* (September 27), 1124–1131
US Government Office of the President (2003). *Budget of the United States government, fiscal year 2004—appendix*. (Washington, DC: US Printing Office)
Walker, F. A. (1874). *Statistical atlas of the United States*. (New York: J. Bien)

Chapter 15
Mapping the Under-Scrutinized: The West German Census Boycott Movement of 1987 and the Dangers of Information-Based Security

Matthew Hannah

Abstract Geospatial technologies have been subjected to critique in geography and other fields over the past ten to fifteen years for their actual or potential complicity in providing knowledge for unjust regimes of control or illegitimate warfare. This chapter argues that in the atmosphere of dramatically intensified concern for 'security' since 9/11, one of the chief dangers of using geospatial technologies lies not in the knowledge they produce but rather in the ways they tend to transform a *lack of knowledge* into grounds for the withdrawal of rights from disadvantaged groups. Using some of Foucault's ideas on 'race war discourses,' I suggest that it thus makes sense to see the 'underscrutinized' as an emerging 'race'. The cultural context for this claim is set via a survey of stigmatizations of groups deemed 'inscrutable' or 'subversive' in US history. The bulk of the chapter is then devoted to setting a second, 'techno-political' context through an account of the nationwide census boycott movement in West Germany in 1987. This controversy from an earlier stage in the history of the information age illustrates one of the ways in which the inevitably uneven geographical coverage of a geospatial data set can lead to stigmatization and discrimination against the unregistered, even in the absence of any intent on the part of experts and state authorities.

Keywords Subversion profiling, census, West Germany, state of exception

15.1 Introduction

Critiques of GIS and other geospatial technologies have taken a number of angles. In his influential early study, J. Brian Harley highlighted the ways in which maps have encoded and reinforced colonial and other relations of domination for centuries (Harley 1989; see also Edney 1997; Clayton 2000). Censuses, another geospatial technology with a long and chequered history, have likewise been examined for the

University of Wales at Aberystwyth

ways in which they have served as instruments of power (Cohn 1998; Hannah 2000). Geographic information systems (GIS) and other digital technologies began to come under scrutiny in the late 1980s (Pickles 1995; Smith 1992). More recently, Crampton has explored the politics of cyberspace as a problem of mapping (Crampton 2003). Some of the main themes to emerge from these critiques have included: (1) the direct role of geographical data in establishing territorially-based administrative control of people, activities and resources ('power/knowledge') (Hannah 2000); (2) the role of maps, censuses, and digital files in constructing images of the world or parts of it that support particular geo- or domestic political or economic aims (e.g. of imperial control over 'India' (Edney 1997)); and (3) the illusion of objective detachment encouraged by abstract technical forms of knowledge-production, which makes it easier for technicians to engage in manipulations up to and including violent warfare (and for viewing audiences to watch over their shoulders) without either group having directly to confront the human destruction they produce (Smith 1992).

All of this work shares one common starting point: that it is the knowledge produced or represented through geospatial technologies that is the (potential) source of harm. Although the critics cited above would not argue that we should do away with GIS systems or censuses, one implication of their critiques is that if these various forms of knowledge were not available, the specific ills and injuries they analyze would not have occurred (or at least, would have been more difficult to cause with impunity). Operating with the same basic assumption, defenders of geospatial technologies have often argued that the data and the technologies are themselves neutral, and that the danger lies with the users and the contexts of use (Goodchild 1995). Geospatial knowledge is potentially beneficial as well as potentially harmful, in this line of thinking, so the important thing is to strive to ensure that it is used properly and ethically. Rather than either attack or defend geospatial knowledge, I propose in this chapter to shift the focus to the specific ways in which geospatial technologies produce a relative *lack of knowledge* about some individuals and groups. Further, I argue that in the present political climate, *not* being registered, *not* making it into certain databases, can have its own potentially quite profound negative consequences. Indeed, we may be witnessing the emergence of a new and very important social distinction between the normally visible and the 'underscrutinized.' In a 'state of exception' (Agamben 1998, 2005) characterized by a new intolerance for the risks associated with the unknown, those about whom less is known become highly vulnerable to discriminatory treatment and social stigmatization.

A number of scholars have begun to explore the impacts of post-9/11 discourses and technologies of 'security' upon inherited notions of citizenship and state power. Davina Bhandar (2004) fleshes out Dick Cheney's claim that North Americans will have to accommodate themselves to the 'new normal,' a form of political and social life reorganized around risk, loss, and trauma, and regulated more closely by technologies of security. In the same general line, Louise Amoore (2006) and Matt Sparke (2006) have both focused on the new border regimes being negotiated by the US and Canada. A central point made by both Amoore and Sparke is that management of cross-border flows will increasingly tend to resolve the tension between geoeconomic imperatives for unimpeded commerce and geopolitical pressures to restrict the movement of dangerous bodies, by establishing divergent mobilities and degrees of access to rights for different classes of people. In this sense Sparke (2006: 168–169) differentiates between 'kinetic elites' and an emerging 'kinetic underclass.' Systems for screening people and goods designed on the EZ-Pass principle (streamlined movement for those willing to be subjected to pre-screening vs. delays and increased manual scrutiny for those not) figure heavily, for example, in the security recommendations of Stephen Flynn, a homeland security specialist much quoted in recent years (Flynn 2004). I would like to suggest that it makes sense to view the underscrutinized as an emerging 'race.' To explain the specific sense in which the term 'race' should be understood here, it is worth a brief detour through some of Michel Foucault's recently published writings on power relations.

15.2 'Race' and Race-War Discourses

In his lectures at the Collège de France from 1975–1976 (published in English as *Society must be defended* (Foucault 2003), Foucault identified a form of discourse winding through European political thought since the end of the 16th century which constituted an entirely new departure from the long-standing 'philosophico-juridical' discourse of politics reaching back to the ancients. This new discourse was fundamentally about war and 'race' as keys to understanding what lay behind the pacification of domestic relations that seemed to characterize the emergence of identifiably modern European societies. Foucault argues that by the late16th century in many parts of Europe, 'the entire social body was cleansed of the [openly] bellicose relations that had permeated it through and through during the Middle Ages' (Foucault 2003: 48). However,

> [w]hen war was expelled to the limits of the State, or was both centralized in practice and confined to the frontier, a certain discourse appeared. A new discourse, a strange discourse. It was new, first, because it was, I think, the first historico-political discourse on society, and it was very different from the philosophico-juridical discourse that had been habitually spoken until then. And the historico-political discourse that appeared at this moment was also a discourse on war, which was understood to be a permanent social relationship, the ineradicable basis of all institutions and relations of power (Foucault 2003: 49).

What made this discourse 'historico-political' rather than 'philosophico-juricidal' was that it understood all proposed political truths (including its own) as partisan, as intimately involved in the history and politics they were seeking to explain, rather than as matters of transhistorical 'law,' as abstract and universal, 'above the fray.' The various writers advancing this discourse were, according to Foucault, 'at once obscure and heterogeneous,' and their writings were usually linked to specific social struggles, whether the English bourgeois revolution, struggles of the French aristocracy against the new absolutist-administrative monarchy at the end of the 17th century, or the eugenics movement in the decades around 1900 (Foucault 2003: 49–50). These political theorists conceived history first and foremost as a matter of 'physical strength, force, energy, the proliferation of one race, the weakness of the other, [...] defeats, victories, the failure or success of rebellions, the failure or success of conspiracies or alliances' (Foucault 2003: 52–54). If history is made thus, any existing, apparently peaceful social order is but a fragile and perhaps transient surface phenomenon. This kind of discourse 'is interested in rediscovering the blood that has dried in the codes. .... It is interested in the battle cries that can be heard beneath the formulas of right, in the dissymmetry of forces that lies beneath the equilibrium of justice' (Foucault 2003: 56).

Foucault argues that the specific form attributed by such discourse, from the 17th century onwards, to the battles silently suspended under the forms of social peace, is that of *race war*, with 'race' understood broadly to encompass 'ethnic differences, differences between languages, different degrees of force, vigor, energy and violence; the differences between savagery and barbarism; the conquest and subjugation of one race by another' (Foucault 2003: 60). From the early 19th century, this fundamentally racial way of understanding social divisions begins to be invested with the vocabulary, on the one hand, of biology, and on the other, of class[1] (Foucault 2003: 60). Whether in the form of theories of inferior vs. superior 'national stock,' of class antagonism between owners and wage laborers, or of the struggle against 'radicals,' this discourse of race war has consistently constructed modern Western societies as internally divided along lines that, on the one hand, clearly separate the normal from the threatening, but on the other hand, blur biological, ethnic, economic and political categories in stigmatizing the internal 'enemy.'[2] Following Foucault, then I will use the term 'race' to refer to a set of (usually negative) characteristics attributed to or imposed upon one group of people by another in the context of a struggle or conflict. 'Race' in other words, is not a 'natural' category but a politically constructed one. To be sure, it has historically been used chiefly to refer to groups distinguishable from other groups by visible

---

[1] The assertion that accounts of class struggle have always *essentially* been versions of 'race war discourse' is of course contestable. My argument here rests only on the more cautious assumption that the theme of class and economic struggle *can* be (and periodically has been) rendered in the race war idiom, in which economic themes have been entangled with biological, political and ethnic/racial themes. Foucault does not systematically pursue the problem of the ultimate nature and status of class discourse in his lecture course.

[2] A number of commentators on Foucault have interpreted his lectures on the 'race war' tradition as comprising an 'experiment' with a particular model of what it is that sets distinctively modern power relations apart from earlier political forms, an experiment he subsequently abandoned, so the account goes, in favor of analyses of liberal governmentality (Dean 1999: 25; Hindess 1996: 98–104). The tendency in these accounts of Foucault's thinking on power is to infer from his turn toward a genealogy of liberalism a devaluation of the race war discourse as a basis for historical analysis. However, it would be a grave mistake to suppose that because he ceased to pursue this strand of political discourse, his 1975/76 lectures are somehow a false or inaccurate representation of the specific writings Foucault surveys, or that a shift in Foucault's interest amounts to a declaration that the race war tradition was in fact historically insignificant. Especially since the renewed focus on terrorism and the resurgence of xenophobia subsequent to 9/11, it seems wiser to think of the race war discourse as one that has waxed and waned in its importance relative to the liberal discourse with which it has co-existed in various constellations for the past two hundred years.

traits. But this has not always been the case, and indeed the construction of certain groups as 'inscrutable' (see below) can be seen in part as an unintended admission of the futility of inferring anything about people from their external appearance.

Why is it necessary to use this particular, Foucaultian definition of 'race' and 'race war discourses,' rather than just a more common-sense, everyday set of concepts? First, Foucault's relatively expansive sense of 'race' is useful precisely because it allows us to recognize the common threads linking seemingly disparate episodes of stigmatization. The theme of *inscrutability* that forms the context for my argument below about geospatial technologies would otherwise be difficult to recognize as a matter of 'race,' because it has been applied variously to politically- as well as 'biologically'-defined groups throughout American history. But this distinction between political and biological is inadequate to capture a way of portraying 'enemies' that often involves stigmatizing political opponents with biological language (disease, infestation, etc.) or characterizing racial others as threats to the 'nation.' Race war discourses, in other words, are inextricably *bio-political*. Second, and no less important, the *partisan* character of race war discourses highlighted by Foucault makes visible another key continuity from the 19th century to the present day: in US race war discourses, as in the European instances reviewed by Foucault, there has always been a sense of *urgency and immediacy* of the threat posed by racialized 'enemies within.' Not at all detached in their attitude, a basic characteristic of these discourses has been the strident call for immediate action *even if there is a risk of trampling on democratic values or constitutional rights otherwise held dear*. Racial enemies in the Foucaultian sense, in other words, have consistently been seen to threaten the very foundations and conditions of possibility of 'free democracy.' Thus the measures to be taken against them, so the argument goes, cannot necessarily be limited by scrupulous attention to rights. Together, these two special features of the Foucaultian account of race, namely its breadth of definition, which allows us to recognize hybrid, 'racial-political' themes like inscrutability, and the partisan association of race war discourses with calls for urgent and often undemocratic remedies, make it possible to outline much more precisely the specific historical context for understanding how digital technologies and sources of data could be involved in constructing something like a race of underscrutinized persons.

15.3 Cultural Contexts: Inscrutability, Deviousness and Subversion in US Political Discourse

Political discussion in the United States has long proven fertile ground for various modulations of race war discourse. In his influential essay 'The Paranoid Style in American Politics,' historian Richard Hofstadter focuses chiefly on conspiracy theories of the 18th and 19th centuries. According to Hofstadter, early nativist tracts against the Bavarian Illuminati, the Masons, and the Jesuits were the forerunners of mid-20th century anti-communist hysteria (Hofstadter 1965). But the paranoid style has also been strongly evidenced in overtly racist discourses, and Foucault's notion of race war makes it possible to tie the religious and political examples surveyed by Hofstadter to very similar racist narratives. Because of its history as a nation of immigrants, the United States has tended to germinate race war discourses heavily focused on immigrants and their descendants, or less specific but potentially subversive 'foreign influences,' as the dangerous 'other' to be eliminated (or assimilated). Chris Philo notes that in the European examples Foucault surveys, race war discourses were chiefly the work of members of groups outside the main corridors of power, of those, in other words, who had a direct interest in re-discovering the 'blood dried in the codes' (Philo 2007: 354–358). But as Foucault makes clear in the later lectures of *Society must be defended*, during the late nineteenth century such race war discourses were adopted by many European and North American states, and were assimilated into dominant discourses of governance (Foucault 2003). In the US, this historical pattern certainly held. While the more extreme demands of race war theorists have only sporadically gained the upper hand in policy circles, a strong current of nativism/anticommunism has coursed through much of mainstream US political culture since the earliest years of the republic, making the periodic ascension to dominance of broadly racist ideas relatively easy.[3] Space is lacking here to attempt even a partial genealogy of the whole range of these discourses, but for the sake of broad context it is worthwhile at least explicitly listing those groups most prominently constructed as 'subversive' and 'inscrutable' in the US since the early 19th century:

- Early Chinese immigrants seen to pose a 'yellow peril' (Daniels 1988: 39–43; Gyory 1998: 10, 18; Saxton 1975; Takaki 1989)
- Immigrants from Eastern and Central Europe suspected of importing communism and socialism (Higham 1994: Chs. 8–10; Morgan 2003: 3–87)
- Japanese-Americans on the West Coast during World War II (Drinnon 1987; Takaki 1989: 379–405)
- Real or imagined communists during the McCarthy era of the early Cold War (Morgan 2003: 361, 371)

Again, although they were occasioned by quite unrelated events, it makes sense to see these various recrudescences of paranoid narratives of national peril as episodes in a single tradition. The themes of inscrutability, deviousness, and subversion lend them substantive coherence around a core link between *lack of knowledge* and

[3] The question of what sorts of texts are the vehicles of a 'race war discourse' is potentially quite complicated. In his lecture course, Foucault refers chiefly to lengthier treatises that might be characterized as 'learned' rather than 'popular', in the sense that they involve an effort to develop an argument at some length. These texts were written, as he stresses, in support of partisan goals, but they were not simply screeds for mass audiences or exercises in sloganeering. The texts that have made up the US race-war discourses sketched here have been much more heterogeneous in terms of genre, encompassing books and other learned works as well as newspaper columns, pamphlets, handbills, flyers and posters. However, like the earlier European examples of concern to Foucault, they are tied together into a coherent discourse by a readily identifiable complex of themes, concepts and biologistic images.

grave threat. This link took the form of *profiling*, which involves a single identifying trait being taken (in the absence of any specific knowledge) as an indicator of other projected or presumed characteristics. In each instance the threat posed by profiled enemies was seen to usher in a 'state of exception' in which fear-mongers felt safe demanding immediate state actions that would have been far more controversial had they been directed at 'normal' members of American society. At the same time it is important to recognize that the specifics of the race war discourse have always mutated and shifted from one period to the next, and that there were very specific constellations of biological, political, and economic rhetoric involved in each instance. In the post-9/11 US, for example, the specific mix has shifted to the cultural referent of (radical) Islam and the specific political threat of terrorism. Traditional, biological notions of race have given way to a more cultural/ethnic narrative of the 'clash of civilizations' pitting Islam against 'the West' (Huntington 1996). The core of my argument in this chapter regards a second key novelty of the new race war discourse: the central importance of *technical means of surveillance* in constructing the subversive racial 'other.' Digital geospatial technologies are in danger of playing an important role in perpetuating (or better, updating) this tradition. In the final section, I will explain this by contrasting profiling to what I call 'no-filing.' But first, to grasp this specifically technical element of the current racialization of the underscrutinized, it will help to shift to a different cultural and political context: West Germany in the 1980s. At an earlier stage of the rapidly advancing information age, some of the technological capabilities we now tend to take for granted were more often the subject of public debate, and not just in West Germany. However, the political ramifications of geospatial technologies were debated at unusual length there in the context of a nationwide boycott movement against the federal census. After explaining how this episode sheds light on the *technological* construction of the underscrutinized as a racial category, I will return to the context of the US to tie the argument together.

15.4 Techno-Political Context: The West German Census Boycott Movement of 1987[4]

The boycott movement of 1987 was actually preceded in 1983 by a smaller but still quite widespread boycott movement. Many West Germans, especially those under 30 years of age, were already sensitized to governmental snooping by the

[4] The account offered here of West German census boycott movements is compiled from media reports in major German newspapers and periodicals of the period (*Franfurter Allgemeine Zeitung, Frankfurter Rundschau, Die Tageszeitung, Die Zeit, Der Spiegel, Handelsblatt, Süddeutsche Zeitung*) as well as other local, regional and organizationally specific publications. Funding for this research was generously provided by the Alexander von Humboldt Foundation (Bonn). The interpretations and conclusions offered here do not reflect its views in any way. Unless otherwise noted, all translations from the German are by the author.

West German state's heavy-handed security response to the domestic terrorism of the 1970s and early 1980s. The peace and environmental movements were both very strong and active in the early 1980s, and as a result many otherwise non-activist citizens had had first-hand experience of clashes with police at demonstrations (Markovitz and Gorski 1993). The political atmosphere was thus generally tense. The more immediate trigger for the first boycott movement was the census-enabling law of 1982, passages of which would have allowed *individual* information to be shared immediately with other local and state agencies, rather than being off-limits for decades, as is the practice in most modern democracies. Provoked by the census law and already distrustful of the State, many West Germans came to see the census as an Orwellian instrument of surveillance (the impending year 1984 provided ready reference). A grass-roots movement sprang up in the early months of 1983, and soon hundreds of citizens' initiatives against the census had formed throughout the Federal Republic. Conferences, information meetings, and publicity stands on sidewalks and in parks brought the message to an ever-larger number of West Germans, and the mainstream press began to debate the pros and cons of the enumeration. Surveys in late March of 1983 showed that approximately one quarter of West Germans were at least contemplating joining the boycott. But in April 1983, before the census could be taken and the degree of resistance among the population put to the test, the Federal Constitutional Court surprised everyone by voiding parts of the 1982 census law and sending the government back to the drawing board. After a long period of revision and inter-party wrangling over dates, the Kohl government was again prepared to count the people in May of 1987. This time the census was actually taken, but it was very controversial, and many tens of thousands initially refused to answer it. The legal fallout lasted for years, and the accuracy of the final totals remained in dispute.

15.4.1 Maps of 'Resistance Potential?'

The lion's share of rhetoric, analysis and activism against the census in both 1983 and 1987 centered on the purportedly sinister uses to which the state might put the information. Not only fears of Orwell's fictional *1984* scenario but also the very real and deadly role played by censuses and other registrations in Nazi attempts to murder Europe's Jews and other 'undesirables' (Aly and Roth 2004 [1983]) were discussed at great length. Additionally, the census was placed by its critics in the more immediate context of state security measures such as video surveillance of cities and electronic data searches (*Rasterfahndungen*). In 1983, coordination between local boycott groups remained incomplete, so no unified boycott strategy emerged. By 1987, the Green Party, which had gained seats in the federal *Bundestag* for the first time right in the midst of the 1983 boycott, had had four years to establish its national presence and build its organizational capacities. Data protection was one of the Greens' signature issues, and in 1987, they played a crucial role in mobilizing and organizing the new boycott.

One of the results of Green Party leadership was a unified national boycott strategy, the so-called 'hard' boycott. The hard boycott called for individuals to leave their census schedules blank, snip off the bar codes that linked the forms to specific addresses, and return the blank, anonymized forms to 'alternative collection points' where volunteers would tally incoming blank forms, report updated numbers to the central Bonn office of the Green Party for publication, and then return the blank forms to local census offices or for 'recycling.' The chief political advantage of this form of boycott, as against the 'soft' approach (filling out forms incompletely or with false information), was that the boycott movement could publicize running national totals of boycotters without exposing them individually to State sanctions. There was, after all, a fine for refusing to fill out the census form. The federal government did not want to give the impression of being too heavy-handed, but, as it turned out, the authorities were nevertheless willing to pursue individual boycotters with fines running into the hundreds, sometimes thousands, of D-Marks. By participating in the hard boycott, individual West Germans retained some deniability, since all the authorities would know was that they had not received a form back from such-and-such an address. If the local census officials were so zealous as to pursue them with determination, boycotters could always relent later, fill out a form, and thereby avoid all penalties.

This scenario hints already at the fundamental vulnerability of boycotters, and raises an issue that went largely unnoticed in 1983 but became central in 1987, when the census was actually taken. It is here also that we begin to see how geospatial technologies can help stigmatize people based not on information but on a *lack of information*. Back in late March of 1983, in the midst of a long feature story in the influential newsweekly *Der Spiegel* on the first census boycott, one of the *Spiegel* editors, Jochen Bölsche, made an interesting suggestion. His reasoning is worth quoting at length:

> Already in 1984 the Federal Republic [will be] the first state in human history to have registered almost the entire protest potential in the population, with names and information as to the degree of individual willingness to resist—not despite but precisely because of the growth of the boycott movement. The more respondents refuse [to provide] answers, the greater the chance that the statistical value of the census will approach zero, but also the more comprehensive will be the registration of political deviants [*Abweichlern*], [and thus] the greater the danger of abuse of this most sensitive of all data sources. If the boycott initiators succeed in totally mobilizing the oppositional potential of the country, the dissident file would be complete. ... No initiative group explains that whoever gives an empty census form back does not escape registration but instead simply lands in another file. (Bölsche 1983)

Bölsche's sinister-sounding scenario was not taken up for sustained discussion in the press in 1983, and because the 1983 census never took place, it remained a moot issue. However, events in 1987, when the census actually was taken, would prove him at least modestly prophetic.

15.4.2 Dissidence Maps by Default

The focus of debate in 1987 remained the danger to privacy and political freedom attendant upon the state's possession of personal information. There was ample evidence, from the perspective of boycotters, that the state's possession of knowledge was indeed the main thing to worry about. Responsibility for carrying out the May 1987 census lay with local officials and state (*Land*) governments, and there was a great deal of geographical variation in the zeal and determination with which these officials pursued missing census forms. In those places and regions where governments insisted strongly on the duty to respond to the census (for example, the southern states of Bavaria and Baden-Württemberg, but also West Berlin), police investigators attended boycott meetings with cameras and tape recorders, and levied fines against organizers and speakers. Eventually it was revealed that some boycott leaders had been entered into a national data file (APIS) created to track known or suspected terrorists. The resulting public outcry, and vigorous objections by some state Data Protection Commissioners, led to most of these names being removed.

But in the longer term, as the number of West Germans not intimidated or fined into returning their forms steadily dwindled, census authorities increasingly treated the residual un-registered as 'hard nuts,' as presumptive boycotters rather than as people who might have been missed due to obsolete address data, moving, or some other innocent circumstance common to census-taking. Some states, for example Bremen, would essentially grant amnesty to those still not counted months after the official close of the census. But in West Berlin or Baden-Württemberg, individual cases dragged on into 1988 and even beyond the fall of the Berlin Wall in 1989. The organizers of the 1987 boycott movement tended to agree in retrospect that they had failed to anticipate this outcome. One of their oft-repeated reassurances to West Germans worried about facing fines had been that the court system would never be able to handle hundreds of thousands or millions of individual appeals and would thus be forced to dismiss all the census cases. But because the census enumeration occurred at such an uneven pace at all scales throughout West Germany, most local and state court systems were able to keep on top of the ballooning case-loads (sometimes by developing standard decision templates, or by shifting personnel and resources from other tasks).[5]

The result of the gradual attrition of the boycott movement was not an official 'dissident file' in the literal sense, but a set of local and state-wide comprehensive address files containing information on the uneven geography of census responses. In the hands of authoritarian officials, these files could indeed serve to pinpoint actual or potential 'trouble spots' or 'problem areas.' As the chief Green Party boycott

---

[5] The court systems in West Germany had a reputation throughout the post-WWII decades for being a stronghold of right-wing, pro-state authoritarianism as a result of continuities of personnel from the Nazi years (von Miquel 2001).

strategist Roland Appel put it in a June 1988 interview in the *Frankfurter Rundschau*, these databases, set up originally only for the temporary logistical purpose of census-taking, could easily be used in other ways: 'With installations like a nuclear re-processing plant, [the state] will want to know at which locations the least resistance is to be expected from the population' (Appel, in Hebel 1988). Though no evidence has since turned up linking subsequent police investigations to the left-over census address files, this scenario was not a wild fantasy. Many state officials were convinced from the beginning of the whole boycott controversy that the organizers and initiators were the same people who threw rocks and Molotov cocktails at their police forces during public demonstrations, or blocked roads and other public facilities to protest atomic energy, war, or destruction of the environment. The Greens, who played such a prominent role in the 1987 boycott, were regularly stigmatized as an 'anti-constitutional' party tainted by communism and bent on destroying the West German political and economic system. Thus, if boycotters were not literally terrorists, they were seen by many authorities as members of a group for whom the spectrum between peaceful civil disobedience and assassinations or bombings was a very slippery slope (Balistier 1996: 253–257). They were, in short, members of the 'race' of radical subversives, a group discursively constructed and treated by conservatives very much in the 'race war' tradition outlined by Foucault. And the possibility that the state would actually disadvantage those individuals perceived as radicals was real: a 'Decree against radicals' had been in force since 1972, and had led in the intervening years to the exclusion of thousands of West Germans from state employment because of their prior affiliations with banned political organizations (Braunthal 1990). The political climate of the 1980s, in other words, meant that *not having been counted* in the census could indeed lead to a form of political stigmatization with potentially real negative consequences.

15.5 Conclusions: The Dangers of Profiling vs. 'No-Filing' with Geospatial Technologies

The story of the West German census boycott movements is of course extremely complicated, and not unproblematically transferable to the post-9/11 United States. But the sketch given above should be sufficient to pinpoint one specific danger of using geospatial technologies in the service of an aggressive security campaign during times of heightened fear of subversion. If we take the census as a stand-in for the wider range of geospatial technologies that in some way allow the gathering, storage, and processing of geographically indexed data about identities, locations, or movements of people, the general problem is one of the stigmatization and unequal treatment of those individuals who for whatever reason are noticeably absent from, or 'under-represented' in, security databases. Over the longer haul, the amount of information held on individuals will explode, and the processing power that can be brought to bear on these ever vaster databases will likewise grow by leaps and bounds. One effect of this explosion will be an increase in the ease with which *differences in the amount of information held on different people can itself become an important piece of information about them.*

The West German boycott controversies were about a deliberate political campaign, whereas the underscrutinized in today's databases are not (yet) generally presumed to be hiding anything. However, the consequences of the lack of knowledge about them may ultimately render the difference irrelevant. If, as Stephen Flynn urges, the 'E-Z Pass' principle becomes a general model of operation (see the introductory section above), those who cannot afford the technology or simply do not want to be subject to background checks will become the focus of much more intensive searches and delays, as the automated passage of the 'kinetic elite' frees up personnel for more searching scrutiny of the 'kinetic underclass' (Sparke 2006). This latter group will, in other words, become suspicious on the merely technical grounds that less is known about them. At present, the principle of pre-emptive surrender of information in exchange for privileges is being applied chiefly at national borders as a system for monitoring movement (Sparke 2006; Amoore 2006). But it could as easily be extended to the internal human geographies of national territories, and refined in the direction of far more complex digitally mediated systems of differential rights of access to a whole range of public and quasi-public spaces than we have today (see Mitchell 2003 for a sense of the issues that would be sharpened in such a scenario, and Weizman 2007 for a chilling analysis of Israeli government innovations in this direction). Sparke is right to caution against lumping together those who are marginally inconvenienced (for example, academics crossing borders to attend conferences) and those who are truly disadvantaged by such technologies (undocumented workers trying to get to places of employment) (Sparke 2006: 169). But I would suggest that in an atmosphere of fear regularly bordering on paranoia and stoked by ongoing race war discourses of 'terrorists in our midst,' the projection of suspicion and disadvantage onto the underscrutinized will tend to widen and deepen. I have argued elsewhere that the discursive and institutional construction of terrorist threats since 9/11 has dramatically expanded the category of unacceptable risks (Hannah 2006). The more the risks posed by terrorism are viewed as unacceptable, the less willing authorities may be to tolerate lingering blind spots in official knowledge, and hence the more likely it becomes that a simple lack of knowledge about a person is seen as grounds for suspicion.

The role of geospatial technologies in all of this will be to lend precision, and perhaps a degree of automation, to the emerging practice of 'no-filing.' The more familiar practice of racial profiling will continue to be a major threat to the civil rights of Muslim-Americans and other groups construed as suspicious, and here, too, geospatial technologies may play a harmful role (Murray 2004). But profiling and 'no-filing' are subtly different. Profiling, again, is a matter of extrapolating from one or a few visible or obvious traits (e.g. skin color, surnames, etc.), to negative stereotypes about the group identified by those traits, and from there to discriminatory actions. 'No-filing,' by contrast, would involve discriminatory action based directly and simply upon a *lack* of information. No-filing, in other words, dispenses with or short-circuits stereotypes altogether, or more precisely, transforms a lack of information into the basis for a new sort of stereotype. Japanese-Americans and suspected communists were persecuted and robbed of rights in the past because inscrutability, deviousness, or subversion was assumed to be a substantive characteristic shared by members of these groups. Now we may be witnessing the emergence of a form of disadvantage that is more purely technical: if we have less information on an individual, that in itself is a reason for suspicion and discriminatory treatment, regardless of who the person is. Users and designers of geospatial technologies should keep this possibility in mind along with the more familiar dangers of providing knowledge (giving truth) to power.

References

Agamben, G. (1998). *Homo sacer: Sovereign power and bare life*. Trans. D. Heller-Roazen. (Palo Alto, CA: Stanford University Press)

Agamben, G. (2005). *State of exception*. Trans. K. Attell. (Chicago, IL: University of Chicago Press)

Aly, G. & Roth, K.-H. (2004 [1983]). *The Nazi census: Identification and control in the Third Reich*. Trans. E. Black. (Philadelphia, PA: Temple University Press)

Amoore, L. (2006). Biometric borders: Governing mobilities in the war on terror. *Political geography*, 25, 336–351

Balistier, T. (1996). *Straßenprotest: Formen oppositioneller politik in der Bundesrepublik Deutschland*. (Münster: Verlag Westfälisches Dampfboot)

Bhandar, D. (2004). Renormalizing citizenship and life in Fortress North America. *Citizenship studies*, 8(3), 261–278

Bölsche, J. (1983, March 28). Der Ge_lerhut des Orwell-Jahrzehnts. *Der Spiegel*, 47

Braunthal, G. (1990). *Political loyalty & public service in West Germany: The 1972 Decree against radicals and its consequences*. (Amherst, MA: University of Massachusetts Press)

Clayton, D. (2000). *Islands of truth: The imperial fashioning of Vancouver Island*. (Vancouver: UBC Press)

Cohn, B. (1998). The census, social structure and objectification in South Asia. In B. Cohn (Ed.), *An anthropologist among the historians and other essays*. (Oxford: Oxford University Press)

Crampton, J. (2003). *The political mapping of cyberspace*. (Chicago, IL: University of Chicago Press)

Daniels, R. (1988). *Asian America: Chinese and Japanese in the United States since 1850*. (Seattle: University of Washington Press)

Dean, M. (1999). *Governmentality: Power and rule in modern society*. (Thousand Oaks, CA: Sage)

Drinnon, R. (1987). *Keeper of concentration camps: Dillon S. Myer and American racism*. (Berkeley, CA: University of California Press)

Edney, M. (1997). *Mapping an empire: The geographical construction of British India, 1765–1843*. (Chicago, IL: University of Chicago Press)

Flynn, S. (2004). *America the vulnerable: How our government is failing to protect us from terrorism*. (New York: HarperCollins)

Foucault, M. (2003 [1997]). *'Society must be defended': Lectures at the Collège de France, 1975–1976*. Trans. D. Macey. (New York: Picador)

Goodchild, M. (1995). GIS and geographic research. In J. Pickles (Ed.), *Ground truth: The social implications of Geographical Information Systems* (pp. 31–50). (New York: Guilford Press)

Gyory, A. (1998). *Closing the gate: Race, politics, and the Chinese Exclusion Act*. (Chapel Hill, NC: University of North Carolina Press)

Hannah, M. (2000). *Governmentality and the Mastery of Territory in Nineteenth Century America.* (New York: Cambridge University Press)

Hannah, M. (2006). Torture and the ticking bomb: the "war on terrorism" as a geographical imagination of power / knowledge. *Annals of the Association of American Geographers 96,* 622–640

Harley, J.B. (1989). Deconstructing the map. *Cartographica,* 26, 1–20

Hebel, S. (1988, June 14). Im Gespräch: Roland Appel. *Frankfurter Rundschau,* 10

Higham, J. (1994). *Strangers in the land: Patterns of American nativism, 1860–1925.* (New Brunswick, NJ: Rutgers University Press)

Hindess, B. (1996). *Discourses of power from Hobbes to Foucault.* (Malden, MA: Blackwell)

Hofstadter, R. (1965). *The paranoid style in American politics, and other essays.* (New York: Alfred A. Knopf)

Huntington, S. (1996). *The clash of civilizations and the remaking of world order.* (New York: Touchstone)

Markovitz, A. & Gorski, P. (1993). *The German left: Red, green and beyond.* (Oxford: Blackwell)

Mitchell, D. (2003). *The Right to the city: Social justice and the fight for public space.* (New York: Guilford Press)

Morgan, T. (2003). *Reds: McCarthyism in twentieth-century America.* (New York: Random House)

Moss, G. (2006). *Vietnam: An American ordeal.* 5th Ed. (Upper Saddle River, NJ: Prentice-Hall)

Murray. N. (2004). Profiled: Arabs, Muslims, and the post-9/11 hunt for the "enemy within". In E. Hagopian (Ed.), *Civil rights in peril: The targeting of Arabs and Muslims* (pp. 27–68). (Chicago, IL: Haymarket Books)

Philo, C. (2007). 'Bellicose history' and 'local discursivities': An archaeological reading of Michel Foucault's *Society must be defended.*' In J. Crampton & S. Elden (Eds.), *Space, knowledge and power: Foucault and geography* (pp. 341–367). (Burlington, VT: Ashgate)

Pickles, J., Ed. (1995). *Ground truth: The social implications of geographic information systems.* (New York: Guilford Press)

Powers, T. (1973). *The war at home: Vietnam and the American people, 1964–1968.* (New York: Grossman Publishers)

Saxton, A. (1975). *The indispensable enemy: Labor and the anti-Chinese movement in California.* (Berkeley, CA: University of California Press)

Smith, N. (1992). History and philosophy of geography – Real wars, theory wars. *Progress in human geography,* 16(2), 257–271

Sparke, M. (2006). A neoliberal nexus: Economy, security and the biopolitics of citizenship on the border. *Political geography,* 25, 151–180

Takaki, R. (1989). *Strangers from a different shore: A history of Asian Americans.* (New York: Penguin Books)

United States Department of the Army (1976). *The My Lai massacre and its cover-up: Beyond the reach of law?* (New York: Free Press)

von Miquel, M. (2001). Juristen: Richter in eigener Sache. In N. Frei (Ed.), *Karrieren im Zwielicht* (pp. 181–240). (Frankfurt: Campus Verlag)

Weizman, E. (2007). *Hollow land: Israel's architecture of occupation.* (New York: Verso)

Chapter 16
The Importance of Spatial Thinking in an Uncertain World

Robert S. Bednarz[1] **and Sarah W. Bednarz**[1]

Abstract We live in uncertain times. Although we cannot eliminate uncertainty and its effects, it is important to minimize the disruption and loss that result from it. Mitigating the negative effects of uncertainty, especially by applying geospatial technologies, requires spatial thinking skills. We argue that teaching students how to use geospatial technologies will not enable them to deal with uncertainty unless they also learn to think spatially. Spatial thinking can be learned and should be taught. Results from classroom-based research provide guidance in developing effective ways to teach spatial thinking and geospatial technologies.

Keywords Geospatial technologies education, GK-12 Program, science education, spatial thinking

16.1 Uncertainty, Spatial Thinking and Geo-Spatial Technologies

Every day we are confronted with new information about climate change, terrorism, armed conflict, and globalization in its many manifestations. Although it is probably true that every generation perceived the times in which they lived as uncertain, a convincing case can be made for the view that we are living in a world with more uncertainty than ever before. Populations feel vulnerable from threats that did not exist in the past, or if they did, went unnoticed by the vast majority of the world's population. While the perception of vulnerability and risk has increased, it is also true that new technologies have arisen to help us manage uncertainty, and the data on which these new technologies depend have never been more accessible. Geographic information systems (GIS), remote sensing (RS), and Global Positioning Systems (GPS) can play a vital role in helping us cope with our uncertain world. It is our contention, however, that in order for individuals to exploit these new tools and databases effectively, they must possess appropriate spatial thinking skills.

[1] Texas A&M University

D.Z. Sui (ed.) *Geospatial Technologies and Homeland Security*,

The educational system can and should consider its role in helping students to understand the patterns and processes that shape global patterns and to cope with increasing uncertainty. If this consideration is to be effective, it must focus on the fundamental role of spatial thinking in geography and other sciences and recent insights into the teaching/learning process gained by the learning sciences. Spatial thinking comprises the knowledge, skills, and habits of mind to use concepts of space, tools of representation, and reasoning processes to structure, solve, and to express solutions to problems. Spatial thinking underlies a significant amount of geographic learning such as the use of maps, graphs, images, diagrams, models, and visualizations. In addition, it supports the description, explanation, and discussion of the functions, structures, relationships, and operations of a wide variety of spatio-temporal processes. Thus the ability to think spatially is a prerequisite for using and understanding the geospatial technologies commonly used in geography, other disciplines, and everyday life.

In this chapter, we argue that teaching students how to use geospatial technologies will not enable them to deal with uncertainty unless they also learn to think spatially. We assert that spatial thinking skills are relevant to a wide variety of subject matter, but that spatial thinking does not develop automatically nor do spatial skills easily transfer from one context to another. We propose an educational system that purposely promotes spatial thinking, supported, when appropriate, by geospatial technologies; one that also supports both the short- and long-term decision-making required of citizens living in an uncertain world.

We use the term 'uncertainty' to indicate what many feel is a decreasing level of stability in their day-to-day lives, an inability to predict the course of events, or a difficulty in understanding the causes and effects of events that often occur at a non-local scale. It is arguable whether we do, in fact, live in more uncertain times or whether people only perceive our times to be more uncertain. Perhaps this perception is a result of advanced technology that has produced very complex systems that are more difficult to understand and control, or because our global inter-relations have made us more dependent on others who live and work halfway around the globe, or because the increased quality and quantity of accessible information have made people more aware of the negative impacts that unpredictable events can produce. The authors of this chapter are not experts on the literature of uncertainty or emergency management. Our contribution to this discussion is from the point of view of geographers interested in how what we know (and are learning) about spatial thinking can help both the young and old cope with the uncertainty they perceive characterizes their lives. We argue that spatial thinking, which has always been important, is even more valuable today because of the widespread use of new geospatial technologies. (Geo)graphical representations are now more common than ever before. To use such tools effectively, citizens, both young and old, must be able to think spatially. To some extent our belief is exemplified by a recent statement made by the executive director of the Association of American Geographers: 'Geography helps us to understand and enhance our own communities as American citizens—and informs our understanding of the challenges facing the United States in an uncertain world' (Richardson 2007: 5).

This article begins by describing aspects of uncertainty and linking this concept to two scenarios from the recent past that illustrate the importance of spatial thinking when coping with uncertainty. Next we describe spatial thinking in greater detail, with emphasis on its relevance to those who would use geospatial technologies to mitigate the negative effects of uncertainty. A short discussion of some preliminary results of *Advancing Geospatial Skills in Science and Social Science* (AGSSS), a program to introduce spatial thinking into science and geography classrooms, precedes a few recommendations that conclude this chapter.

16.1.1 Scenarios of Uncertainty and Spatial Thinking

All of us face uncertainty at a local, personal scale frequently during our daily lives. We deal with uncertainty when we deal with the weather. For example, we must decide whether to take our umbrella with us when we leave our house or workplace. The effects of the uncertainty we face can be short-term and relatively minor—Will traffic congestion delay us, making us late for an appointment?—or long-term and more significant—In what sort of global environment will my grandchildren live? Some of this uncertainty can be dealt with directly and falls within the individual's locus of control—Should I leave for my appointment a bit earlier to make sure I arrive on time? Other types of uncertainty lie beyond an individual's control—Will reckless behavior by another have a negative impact on me? Often it is difficult to understand the location or the origin of uncertainties. Historical events are inherently uncertain, and current events occurring in our rapidly changing, highly-volatile, tightly interconnected world seem to leave most people more uncertain than ever. As a result, people can imagine the future in many different ways. Although uncertainty can never be eliminated, it can be managed through data analysis, planning, and logical decision-making. Only by taking these actions can we behave rationally (Pollack 2003). We argue that one can manage uncertainty by applying appropriate spatial thinking processes.

How can spatial thinking play a vital role in minimizing uncertainty? Consider the importance of spatial thinking in each of following two situations that occurred recently.

16.1.1.1 London

Mid-morning on 7 July 2005, three explosions rocked the London Underground, and a fourth explosion destroyed a bus. Londoners were horrified at the loss of life and confusion quickly spread. The entire transportation system in Zone 1, comprising central London, was shut down. Mobile phone lines were jammed, not working, or, perhaps as some suspected, shut off by concerned security agencies. Hundreds of thousands of people who commute to and from London via the public transport system chose to walk home that day. But many of these people had difficulty navigating to their home. The ability of Underground riders to find their way on the surface often proved inadequate. Bus and train passengers also found it difficult to navigate on their own. Quickly, newsstands and book stores sold their entire stock of maps; people without spatial guidance attempted to navigate through unfamiliar streets and roads, asking directions and moving from landmark to landmark. Obviously a lack of spatial thinking skills significantly contributed to the difficulties people encountered that day.

16.1.1.2 Houston

About two months later, Hurricane Rita approached the upper Texas coast and eventually made landfall just east of Houston, Texas, the United States' fourth largest city, with a population of three million. Because the storm struck shortly after Hurricane Katrina had devastated New Orleans, most residents, especially those living in the vulnerable areas near the Gulf Coast, followed the advice of emergency planners and began a mass evacuation of the greater Houston region when advised to do so. Within hours, however, the evacuees faced a highway gridlock of unprecedented dimensions. Thousands of people poured onto the three main highway systems in an effort to move inland. The roadways' carrying capacity was quickly exceeded; car-loads of people sat in traffic jams as temperatures rose to the mid-90s. Many of their vehicles ran out of gas, and some of these abandoned cars blocked highway exits. It was apparent that very few people sought alternate evacuation routes or possessed the spatial skill to use a map to find one.

These two examples, chosen because of their scale and significance, demonstrate the value of spatial skills and competencies in managing uncertainty. A large part of the populations of both London and Houston did not have critically important spatial information to help them make geographic decisions. Furthermore, even when information was made available, many people could not use it to locate themselves and to evaluate choices or identify alternative routes because their mental map of their environment was inadequate. Only a few people had the habit of mind or disposition to seek spatial information or to consult a map. In fact, an undergraduate geographer traveled to College Station, Texas, a designated evacuation zone, from a southeastern Houston suburb while thousands of evacuees were trapped on clogged highways. He reported that his trip took only 20 minutes more than usual because he applied the spatial thinking skills he possessed and took advantage of them to avoid the gridlocked roads. Lacking a mental map and complete information, many residents were forced to rely on less efficient methods of finding their way to a safe location in a time of crisis. Of course, the lack of effective spatial thinking was not confined to the general public trying to flee from danger. For example, emergency planners in Houston advised evacuees to use the very routes that clogged and brought the evacuation to a halt. And, although evacuation planning and execution is a complex process, in this case, similar to those stuck in the massive traffic jam, the authorities proved unable to deal with the problem once it occurred.

Besides hurricanes and terrorist attacks, a large number of other threats or hazards exist, and most put a population without adequate spatial skills and competencies at increased risk. We are not the first to recognize this problem. A 2006 National Geographic Society Roper Poll that surveyed 18 to 24 year-old Americans concluded, '...some young Americans lack the basic skills necessary for safety (and employment) in today's world' (Roper Public Affairs 2006: 8). One third (34 percent) of respondents did not possess sufficient spatial thinking skills and knowledge to know which locations relative to a hurricane's eye were most likely to suffer heaviest impact of the storm. The same proportion of respondents could not select the appropriate route to evacuate a city after being given the route's cardinal direction. These findings are sobering given the potential hazards presented by chemical leaks, dirty-bombs, wildfires, or the plethora of other dangers that, at one time or another, threaten the property or personal safety of almost every citizen. It is not difficult to identify other hazards or threats that put a population without adequate spatial skills and competencies at increased risk.

Additional evidence linking spatial thinking and public safety is contained in a recent National Academies of Science (NAS) report concerning geospatial support for disaster management. *Successful Response Starts with a Map*, notes the importance of

> ...the maps that are an essential part of search-and-rescue operations...the GPS receivers that allow first responders to locate damaged buildings or injured residents,...images that are captured from aircraft to provide the first comprehensive picture of an event's impact... road maps that form the basis of evacuation planning, and...all of the other information connected to a location that can be used in emergency management (Committee on Planning for Catastrophe 2007: 1).

Given the examples of recent disaster scenarios, the results of the National Geographic Society Roper Poll, and the NAS report, it is clear that spatial thinking, supported by geospatial technologies, can, at least to some degree, mitigate uncertainty.

Having made the case for the importance of spatial thinking in the context of uncertainty and the threats it creates for today's citizens, it is important to note that the use of geospatial technology is expanding in a wide variety of other contexts that are quickly making spatial thinking a requirement to participate effectively as a citizen in modern society (Bednarz and Acheson 2003). A large number of governments, agencies, and organizations are making their data and findings available to the public via online mapping (iGIS, Google mashups) systems. Thus, it is becoming increasingly necessary for citizens to use these products, or to manipulate them or create their own, in order to make the informed decisions required to participate in democratic government (Bednarz and Acheson 2003). In fact, geospatial technologies are commonly used to engage stakeholders by giving them access to information and supporting their decision-making through participatory GIS and similar systems (Elwood 2006; Nyerges et al. 2006; Kirschner et al. 2003). Understanding and interpreting spatial data and maps are becoming increasingly important for 21st century citizens.

The diffusion of geospatial technology has also made spatial thinking more important in the workplace. According to the US Department of Labor (2007) geospatial technology jobs are among the fastest growing employment areas. Both the public and private sector are seeking employees with spatial-thinking knowledge and skills. They seek not only those with the technical ability to operate modern software but also those who understand spatial concepts and know how and when to apply them to solve problems.

16.2 The Nature and Importance of Spatial Thinking

As stated previously, spatial thinking is defined as the knowledge, skills, and habits of mind to use concepts of space (such as distance, direction, distribution, and association), tools of representation (such as maps, graphs, and diagrams), and processes of reasoning (such as cognitive strategies to facilitate problem-solving and decision-making) to structure problems, find answers, and express solutions to these problems (Committee on Support for Thinking Spatially 2006). It is a collection of cognitive skills that allow individuals to use space to model the world, real and imagined, in powerful ways. Spatial thinking permits individuals to comprehend relationships and structures in multiple ways and to remember them, through a range of representations, in both static and dynamic forms. To paraphrase the National Geography Standards (Geography Education Standards Project 1994), spatial thinking can be framed through two questions: what do students know (about space and tools of representation) and what can they do (processes of reasoning) with what they know. Expanding on the examples provided in the previous section, a spatially literate resident of the Gulf Coast of the United States would have understood a physical phenomenon such as a hurricane well enough to have identified the best direction to evacuate when shown an image of the approaching storm. She would also have known how to reach her destination via more than a single route. A spatially literate Londoner would have had a well-developed mental map of the city and his home location in it (concepts of space), and would have used a map (tool of representation) to devise an efficient route home on July 7 (processes of reasoning).

Spatial thinking has traditionally been a subject of interest to cognitive scientists, psychologists, and behavioral geographers. Before there was 'spatial thinking' researchers defined and studied spatial skills and spatial abilities. Much of the work done to determine the nature and importance of spatial ability was done by psychologists (McGee 1979a,b; Gardner 1983; Newcombe and Dubas 1992). In general, these researchers identified two spatial abilities, visualization (the ability to picture and mentally rotate objects) and orientation (the ability to see objects from a different perspective). A review of that research shows a concern with replicability and experimental design, performance tasks that were confined to small scales using paper and pencil, tests that were timed, and the use of small samples or samples of convenience.

Geographers have argued that the psychologist's conceptualization of spatial ability overlooked some important aspects of spatial phenomena such as distribution, process, association, and structure which are important elements used in spatial activities (Golledge 1993; Self and Golledge 1994), and that the term 'spatial' referred only to small-scale (table-top) spaces (Liben 1981). Golledge and Stimson, among others, argued for addition of an ability termed 'spatial relations' composed of a large collection of skills such as the

> ability to recognize spatial distributions and spatial patterns, to connect locations, to associate and correlate spatially distributed phenomena, to comprehend and use spatial hierarchies, to regionalize, to orientate to real-world frames of reference, to imagine maps from verbal descriptions, to sketch maps, to compare maps, and to overlay and dissolve maps (Golledge and Stimson 1997: 158).

To some extent, when the National Research Council (NRC) published *Learning to Think Spatially* (Committee on Support for Thinking Spatially 2006), they avoided the controversy about the nature of spatial abilities by concentrating on spatial thinking, the broader concept defined in the opening sentence of this section. To think spatially entails (1) knowing about space (e.g., different ways of calculating distance, the basis of coordinate systems, and the nature of spaces); (2) understanding representation (e.g., the relationships among views, the effect of projections, and the principles of graphic design); and (3) reasoning in and about space (e.g., the different ways of thinking about shortest distances, the ability to extrapolate and interpolate, and making decisions).

Learning to Think Spatially (Committee on Support for Thinking Spatially 2006) identifies three types of spatial thinking. Cognition **in** space involves thinking about the world in which we live. It is exemplified by wayfinding and navigation, actions that we perform in space. Walking to school, taking a shortcut to avoid a traffic jam, playing a team sport such as football, evacuating after a chemical spill, or packing a suitcase—actions that are performed *in space*—all require spatial thinking in a real-world context. This type of thinking also extends to other everyday activities: assembling a piece of furniture or a bicycle; packing a storage container; and building a shed or other structure. Because many of these actions are frequently engaged in by most of us, they make a strong case for the relevance of spatial thinking. Success in a wide variety of vocational and academic pursuits such as air traffic control, medicine, urban planning, engineering, and the geosciences also depend on this skill. Thinking in space is relevant to safety as well—for example, by knowing which way to flee in order to avoid an approaching hazard.

The second type of spatial cognition, thinking **about** space, helps individuals understand how the world works, that is, the nature, structure, and function of phenomena that range from microscopic to astronomical scales. Thinking about space supports a significant amount and variety of knowledge and performance. It enables individuals to use maps, graphs, images, diagrams, models, and visualizations that describe and explain the functions, structures, relationships, and operations of all sorts of phenomena (Bednarz et al. 2006). Thus, spatial thinking is important to most, if not all, of the natural sciences, social sciences, and humanities where

Cosgrove (2004) argues 'the spatial turn' is evident, particularly in sociology, economics, and history. The centrality of spatial thinking in a range of disciplines reinforces its worth. The value of thinking about space was exemplified during the Indian Ocean tsunami of 2005 when a young British girl on holiday with her family observed water moving away from the shore. She had learned about this phenomenon in geography class and by thinking spatially about its implication, was able to warn those around her and save their lives.

The third type of spatial thinking, thinking **with** or through the medium of space, is more abstract but perhaps the most powerful form of spatial thinking. Spatializing non-spatial data or using space as an organizing framework to conceptualize problems and make decisions is a very effective cognitive strategy. This third context is the least understood yet perhaps most generative context for spatial thinking. Often, inherently non-spatial data are spatialized and expressed graphically to aid in analysis and comprehension. As society's access to data increases and the analytical abilities to process data increase, graphic representations become increasingly more necessary. For example, although it may be difficult to detect trends and patterns in demographic data, when the data are graphically represented as a population pyramid, the structure of the population becomes much easier to understand and drawing inferences becomes less difficult. Thinking with space supports the acquisition of knowledge, provides students with the skills they need to become more independent and successful learners, and meets society's needs for individuals with the habit of mind to use space as an organizing concept.

Another important conclusion contained in *Learning to Think Spatially* (Committee on Support for Thinking Spatially 2006) is that the benefits of practicing spatial thinking initially tend to be domain specific; and as is the case for developing other forms of expertise, learning to think spatially is best conducted in the context of the materials and situations an individual is seeking to learn and understand. It is unlikely that there will be instant transfer of some skill to a problem in another domain of knowledge, yet some components of existing spatial skills can be drawn upon to tackle new problems. Thus, practicing spatial skills is most effective if it is contextualized within a domain of knowledge, and it might be necessary to develop expertise in particular contexts before one can see the connections to a more general spatial skill (e.g., one might become expert at seeing things in three dimensions in biology, but still need considerable practice to learn to apply the skill to see new kinds of forms, shapes, and positions in another domain). Finally, although the research also confirmed that spatial thinking develops uniquely in individuals, the good news is that spatial thinking can, and should, be learned (Committee on Support for Thinking Spatially 2006). This is especially significant if we expect citizens to exploit the new geospatial technologies to manage the increasingly complex, uncertain world in which they live.

Although three types of spatial thinking can be identified, the research suggests that they are linked: thinking in space promotes thinking about space, while the use of space as a cognitive strategy adds power and aids in thinking in and about space. We also now know that instruction that explains general principles supports transfer between domains better than does instruction that is more specific and focused.

Second, using multiple examples during initial learning and/or varying the conditions of practice also facilitates far transfer.

The NRC study makes a strong argument for the benefit to students—and society—from thinking spatially. The findings of *Learning to Think Spatially* (Committee on Support for Thinking Spatially 2006), however, are not new or isolated. Other researchers have found that success in spatial thinking correlates with success in school. Spatial thinking plays a major role in a variety of cognitive processes: generic learning, remembering, and problem-solving (Golledge and Stimson 1997). Spatial thinking also facilitates encoding and recalling information and communicating both spatial and non-spatial relationships (Kulhavey and Stock 1996; Kitchin and Blades 2002; Verdi 2002; Committee on Support for Thinking Spatially 2006). For example, readers remember information both verbally and visually. This 'dual encoding' or conjoint retention is enhanced when readers are encouraged to link what they read to graphic representations such as maps, graphs, flowcharts, or concept maps (Paivio 1986). The process of producing spatial representations creates schemas that link related items and provide an efficient means to search one's memory. Linking *what* with *where* (i.e., reading and thinking *with* and *through* visual representations), makes content easier to understand and remember (Liben 2001).

16.3 Spatial Thinking and Geospatial Technologies

With the advent of sophisticated visualization tools and geospatial technologies and the concomitant increased need for a spatially literate citizenry and workforce, interest in spatial thinking has exploded. This growth is evidenced by the previously mentioned NRC's spatial thinking study, an NSF-funded Spatial Intelligence and Learning Center (http://www.spatiallearning.org), the United Kingdom's government-funded Spatial Literacy in Teaching Centre of Excellence (http://www.spatial-literacy.org) and calls for spatial literacy in the United States (Goodchild 2006). In addition, awareness of the importance of spatial literacy as a matter of public safety, security, and personal empowerment is increasing.

It is not surprising that the call for more and better spatial thinking and the development of geospatial technology have been, and continue to be, connected. As early as 1995, Golledge and Bell (1995) noted that:

> The rapid development of GIS as a means for coding, storing, accessing, analyzing, representing and using spatial data creates an obvious parallel with the cognitive mapping process—which also encodes, stores, internally manipulates, decodes and represents spatial information. We offer a metaphor that may be useful when dealing with both concepts—i.e., that the cognitive map is an internalized GIS.

These research results lead to two obvious questions: What is the state of spatial-thinking instruction, especially instruction supported by geospatial technologies, in today's educational system; and What are the results of efforts to teach students to think spatially?

Despite its growing importance, spatial thinking and geospatial technologies are not an explicit part of curricula at any level in the United States, and the evidence suggests, not surprisingly, that most students are not proficient spatial thinkers. An analysis of the 2001 National Assessment of Educational Progress (NAEP) geography exam revealed that at every level (grades 4, 8, and 12) students scored low on test items that required them to use and interpret maps (that is, to think spatially). Baker (2002) found that applying spatial thinking is the most difficult component of GIS for students to master, that is, even when they could produce maps, students could not interpret them effectively or use them to solve problems or make generalizations.

Even if spatial thinking and geospatial technologies are not part of any particular curriculum, given its importance, it is fair to ask why so little instruction is offered to students. At the pre-collegiate level, the reasons are many and varied. Our experience with middle and high school teachers through an NSF-funded project, *Advancing Geospatial Skills in Science and Social Science* (AGSSS), offers some relevant information. AGSSS connects geospatially skilled graduate students (AGSSS Fellows) with science and social studies teachers, grades 6–12, in a collaborative two-year cycle to enhance teachers' and students' spatial-thinking knowledge and skills. The AGSSS program has five goals: (1) to develop and test a set of training experiences for geospatially skilled graduate students to prepare them to enhance their own spatial thinking and that of teachers and students; (2) to prepare graduate students with the knowledge and skills to work collaboratively with teachers to enhance spatial thinking as supported by geospatial technologies; (3) to develop and test a set of training experiences for teachers to prepare them to use geospatial technologies; (4) to develop and test a set of training experiences to assist teachers in conducting research in their classrooms about student spatial thinking; and (5) to share research results within the AGSSS partnership and other communities thereby building a foundation of practical, classroom-based findings regarding spatial thinking. As the program has evolved, the goals related to spatial thinking have focused on three questions: (1) What is the nature of spatial thinking in classroom settings? (2) What practical, classroom-based strategies can be used to develop spatial thinking? and (3) What is the role of spatial thinking in the implementation of geospatial technologies?

As we have reported elsewhere (Bednarz and Bednarz 2008) one important barrier to formal instruction in spatial thinking is educators' lack of understanding of spatial thinking and the fundamental concepts and cognitive processes that support it. One reason this barrier exists is that teacher-education curricula, like those in the pre-collegiate system, do not include instruction in spatial thinking. Teachers' lack of knowledge of, and appreciation for, spatial thinking also deters the introduction of geospatial technology into classrooms. Of course, the introduction of GIS and other technologies face other impediments, such as the cost of appropriate hardware and software, the lack of guidance to assist educators in using technology to teach specific content, and insufficient technical support (Alibrandi 2001; Bednarz 2001; Kerski 2003; Wallace 2004).

At the university level, where instruction in geospatial technologies is more common, instructors who endeavor to teach spatial thinking often face resistance from students. These faculty report that a significant portion of their students are interested in only the technical aspects of GIS and other geospatial technologies. The goal of such students is to acquire the skills necessary to secure employment after graduation; few appreciate the importance of learning more than how to use the software and hardware.

Although the state of spatial-thinking instruction in pre-collegiate and university instruction leaves much to be desired, our research findings (and those of our graduate students) lead us to believe that teaching spatial thinking along with geospatial technologies is possible and worthwhile. At the university level, students who completed GIS and remote sensing courses improved their spatial-thinking skills. Furthermore, the improvement was greatest for students who undertook term projects that required them to apply the technology and the spatial theories and concepts to solve an authentic problem (Vincent 2004; Lee 2005; Lee and Bednarz 2005).

16.3.1 Lessons We are Learning

At the pre-collegiate level, the situation is more complicated, but the preliminary results generated by the AGSSS project suggest that, with appropriate training and provisions, spatial thinking, supported by geospatial technologies, can be successfully incorporated into science and geography/social studies courses at the middle and high school grades, respectively. After struggling for the first year of the project to explain spatial thinking and its importance to the teachers participating in the project, they reported 'finally beginning to get it' when they completed a week-long training session in geospatial technology and spatial thinking that was held during the summer at the end of the first year. Both teachers and graduate Fellows participated in this co-learning experience. Nevertheless, teacher awareness of spatial thinking and understanding of its importance (especially among geography teachers) was (and remains) slow to develop. Only gradually and incrementally, as Fellows work with them to spatialize their curriculum, have the teachers begun to understand how spatial thinking can organize and enhance their students' subject-matter learning.

Of course teachers must understand not only the importance of spatial thinking; they must also be aware of the techniques and strategies that can best be used to support students' learning. Teachers' understandings of the connections between spatial thinking, subject-matter content, and appropriate instructional methods are also developing at a leisurely pace. This slow progress is not surprising, given the struggle teachers experienced in understanding the importance and relevance of spatial thinking. If they have any doubts about the value of spatial thinking, they are unlikely to invest time and energy in finding effective ways to use it in their teaching, let alone to determine when, where, and how geospatial technologies could help students learn particular concepts and skills. Our objective is to explain how spatial thinking and geospatial technology supports learning more clearly and more powerfully so that teachers will enthusiastically embrace and use them to develop new skills and master their subject matter content more easily.

Somewhat surprisingly, teachers have developed strategies to support spatial thinking with technology more quickly and successfully than expected. With the assistance of the AGSSS Fellows, the participating teachers have incorporated computers and related technologies such as whiteboards, computer projectors, and the Internet into their courses. Although teachers were reluctant to adopt new methods at first, they quickly began to innovate and adapt technology-based lessons and materials to accommodate their learners. They also began experimenting with different student work schemes, such as letting students work on their own or in pairs. Thus, we perceive success in effecting change in teachers' willingness and ability to use technology to support spatial thinking in their courses.

Through the collaboration of teachers, Fellows, and university faculty, initial observations have begun to address the nature of spatial thinking in the classroom. Among the things we have learned is that in high school World Geography classes, students have difficulty using geospatial technology—GIS—to learn geography because of their inadequate vocabulary. When asked to describe world regions, spatial patterns, or spatial relationships, they experienced more difficulty than we expected. During a Google Earth activity when students were asked to 'zoom in to' a location 'at the edge' of Cairo and the surrounding desert, many were puzzled. We expected students to interpret the distinctive border between urban land use and desert as an 'edge' with little or no problem. Although students understood 'zoom in,' 'edge' confused them. 'What do you mean by the edge?' they asked. We suddenly realized that we were asking students to perform tasks (pattern seeking, description, navigation from place to place) that were completely new to them. Not surprisingly, students were unable to accomplish these tasks until we introduced and defined the appropriate spatial concepts, supplied them with support (i.e., illustrated vocabulary booklets). They also needed guided practice to help them understand how these new ideas related to the geographic representations they were using, and how they should proceed to use the technology (zooming in; zooming out; maintaining a frame of reference) appropriately and effectively. We consider this finding as one of the most important outcomes of the AGSSS project. Before this experience, the teachers, fellows, and university faculty did not fully understand the necessity of providing students with explicit vocabulary training, appropriate support materials, and guided practice using geospatial technology to help students to learn to think spatially.

16.4 Conclusions

We live in uncertain times. The uncertainty we face occurs at a variety of scales. Some can be managed relatively easily, but some lie virtually beyond the control of humans. Although we cannot eliminate uncertainty and its effects, it is important to minimize the disruption and loss that results from it. Mitigating the negative effects of uncertainty, especially by applying geospatial technologies, requires spatial-thinking skills. *Learning to Think Spatially* (Committee on Support for Thinking Spatially 2006) makes a convincing argument that spatial thinking can and should be taught. It bases this recommendation on the best available evidence from the cognitive sciences and disciplines ranging across the human-natural science spectrum. A strong connection between geospatial technology, especially GIS, and spatial thinking has been proposed and supported by spatial cognition researchers. *Learning to Think Spatially*, in fact, states that '[t]he key to spatial thinking is a constructive amalgam of three elements: concepts of space, tools of representation, and processes of reasoning' (Committee on Support for Thinking Spatially 2006: 5).

Of course, for the same reasons that spatial thinking is important in managing uncertainty, it is valuable for other types of decision making required of citizens who wish to participate fully in their democracy. Spatial thinking has also become an important skill in the workplace as the diffusion of geospatial technology reaches large parts of the public and private sectors.

Initial results from the AGSSS project suggest that, although introducing instruction in spatial thinking into the classroom is not a simple or easy process, it can be accomplished. Anecdotal responses from teachers and from classroom observation by graduate Fellows and university faculty indicate that in addition to the amount of spatial thinking occurring in the classroom, spatial thinking is having a positive effect on student learning in science and geography.

Teachers, who were skeptical at first, have recognized the advantages of explicitly teaching spatial-thinking skills. They have enthusiastically sought out the assistance of the graduate Fellows and have successfully incorporated a significant quantity of new learning activities into their courses. Both the teachers and the students themselves report that these new hands-on and technology-based lessons are worthwhile, interesting, and effective.

Baseline data gathered from more than 940 students provide an interesting portrait of the current status of their spatial literacy. Nearly two-thirds (65.7 percent) of all students agree that they are 'good at reading and interpreting maps.' Approximately three-fourths find that following illustrated directions is easy and that 'graphs, maps, and charts help me learn.' Although students are confident in their ability, they have not developed the habit of mind to use spatial representations. Only about 28 percent report using maps frequently and just 35 percent use maps and diagrams to help them think and communicate. These results support our assertion that many individuals employ spatial thinking only passively. Thus, if we expect citizens to fully exploit the capabilities offered by the information available and the technology to display and analyze it, more direct instruction in spatial thinking seems warranted.

Our data also reveal that attitudes and behaviors only sometimes vary significantly by gender, age or both. For example, frequency of map use increases dramatically from middle to high school, perhaps as a result of high school students' higher level of mobility. Significant gender differences exist for respondents' inclination to visualize, propensity to take short cuts, frequency of getting lost, and preference for written directions rather than a map to find one's way. These age and gender patterns for spatial abilities and preferences provide important information about how spatial thinking can best be taught to diverse learners. Sequencing instruction appropriately and providing effective learning opportunities for both males and females will improve the outcomes of any spatial-learning program that may be established.

In the spring of 2007, more rigorous assessments of student spatial learning were conducted. While they are still being analyzed, we are confident that the results of that research will confirm the value of spatial-thinking instruction. Although we are convinced that spatial thinking is an important 21st century skill and that it must be taught, we recognize that only the first steps have been taken. We hope that demonstration of the importance of spatial thinking will convince others to implement programs of their own. We also know that the AGSSS project will generate findings that will assist in the development and refinement of effective teaching and learning programs. Perhaps this chapter will serve as one of the first vehicles for the dissemination of information that will catalyze educators to recognize the value of spatial thinking and to incorporate explicit instruction in it into their classrooms.

Note: Much of the material in this chapter was presented at the 2006 ESRI Education Users Conference; the Texas A&M symposium *Geospatial Technologies and Homeland Security* (2006); the conference *Changing Geographies: Innovative Curricula* sponsored by the Institute of Education, University of London (2007); the National Council for Geographic Education (2006); and at the Seventh National Science Foundation Graduate Teaching Fellows in K-12 Education Project Meeting. Related discussions of spatial thinking and its use to manage uncertainty appear in Bednarz, S. W. (2007) and Bednarz S. W. and R. S. Bednarz (2008) cited below.

References

Alibrandi, M. (2001). Making a place for technology in teacher education with geographic information systems (GIS). *Contemporary Issues in Technology & Teacher Education*, 1(4), 483–500

Baker, T. R. (2002). *The effects of geographic information system (GIS) technologies on students' attitudes, self-efficacy, and achievement in middle school science classrooms*. Ph. D. dissertation. (Lawrence, KS: The University of Kansas)

Bednarz, S. W. (2001). Thinking spatially: Incorporating geographic information science in pre and post secondary education. (In L. Houtsonen et al. (Eds.), *Innovative practices in geographical education*. Helsinki: Proceedings of Helsinki Symposium, IGU Commission on Geographical Education, 3–7)

Bednarz, S. W. & Acheson, G. (2003). Learning to be a citizen in post 9/11 United States: What role for geography? *Proceedings of the commission on geographic education, geography and citizenship education*. (London: Institute of Education)

Bednarz, S. W. & Bednarz, R. S. (2008). Spatial thinking: The key to success in using geospatial technologies in the social studies classroom. (In A. J. Milson & M. Alibrandi (Eds.), *Digital geography: Geo-spatial technologies in the social studies classroom* (pp. 249–270), New York: Information Age Publishing)

Bednarz, S. W., Acheson, G. & Bednarz, R. S. (2006). Maps and map learning in social studies. *Social Education*, 70(7), 398–404.

Committee on Planning for Catastrophe. (2007). *Successful response starts with a map: Improving geospatial support for disaster management.* (Washington DC: The National Academies Press)

Committee on Support for Thinking Spatially. (2006). *Learning to think spatially.* (Washington DC: National Academies Press)

Cosgrove, D. (2004). Landscape and landschaft. *German Historical Institute Bulletin*, 35(Fall), 57–71

Elwood, S. (2006). Beyond cooptation or resistance: Urban spatial politics, community organizations, and GIS-based spatial narratives. *Annals of the Association of American Geographers*, 96(2), 323–341

Gardner, H. (1983). *Frames of mind: The theory of multiple intelligence.* (New York: Basic Books)

Golledge, R. G. (1993). Geographical perspectives on spatial cognition. (In T. Gärling & R. G. Golledge (Eds.), *Behavior and environment: Psychological and geographical approaches* (pp. 16–46). Amsterdam: Elsevier Science Publishers)

Golledge, R. G. & Bell, S. M. (1995). *Reasoning and inference in spatial knowledge acquisition: The cognitive map as an internalized geographic information system.* (Santa Barbara, CA: Santa Barbara Geographical Press)

Golledge, R. G. & Stimson, R. J. (1997). *Spatial behavior: A geographic perspective.* (New York: Guilford)

Kerski, J. J. (2003). The implementation and effectiveness of geographic information systems technology and methods in secondary education. *Journal of Geography*, 102(4), 128–137

Kirschner, P. A., Shum, S. J. & Carr, C. S. (Eds.). (2003). *Visualizing argumentation: Software tools for collaborative and educational sense making.* (London: Springer)

Kitchin, R. & Blades, M. (2002). *The cognition of geographic space.* (London: I.B. Taurus)

Kulhavey, R. W. & Stock,W. A. (1996). How cognitive maps are learned and remembered. *Annals of the Association of American Geographers*, 86(1), 123–145

Lee, J. (2005). *Effect of GIS learning on spatial ability.* Ph.D. dissertation, College Station, TX: Texas A&M University.

Lee, J & Bednarz, R. S. (2005). Video analysis of map-drawing strategies. *Journal of Geography*, 104(5), 211–221

Liben, L. S. (1981). Spatial representation and behavior: Multiple perspectives. (In L. Liben, A. H. Patterson & N. Newcombe (Eds.), *Spatial representation and behavior across the life span: Theory and application* (pp. 3–36). New York: Academic)

Liben, L. (2001). Thinking through maps. (In M. Gattis (Ed.), *Spatial schemas and abstract thought* (pp. 45–78). Cambridge, MA: MIT press)

McGee, M. G. (1979a). *Human spatial abilities: Sources of sex differences.* (New York: Praeger)

McGee, M. G. (1979b). Human spatial abilities: Psychometric studies and environmental, genetic, hormonal, and neurological influences. *Psychological Bulletin*, 86(5), 889–918

National Education Standards Project. (1994). *National geography standards 1994: Geography for life.* (Washington DC: National Geographic Society Committee on Research and Exploration)

Newcombe, N. & Dubas, J. S. (1992). A longitudinal study of predictors of spatial ability in adolescent females. *Child Development*, 63, 37–46

Nyerges, T., Jankowski, P., Tuthill, D. & Ramsey, K. (2006). Collaborative water resource decision support: Results of a field experiment. *Annals of the Association of American Geographers*, 96(4), 699–725

Paivio, A. (1986). *Mental representations.* (New York: Oxford University Press)

Pollack, H. N. (2003). *Uncertain science...uncertain world.* (New York: Cambridge University Press)

Richardson, D. (2007). Educating Congress on geography education. *AAG Newsletter*, 42(7), 2–5

Roper Public Affairs. (2006). *National Geographic-Roper public affairs 2006 geographic literacy study* [electronic version]. Retrieved from http://www.nationalgeographic.com/roper2006/findings.html

Self, C. M. & Golledge, R. G. (1994). Sex-related differences in spatial ability: What every geography educator should know. *Journal of Geography*, 93(5), 234–243

US Department of Labor. (2007). Local solutions with national applications to address geospatial technology industry workforce needs [electronic version]. Retrieved from http://www.doleta.gov/BRG/indproof/geospatial.cfm

Verdi, M. P. (2002). Learning effects of print and digital maps. *Journal of Research on Technology in Education*, 35(2), 290–303

Vincent, P. (2004). *Using cognitive measures to predict the achievement of students enrolled in an introductory course of geographic information systems*. Ph.D. dissertation, College Station, TX: Texas A&M University

Wallace, R. M. (2004). A framework for understanding teaching with the Internet. *American Educational Research Journal*, 41(2), 447–488

Chapter 17
GIS and Homeland Security Education: Creating a Better Tomorrow in our Classrooms Today

David H. McIntyre[1] and Andrew G. Klein[1]

Abstract The future stacks up to be a very dangerous place. To meet homeland security challenges, integrated solutions that cross discipline boundaries and incorporate new technologies like Geographic Information Systems (GIS) are required. Constructing an integrated homeland security curriculum in our classrooms today can help to shape a safer future. Developing such a curriculum that successfully incorporates the full potential of geospatial solutions including Geographic Information Systems (GIS) requires efforts on the part of GIS experts. These professionals must learn the challenges and the components of the solutions, and find ways to incorporate their expertise into those solutions and then convince experts in other disciplines who remain skeptical of unproven programs and unfamiliar technologies, of the potential benefits of integrating GIS into homeland security education.

Keywords CBERNN threats (chemical, biological, explosive, radiological, nuclear and natural), critical infrastructure, geographic information systems, graduate education, national scenarios, taxonomy for homeland security

17.1 Introduction

As we look to the future and to how GIS might be applied to that future, one clear if unpleasant fact is that tomorrow is going to very unlike today. People who are happy, comfortable, satisfied, and safe will be unhappy to hear this, but the truth is that in many fundamental ways the future is going to be less comfortable, less safe, and more dangerous, making homeland security more important in people's lives. The good news is that geographic information systems (GIS) can help us change the way leaders think about homeland security. The technology is important to

[1] Texas A&M University

homeland security and it can help achieve safety and security in an uncertain age—if we can figure out how to successfully incorporate it into our homeland security education programs.

17.2 Challenges

The new security challenges we will face arise from several related causes. First, the new security challenge springs not just from terrorism, but from the rise of technology, because it is changing what political scientists call the offense/defense balance.

Military strategists, and historians in particular, but also anyone familiar with how the military operates, understands that over time, advances in technology have changed the way wars are fought. One side develops a better sword so the other side counters with an improved shield. So the first side gets arrows, so the second side creates armor. The first puts their armored people on horseback, and the second constructs walls around its cities. And this cycle has continually repeated itself, with technological advances constantly changing the ability to attack and defend.

The First World War was a costly example of this technological cycle. During World War I, the rifle, machine gun, and improved artillery made it much more difficult for people to attack than to defend, and changed the whole balance on the battlefield. Old techniques of massed attacks that worked in the French Revolution suddenly failed, and the world reeled from the mass slaughter that occurred as the old tactics failed in the face of new technology. The best military minds of the day were immobilized by new strong defenses. Mobility was restored only when new technological advances appeared in the form of the tank and the radio. The Germans combined them with astonishing success in their Blitzkrieg during the Second World War.

For the moment and into the foreseeable future, the world is, and will be, victimized by a similar shift in power, created by a similar shift in technology. Developing technologies make it easier to attack a civilian society than it is to defend it. For example, it is easier to attack a subway than to defend it and, unfortunately, easier to create a new virus or disease than to create an antiviral or a vaccine against it. The logical result is a situation where big weapons are now available to small people, even non-nations or criminal organizations. This is the situation we are faced with in the foreseeable future. Even if there were no radical fundamentalist terrorist threat, modern societies remain increasingly vulnerable to sophisticated attack by small organizations.

Of course, the second reason the world is facing a more dangerous future is that we are dealing with a radical fundamentalist terrorist threat. Independent of political ideology, it remains clear that the world is facing a different type of enemy than we have faced in the past. The bipartisan 9/11 Commission made this very clear, explaining the origins of this attractive ideology, and the fact that it is growing and gaining adherents worldwide. The Commission's report stressed that the fundamentalists who adhere to that ideology have a new approach to the old idea of terrorism (National Commission on Terrorist Attacks Upon the United States 2004).

The 9/11 attacks provided an unfortunate example of how changing tactics have increased the national security threat. For many years, terrorism experts advised the pilots of hijacked aircraft to cede control of the aircraft because the fundamental reason terrorists hijacked aircraft was to generate publicity for their cause. The buzz word at the time was, 'They want a lot of people watching, not a lot of people dead.' Well, that has changed as the era of the Red Army Brigade, the Shining Path, and the other localized, politically-driven terrorists has passed. The world now faces an enemy with an ideological ax to grind and who, unfortunately, is seeking 'a lot of people dead.' We are facing, in author Ralph Peter's words, 'apocalyptic terrorism.' (Peters 2002). This conclusion is not our own, it is that of the bipartisan 9/11 Commission, who stated that a dangerous new and attractive ideology is producing a new type of enemy—empowered by new technology—with whom there is no hope of negotiation and no alternative to force.

Advances in technology have not only aided individuals and organizations aiming for a different type of terrorism, they have also increased society's vulnerability. This increased vulnerability is a result of the world's ever increasing technological complexity. Just in time delivery while boosting our productivity and creating a new economy and prosperity more broadly shared than at any time in the history of the world, has also created a web of international dependencies. If that just in time system is interrupted, consequences cascade rapidly, and problems unimaginable in a less interdependent world quickly emerge.

One example of a vulnerable just in time system is gasoline tank farms, familiar to anyone driving down an American highway. In the past, these tanks contained perhaps three or four weeks of gasoline. Today, many refineries now keep less than two weeks of supply on hand, because technological advances have made the delivery system much more efficient, and there is no reason to bear the cost of additional stored inventory. This efficient delivery system, however, is much more vulnerable than the old to any interruption, whether it is by terrorist attack, flu pandemic, or simply a hurricane. Changes in one location immediately ripple through the entire system. In today's world, redundancy is a luxury only military forces can afford. There exists very little cushion for the civilian population.

Increasing efficiencies have also led to increased vulnerabilities in our food distribution networks. Most supermarkets have only a four day supply on the shelf. The same holds true for medicines where hospitals no longer store two or three week's worth of supplies, but frequently expect to be re-supplied twice a week. Such changes are true throughout our society because of the increasing specialization and the resulting complexity of our economic system. These changes together with rising expectations of the population make us more dependent on government than we were in previous generations.

At the same time the combination of global climate change, expanding populations, and increased development in riskier areas, including the world's coastal zones, has the potential to make societies more vulnerable to economic disruption from natural disasters than ever before.

Finally, in the homeland security realm there is the new reality of growing divisions between individuals and organizations who are concerned first of all about prevention and protection against terrorism, and those who are concerned with mitigation against natural disaster. The emergency management community, in particular, currently believes that the priority placed on terrorism since 2001 has diverted resources and attention better devoted to improving response capabilities.

Drawing everything together suggests that if we do not think carefully and plan ahead to prepare for these new challenges we face, we could end up losing the fight. By 'the fight,' we do not mean against an enemy who invades the United States of America and occupies the White House. Rather, we mean our society could lose the advantages that modern civilization currently provides to us as citizens of this modern nation. Similarly, we could lose the race against proliferation of weapons of mass destruction. We could lose common cause with our allies, resulting in the United States of America having fewer friends abroad. We could lose the struggle for the soul of moderate states with whom we share common values; thus losing the struggle against a wider war.

We could also lose national power here at home as our country is required to expend more of our finite resources, economic and other, on domestic security. Or we could even see our nation's coherence or our national character challenged at home as it was in the days immediately following Hurricane Katrina. In other words, loss of national security means we could lose the character of the modern world—the free flow of people, things and ideas—if we do not think, organize, and use our national resources properly in the future.

17.3 The Role of GIS

Geographic information systems (GIS) can be an important tool in this struggle for a secure future. One area of homeland security where there is an immediate opportunity for application of GIS is the area of critical infrastructure.

The nation's critical infrastructure is not simply what a citizen views as critical, nor is it what local authorities, such as the nation's mayors, view as critical. Our Critical Infrastructure is clearly defined in the *National Strategy for the Physical Protection of Critical Infrastructure* as those resources whose loss would: (1) impair the federal Government's ability to perform essential national security missions and ensure the general public's health and safety; (2) undermine state and local government capacities to maintain order and to deliver minimum essential public services; (3) damage the private sector's capability to ensure the orderly functioning of the economy and the delivery of essential services; and (4) undermine the public's morale and confidence in our national economic and political institutions (Bush 2003).

At the national level, the nation's critical infrastructure includes elements like agriculture and food production, energy, transportation, banking, and finance. Each of these areas stands to benefit immediately and directly by improved application of the specialized knowledge and predictive capabilities provided by GIS methods and technologies. In fact, GIS is already heavily, if not always imaginatively, used by experts and managers alike in these fields.

But one thing that has not yet been considered at the national level is that the key components of local government also constitute Critical Infrastructure, starting with public administration, management, and leadership. Hurricane Katrina clearly demonstrated that the presence of legal and tax records, the availability of social services to families and vulnerable populations, the ability to support business and jobs and job programs, continued access to schools and libraries, jails and prisons, and even recreation facilities—all these are aspects of critical infrastructure that perhaps our nation has not paid adequate attention to in the past. But we should.

To take one example, parks and recreation facilities may well serve a dual use in times of emergency; many of them can be successfully be pressed into service in secondary roles as dormitories, hospitals, or for evacuation shelters. However, knowing their geographic location relative to an area's populations and terrain, and being able to spatially relate these locations to other support systems like water, electricity, and roadways are important. This useful knowledge can be created through GIS overlay operations, but only if there has been successful integration of spatial databases held by numerous entities and organizations that use and maintain the required spatial databases and GIS technology skills.

Examining the list of local services, it becomes clear that many of the functions of local government—public works, waste water and sanitation, public safety, animal control, electricity, emergency medical care—should be considered critical public infrastructure. And every single one of these items, just like every aspect of national critical infrastructure, could be influenced, supported, promoted and secured by GIS.

If it is not, numerous opportunities will be missed for bringing the current and emerging capabilities of both geographic information systems (GIS) and geographic information science (GISci) to the service of homeland security in imaginative ways. Our concern is that these opportunities will be missed—only to see lives lost and the legitimacy of the government suffer—because of a too narrow and constrained vision for how boldly and completely GIS should be incorporated into homeland security efforts.

At the national level, the federal government has established fifteen national planning scenarios, ranging from a nuclear attack to a volcanic eruption. The idea behind the scenarios is not to identify the most likely disasters, but rather to examine the potential situations that would most severely stress the federal government and its interaction with state and local responders. In the near future, these situations will be explored through a national scenario exercise program. Although knowledge of the National Planning Scenarios is widespread in the homeland security community, the scenarios themselves have not been released to the general public. However, discussion of the scenarios and their role in promoting national preparedness may be found at http://www.dhs.gov/xnews/testimony/testimony_1166567952901.shtm.

In every single one of these situations, GIS offers entirely new ways of viewing cross-cutting data and providing new solutions for command and control, decision making, and logistics. Unfortunately, none of the planning scenarios have been significantly influenced by GIS or its presence thus far. That is because too many responders and homeland security officials still see GIS simply as a quick method of gaining access to better maps. We continue to viewing this relatively new technology and new opportunities through our old perceptions and perspectives.

17.4 Solutions with GIS

It is informative to examine two of the planning scenarios and consider what the impact might be if GIS were used to its fullest potential. One of the planning scenarios is a pandemic influenza scenario. This scenario considers that waves of infection and future pandemics could originate anywhere on Earth, but are especially likely where human or animal populations are dense. Access to relevant spatial databases and GIS-based models of the geographic spread of a pandemic are key to early informed decision making. Numerous case studies using GIS to study past pandemics have been undertaken. Since the 1990s, concerted efforts have been made to make much better use of GIS for the planning, analysis, and monitoring of the spread of infectious diseases (Martinez 2000).

Based on experience, waves of infection are expected to arrive six to eight weeks apart, but it could be six months to a year before an effective vaccine becomes available. Thus countering the spread of the pandemic requires mobilization of key resources to the most effective and efficient locations. Such positioning and moving of medical care and medicine distribution points presents a big challenge for law enforcement and the military. Successful deployment also requires a system capable of simultaneously monitoring both the evolving pandemic and the distribution of our response and recovery assets. Again, integration of GIS information and models would be a key component of the homeland security response.

Estimating the overall progress and impact of the disease would fall to government leaders—informed, we would hope, by bold integration of GIS. The 1918 Pandemic serves as a rough guide to the possible impacts of such a pandemic. During the 1918 pandemic, 35 percent of the population in the United States fell ill and two percent of those subsequently died. Given the population of the United States today, these figures would translate into approximately 200,000 deaths. Understanding and responding to the distribution of sick and dying, as well as tracking the healthy and immune, and projecting the growth of both groups, would be greatly facilitated by integrating GIS into the reporting system. Understanding the impact on our economy and our national power would offer a similar opportunity.

A second planning scenario focuses on the effects of a moderate earthquake, which would lead to 1,000 fatalities and 5,000 hospitalizations. In this scenario, more than a million people would be evacuated and perhaps two hundred thousand would seek long-term shelter from destroyed homes. Hurricane Katrina saw the early use of GIS to track such population movements. Emerging GIS technologies and their application should provide immensely improved geographically-oriented visualizations of both the impacted areas and relief sites, and improve the decision making process—leading to better services for our impacted citizens. FEMA and other government agencies should provide tremendously improved support to those impacted by the earthquake *if* full advantage is made of GIS analysis and mapping capabilities.

And of course, we expect to see the integration and exchange of GIS information improve in the face of the ever growing hurricane menace. It is important to remember the point made by Max Mayfield, Director of the National Hurricane Center, at his retirement. 'Katrina,' he said, 'was not the big one. That is yet to come.' (Mayfield 2007).

As these scenarios make clear, there are many potential challenges short of a terrorist attack on the US that would require an integrated homeland security response. These scenarios highlight the benefit that would come from the improved incorporation of GIS concepts and experts into our disaster preparedness and response efforts.

Another scenario is that of a volcanic eruption which is likely to impact populations on the tectonically-active western coast of the United States, which is part of the Ring of Fire surrounding the Pacific Ocean. Experts suggest that a major volcanic eruption might create an undersea earthquake that would cause a tsunami in Alaska or along the west coast that might be as much as 18 feet high. Only GIS technology and analysis would allow us to rapidly anticipate the impact of such an event, and begin all important mitigation actions now.

Finally it is necessary to consider the scenario of a major terrorist attack upon a United States city with a nuclear weapon. Virtually every major intelligence and homeland security official in both in the United States and Great Britain believes that our enemy is actively seeking to mount such an attack—that the question is 'when, not if.' While it is not possible to anticipate the specifics of such an attack ahead of time, the United States attacks on Hiroshima and Nagasaki provide a basis to estimate the scope and degree of destruction. Again, GIS technologies can play a vital role in investigating and visualizing how shock patterns interact with the ground, how radiation will be distributed among the impacted population, as well as longer-term impacts, such as potential impacts on groundwater.

Actually, GIS can serve an important role in providing a template for disaster relief efforts prior to such a disaster. And the amount of relief required would be large. The RAND Corporation issued a report in August of 2006 in which they evaluated the impact of a nuclear attack on the port of Los Angeles. The answer was catastrophic (Meade and Molander 2006), resulting in more than 60,000 people dead, 150,000 requiring emergency medical treatment, and a tremendous challenge to response and recovery efforts. GIS can serve a valuable role in anticipating what would be needed, where it would be needed, and how the response and recovery efforts could be coordinated.

So in summary, while the future poses many serious challenges in the field of homeland security, the nation does have some advantages in shaping our responses to these challenges. GIS is at the top of the list of those new advantages.

To understand how GIS can be better integrated into homeland security, it is advantageous to briefly introduce the conceptual model that describes the four temporal phases of managing before, during and after a disaster which are mitigation, prevention, response, and recovery (e.g., Alexander 2002). GIS can be used effectively in many different ways during different phases of a disaster (e.g. Tierney et al. 2001).

Mitigation consists of actions taken before catastrophic events in order to reduce their impact. It is typically more tied to natural disasters and includes a long-term vision and planning. Prevention, sometimes referred to as preparedness, refers to actions taken before an event to actually avoid its occurrence. While unlikely for natural disasters, it is the preferred way to address terrorist attacks, because actions against active targets can be more focused on time and place than mitigation. Protection is a form of preparedness that involves the hardening of potential targets, and may include either long-term analysis or very short term activities such as evacuation before a storm. Response includes actions taken immediately after the event (generally within 72 hours) to save lives and property. Recovery includes actions taken to restore people, places, and things as much as possible to their pre-disaster condition. This will usually take months, and may take years. Obviously, GIS and other spatial technologies can play a role in each of these temporal phases of a disaster.

Most emergency managers use the term ‘all hazards’ to describe the way in which preparation for one type of attack or incident (i.e., chemical, biological, explosive, radiological, nuclear, cyber, electro magnetic pulse, or natural disaster) can apply to all. We have already established that GIS data and analysis can be critical in understanding the actions required, and the similarities and differences between these different threats.

If we align homeland security threats with the phases of the disaster cycle, it is possible to see the range of GIS products and efforts that can support homeland security as a whole. For example, mitigation, prevention, protection, response, and recovery against a chemical event, against a radiological event, against a hurricane, against a cyber attack, etc., as each of these threats has its own unique attributes.

And suddenly, the huge range of needs and opportunities for GIS to be incorporated into homeland security becomes evident. But these challenges and opportunities can only be met if the security and emergency experts learn to use GIS—and if the academic experts learn to teach it in ways that operators and policy makers find practical.

If we are going to use and teach GIS to its maximum effectiveness in the emerging homeland security, some serious questions must be asked. These questions are raised in small communities and municipalities just as routinely as in big state and federal agencies. As GIS increasingly becomes an essential tool for security and response, a whole range of issues come into view and must be addressed. Fortunately, the larger geospatial community has begun to address the questions we raise here. Following the aftermath of 9/11 and Hurricane Katrina, in 2007 the National Academy of Sciences published *Successful Response Starts with a Map: Improving Geospatial Support for Disaster Management,* which examines how geospatial data can be better utilized for disaster management, and makes specific recommendations on how this can be accomplished. It is recognized that the effectiveness of any technology is only as good as the human system that encompasses it (Committee on Planning for Catastrophe 2007).

The first question is, 'Who owns the information?' If information has been collected specific to someone's particular needs or geographic location and it has not been passed on to larger GIS data providers, or has inadequate metadata, is it reliable enough to use for homeland security purposes? If the data has been shared, who now owns that information and who is responsible for updates and changes? Who is responsible if the database contains errors or is out of date? Who is liable? Small errors, inaccuracies, or out-of-date information that may be no more than minor annoyances for typical applications, may cause serious problems in an emergency situation.

Numerous federal, state, and local agencies will all respond to an emergency, and in such situations data sharing can be an issue. The midst of a major emergency is not the time to be sorting out these issues. Because multiple agencies respond to disasters, sharing of spatial information can be key and brings up a litany of other issues. Can anybody in the organization use the information or is it restricted to certain individuals or entities? Can this information be shared with other organizations? Does data sharing require interagency agreements in order to be shared? In such a data sharing situation, who is responsible for changes and updates, and who is responsible and liable if these do not take place? From a homeland security standpoint, these issues must be addressed—Who is going to be responsible for updating GIS and where are they going to get the budget?

These concerns are addressed in the NAS report as well, which identifies in it's recommendations the special needs of the emergency management community. It is suggested that the existing National Spatial Data Infrastructure (NDSI) can provide the necessary framework for sharing information, and recommends that geospatial data plans and procedures be put in place for all but the smallest emergencies and involve all levels of government agencies and appropriate nongovernmental agencies (NGOs) as well. The special needs of emergency management for effective data sharing and collaboration are also recognized, including concerns over data quality, ownership, and interoperability (Committee on Planning for Catastrophe 2007).

The 9/11 tragedy has had serious ramifications on data access as well. Who controls access? Who is able to say how much GIS information we want to collect and release? Who is able to say, 'You know I don't want to see the entrance to the layout of the police station or the court house from space. I want you to take that information and blur that. I don't think we ought to have people clearly looking at the ground around our water towers so let's blur that.'

Who has the authority to do that? And how do you notify the population that it has happened? We criticized the Soviet Union for their lack of an open society when they routinely camouflaged the location of buildings in public documents and in encyclopedias. But now we find ourselves trying to protect our own people and our own facilities, and need to decide whether somebody ought to control and withhold basic information about our government agencies and private infrastructure.

The NAS report also examines issues surrounding the necessity of restricting access to certain types of geospatial data. The report does also note, however, that overly restricting access to information can lead to skepticism, even hostility, from the media and the public (Committee on Planning for Catastrophe 2007).

Actually, authority at the moment is probably driven in most locations by the question, 'Who pays?' When you call for assistance from a GIS organization or a team, they are likely to ask in return, 'Who is going to pay for that information and analysis?' This is frankly one of the first order questions any politician is going to ask about GIS capabilities. So we would suggest that what we need for the future is not just operational expertise, but organizational and financial expertise to anticipate questions about GIS availability and use. The midst of a crisis is not the time to be sorting out this question. And neither is after the event, when somebody is going to be stuck with the bill. While there certainly is much work to be done in answering these questions, there is a recognized need and some guidelines to better utilization of geospatial data within the GIS community.

17.5 Getting to a GIS Informed Homeland Security Future Through the Classroom

We think educational organizations and GIS should have a major role in anticipating such questions and developing standard answers. An example of how GIS could be incorporated into homeland security educational programs is offered by Texas A&M University, not as an example of how everyone should do it, but as an example of how difficult it is to accomplish. Despite three years of effort, the truth is GIS is not yet integrated into the Texas A&M homeland security program, despite it being a national leader in the field. Progress is being made, but it is hard, slow work, and we are just beginning to make inroads.

Over the last ten years, Texas A&M has established a number of very good independent offerings in homeland security, some of which incorporate GIS capabilities. The National Emergency Response and Rescue Training Center—a component of the Texas Engineering Extension Service (http://www.teex.com) is a world-class facility for fire and disaster training. Its educational outreach is large with more than 70,000 students a year using the facility to learn basic and advanced responder techniques. Our university also offers a broad range of homeland security courses at the undergraduate and graduate levels by individual departments, both in residence and online. An overview of the homeland security courses offered in the program can be found at http://homelandsecurity.tamu.edu/education. These courses range from agriculture, foreign animal, and zoonotic disease issues, to computer security and rural and public health. Texas A&M University also offers a specialized graduate program in hazard mitigation and urban planning. Many of these individual classes include GIS learning objectives, and yet it has been difficult to pull these varied approaches together in a coordinated fashion, because to date, no one has been quite sure of what to use as an organizing concept around which to build a homeland security educational program.

Specifically, it is possible to build a world-class program in homeland security from a security perspective—which examines protection first, or from an emergency management perspective, which is less interested in security or terrorism and more interested in disaster mitigation. In the wake of Katrina, Texas A&M social scientists are starting to examine the influence, judgment, and the nature of different populations in disasters, which is an area in which GIS has been applied to before (e.g. Cutter et al. 1997, 2000) and at Texas A&M, specifically (Burns 2007). There is also merit in studying governmental issues having to do with organizational policy as related to homeland security. Faculty working in the Business School has an entirely different perspective.

The bottom line is that this field of homeland security is ripe for individuals to begin doing world-class research on specific issues of interest to their disciplines; however, individualized research and study does not always lead to a comprehensive view of larger, overarching issues within homeland security, nor of the overarching role that GIS could play in such research and teaching in the field.

The Integrative Center for Homeland Security (ICHS) at Texas A&M is attempting such a coordinated, inter-disciplinary approach to distance learning on homeland security education through the Bush School of Government and Public Service. While the program has a wide range of individual courses and an excellent integrated graduate program capable of reaching a large audience through distance learning, it does not yet have a broad, integrated effort to include GIS into the curriculum even though the customers who will hire our graduates want to see those skills taught. The question is, 'Why not?'

The answer lies in the fact that the faculty are simply unaware of the capabilities of GIS and how it can be applied to their program. Remedying this shortcoming requires the help of GIS experts. To explain this conundrum requires more exploration of ICHS as an example of a detailed, extensive homeland security program that does not yet take full advantage of GIS knowledge and offerings.

The mission of ICHS to examine the entire range of homeland security issues, ranging from overarching issues like the Patriot Act to small details such as the four fluid ounces that can be used to make a bomb onboard an airplane. This makes the program somewhat unique. Our program's approach has been to pull all these issues together in a framework, or in our terminology—a single taxonomy—that allows a student, faculty, or researcher to examine the entire range of homeland security issues and say, 'What does this discipline look like?' and 'How are different homeland security issues connected?' This is accomplished via a website (http://homelandsecurity.tamu.edu) that enables individuals to walk through the major issues in homeland security and examine the major documents in each. ICHS also produces newsletters, a daily RSS Feed, and a lecture series through the George Bush Presidential Library. All of these efforts are structured around the taxonomy, so they can be offered to individuals as one integrated holistic interdisciplinary framework.

But GIS is not yet a major part of this effort. This is because GIS experts have not yet taken the lead in figuring out how to incorporate themselves into the different areas comprising homeland security. GIS experts cannot simply wait for non-experts to request help. To overcome this problem, GIS educators and researchers must be more aggressive in offering their help to homeland security experts.

The ICHS curriculum includes a fundamentals course that touches the major issues in the homeland security taxonomy, and nine additional courses that focus on various aspects of the taxonomy. For example, terrorism, critical infrastructure, business, state, and local government are all headings in the taxonomy and are courses offered in the graduate certificate program.

Yet GIS is hardly mentioned in any of the most carefully, holistic, homeland security courses in the United States. The reason, again, is that the subject matter experts do not know enough about GIS to include it in their courses, and GIS experts have not been aggressive in finding ways to incorporate their expertise into the curriculum. We are not yet working together.

ICHS is working with a large number of people and organizations outside the Texas A&M community—individuals and organizations who have a direct interest in homeland security and some aspect of homeland security education. These range from the Department of Homeland Security, to the United States Northern Command, to the Naval Post Graduate School, the FEMA Higher Education project, and the Homeland Security and Defense Education Consortium, which is the most active collection of homeland security related organizations of higher learning in the world. Yet almost no individual in these groups seems to be recommending that GIS be wrapped into the education materials they are developing. They are not hostile to GIS—they are simply blind to its possibilities. We feel that GIS should to be integrated into almost every class, from policy, to science, to engineering. But hardly anyone is thinking in this direction, nor will they awaken to the opportunities GIS offers in the future, unless the GIS community takes the initiative.

Texas A&M is attempting to address the underrepresentation of GIS in homeland security as we develop a new master's degree. However, our current plan still underrepresents GIS. Better integration of GIS into the curriculum requires answering such fundamental questions as 'Where does GIS fit in?' 'Should it be integrated into the core courses of the program?' or 'Into the broad range of classes offered by different departments in the program' or 'Into the in-depth specialty areas?'

The answer lies in GIS experts stepping forward with options and ideas. The GIS field must fend for itself and represent itself as a solution to problems in every discipline within the homeland security arena. Until that happens, we will not see an integrated GIS program in homeland security take shape.

Anyone who has created an interdisciplinary educational program before knows what a huge challenge this is. Once a successful organization has been created, it is very difficult to change that organization. Integrating a new technology requires that the practitioners of this new technology demonstrate its benefit before individuals in the organization will accept it. Individuals want to see new technology work, not help make it work. Individuals in a successful program want to apply the first rule of mountain climbing: 'Never let go of your current handhold until you have a firm grasp of the next.' They will want to see GIS work in the classroom before they include it in their class.

GIS professionals can crack this resistance, and can help create the future homeland security programs our country needs by taking it upon themselves to incorporate GIS into our classrooms in a variety of disciplines today. Accomplishing this requires:

First, learning what is really of overarching importance in homeland security. This will take a bit of study, but the subject matter is finite. It includes a bit on all the major threats: chemical, biological, explosives, radiological, nuclear, and natural. It is also important to understand the major national strategies for homeland security and what constitutes critical infrastructure. GIS professionals need to become familiar with the major national policies, starting with the twenty Homeland Security Presidential Directives and the roles of the major federal players in homeland security: including the Department of Homeland Security (DHS), the Department of Defense (DOD), the National Guard, the Department of Energy (DOE), the Department of Transportation (DOT), the Department of Agriculture, the Centers for Disease Control and Prevention (CDC), and a handful of others. And finally, GIS professionals must learn something about the National Response Plan and the National Incident Management System. This knowledge is the coin of the realm in homeland security and is required to bargain with the locals.

To incorporate GIS into homeland security education, GIS practitioners must approach homeland security educators now, and ask to be included in their work. They should also seek partners looking to develop homeland security classes, but who are not entirely sure exactly what to teach. Help them flesh out their curriculum with GIS capabilities from the beginning.

Enlist students both as an attractor and a driver for classroom offering. Homeland security students are eager for technological applications in their classes, as they see the future and know they will need technical credibility which includes GIS. Use their demand to leverage your participation in classes.

Find employers who understand the opportunities provided by GIS in the federal, state, local governments, and in the private sector. Use their needs for employees conversant in GIS solutions to help shape homeland security curriculum as it is developed.

It is interesting to note that among the major recommendations in *Successful Response Starts with a Map: Improving Geospatial Support for Disaster Management* are some that deal explicitly with education and align well with our observations. One recommendation is that emergency management programs should increase the emphasis given to geospatial data and tools. As we highlight, geospatial professionals should received increased training in emergency management practices (Committee on Planning for Catastrophe 2007). So while work remains to be done, at least some groundwork for better incorporating GIS into homeland security has already been laid.

The bottom-line is that the future stacks up to be a very dangerous place. We need to be looking at integrated solutions to those challenges—solutions that will cross disciplines and incorporate new technologies like GIS. We can shape a safer future if we build that sort of integrated curriculum in our classrooms today, but that will require an effort on the part of GIS experts. GIS professionals will have to learn the challenges, learn the components of the solutions, and find ways to incorporate their expertise into those solutions—and sell the whole package to experts confident in their own discipline but skeptical of unproven programs. This is a tall order, but one that must be met it if we are to prosper in the dangerous future that looms ahead of us.

References

Alexander, D. (2002). *Principles of emergency planning and management.* (New York: Oxford University Press)

Burns, G. (2007). *A case study of social vulnerability mapping: Issues of scale and aggregation.* M.S. thesis, College Station, TX: Texas A&M University

Bush, G.W. (2003). *National strategy for physical protection of critical infrastructure and key assets* [Electronic version]. Retrieved from http://www.whitehouse.gov/pcipb/physical.html

Committee on Planning for Catastrophe: A Blueprint for Improving Geospatial Data, Tools, and Infrastructure, National Research Council. (2007). *Successful response starts with a map: Improving geospatial support for disaster management* [Electronic version]. (Washington, DC: The National Academies Press). Retrieved from www.nap.edu/catalog/11793.html

Cutter, S., Mitchell, J. & Scott, S. (1997). *Handbook for conducting a GIS based hazard assessment at the county level.* (South Carolina: Hazards Research Lab, Department of Geography, University of South Carolina)

Cutter, S., Mitchell, J. & Scott, S. (2000). Revealing the vulnerability of people and places: A case study of Georgetown County, South Carolina. *Annals of the Association of American Geographers*, 90(4), 713–737

Martinez, L. (2000). Global infectious disease surveillance. *International Journal of Infectious Diseases*, 4(4), 222–228

Mayfield, M. (2007). Interview. *Homeland security inside & out* 6 June 2007 [Electronic version]. Retrieved from http://hlsinsideandout.org

Meade, C. & Molander, R. (2006). *Considering the effects of a catastrophic terrorist attack* [Electronic version]. Santa Monica, CA: RAND Center for Terrorism and Risk Management Policy. Retrieved from http://www.rand.org/pubs/technical_reports/2006/RAND_TR391.sum.pdf

National Commission on Terrorist Attacks Upon the United States. (2004). *Final report of the national commission on terrorist attacks upon the United States* (Authorized Edition). (New York: W. W. Norton)

Peters, R. (2002). *Beyond terror.* (Mechanicsburg, PA: Stackpole Books)

Tierney, K., Lindell, M. & R. Perry. (2001). *Facing the unexpected: Disaster preparedness and response in the United States.* (Washington DC: Joseph Henry Press)

Chapter 18
Geospatial Technologies and Homeland Security: Challenges and Opportunities

Michael F. Goodchild

Abstract The acronym GIS can be decoded in three distinct ways: GISystems, GIScience, and GIStudies. This framework is used to provide an overarching synthesis, and to ask whether the chapters of the book provide a complete picture of geospatial issues and applications to homeland security. Three characteristics distinguish homeland security applications from other domains: the need for speed; the difficult environments in which technology must operate; and the impossibility of anticipating many relevant kinds of events in either space or time. One of the strongest factors impeding the effective use of geospatial technologies is lack of collaboration between institutions and the cultures of emergency response and GIS. The final section identifies four themes that are largely missing from this book, but nevertheless represent opportunities and challenges for the future.

Keywords Geocollaboration, geographic dynamics, sensor networks, volunteered geographic information

18.1 Introduction

Secure borders, freedom from fear, effective law enforcement, and well-laid plans for emergencies are just a selection of the many themes that are encompassed by the notion of homeland security, and that emerge in the chapters of this book. As the chapters have clearly shown, all of these themes have geospatial dimensions, and can be addressed by the kinds of tools that we now lump together under the heading of geospatial technologies, and that range from the lowly paper map to the most advanced gadgets and simulation models, within the broader compass of abstract concepts that Bednarz and Bednarz (this volume) term spatial thinking. In this final chapter I will risk an attempt to bring the great diversity of what has

University of California, Santa Barbara, CA

D.Z. Sui (ed.) *Geospatial Technologies and Homeland Security*,

gone before to some kind of concluding synthesis, and to offer some ideas about future directions, for both research and practice.

The computer application known as GIS began to emerge in the 1960s, but not until the 1980s was there a consensus on its definition and domain, and a substantial commercial software industry. While the acronym originally stood for geographic information system, I suggested in the early 1990s (Goodchild 1992) that it could also be decoded as geographic information science, and could stand for the many fundamental ideas that lie behind the software, or belong to the body of scientific knowledge that the software implements. This body of knowledge is extensive, encompassing not only the more traditional fields of cartography and surveying, but also more recent interest in photogrammetry, remote sensing, spatial databases, spatial cognition, and spatial statistics. Later (see, for example, Longley et al. 2005) the decoding geographic information studies came to be associated with research on the societal context and social impacts of the technology. Today other terms such as geomatics have become popular in some countries and communities, but this three-part structure still provides a useful framework.

On this basis the chapters of this book might align as follows. In the first category of applications of GISystems, one could place the work of Ashby, Chainey, and Longley (this volume) on geodemographics and policing; Brody and Zahran (this volume) on floods and wetlands; Huang et al. (this volume) on pest management; Mesev et al. (this volume) on demographic segregation; Pan et al. (this volume) on the economic impacts of disasters; Shroder (this volume) on Afghanistan; Ward (this volume) on medical geography; Wunneburger, Olivares, and Maghelal (this volume) on sex offenders; and Zhan and Chen (this volume) on evacuation. All in their various ways demonstrate the power of geospatial data and tools in tackling specific problems within the broad rubric of homeland security; and together they account for almost half the book.

The second category of GIScience would emphasize the principles underlying geospatial technologies, research on their development, and fundamental questions raised by their use. In this category one might place the chapter by Castle and Longley (this volume) on software for the analysis and simulation of pedestrian movement; Filippi (this volume) on the technology of remote sensing; and Monmonier (this volume) on the acquisition of detailed elevation data. Finally, the third category of GIStudies could include the Bednarz and Bednarz (this volume) chapter on spatial thinking and its importance to homeland security; the Crampton (this volume) chapter on the politics of fear and its implications for geospatial perspectives; Hannah (this volume) on Foucault and the future of the German census; and McIntyre and Klein (this volume) on the importance of education in GIS.

While this categorization is by no means perfect, it does allow one to ask two important questions: how well do the chapters of this book cover the ground in each of the three areas, compared with the literature and the work of others; and what gaps exist that might present challenges or opportunities for the future? In the first category of GISystems, it is easy to think of application areas important to homeland security that are not addressed in the chapters of this book. For example, GISystems might be used to study the resilience of conventional infrastructure networks (road, rail, pipelines, etc.), and the beginnings of such a literature can be found in the work of Grubesic and Murray (2006), Church and Scaparra (2007), and others. GISystems are crucial in the study of risk from hazards such as earthquakes, severe storms, wildfires, and in risk mitigation.

In the second category, several reports (e.g. NRC 2003, 2007) have drawn attention to the need for research on the adaptation of geospatial technology to the extreme conditions of emergency management when smoke, darkness, and noise may make conventional tools useless.

Finally, there are challenges and opportunities in the area of GIStudies, dealing for example with the economics of GIS and returns on investment (GITA 2007), with the conflict between surveillance and the right to locational privacy, and with the need to build a secure, resilient national spatial data infrastructure. In this last area, it would be good to know the views of the US Department of Homeland Security, and to learn more about its specific activities and how it organizes its own commitments to GIS.

In the following sections I expand on these ideas, and point to specific challenges and opportunities. The perspective is essentially personal, reflecting my own experience and interests, and ideas that seem particularly current within the broad framework of GIS.

18.2 What is Special about Homeland Security?

In 2005–2006 I had the pleasure of chairing a committee of the National Research Council charged to address the role of geospatial data and tools in emergency management. The report (NRC 2007) first concludes that GIS is essential in all four phases of emergency management—preparedness, response, recovery, and mitigation—and moreover that the case for GIS is extremely powerful and compelling. It identifies three ways in which emergency management is distinct from other applications of GIS.

First, while traditional applications of GIS have often proceeded at a fairly leisurely pace, and it is not uncommon for GIS projects to last several years from initial conception through problem formulation, data collection, analysis, and final publication, applications in emergency management almost always stress the need for speed. The term golden hour is often used by responders to denote the first sixty minutes following the event, when there are the greatest opportunities for saving lives. In the case of the bombing of the Murrah Federal Building in downtown Oklahoma City in 1995, for example, the explosion which occurred at 9:02 am local time on April 19 caused an immediate need for all relevant geospatial data, including evacuation routes, locations of hospitals, and building plans. Yet in this and many other recent disasters it has taken hours, and in some cases days, to assemble the trained personnel, hardware, software, data, and network communications of an operational geospatial capability. Paradoxically, people all over the world were able to access geospatial data about New Orleans in the immediate aftermath of Hurricane Katrina through services such as Google Earth—but those on the ground in the impacted area had to wait days for the necessary power and Internet connectivity.

The report has several recommendations designed to get the geospatial operation up and running faster. Central to these is the need for overhead imagery, whether from satellite, unmanned aircraft, or helicopter. In some recent events it has been people with cameras in helicopters that have provided the first usable images of damage, and the basis for planning organized response. In many cases the lack of prior arrangements for acquisition of imagery from commercial sources has been a major impediment to progress, as has the lack of arrangements for distributing such imagery to those who need it most.

Second, emergency responders often work in difficult environments that are in sharp contrast to the highly functional office environments of desk-top GIS. The New York firemen climbing the stairwells of the World Trade Center on September 11, 2001 were working in environments that were often dark, full of smoke and dust, and noisy, and were often unable to communicate with their peers. This was no place for laptops, or even PDAs. In the evidence presented to the committee, it was often clear that the most useful geospatial product in such situations is a paper map, and indeed that there was often a strong negative reaction on the part of emergency responders to sophisticated electronics. As a result, the report calls for research on how to adapt geospatial data to the particular environments in which responders must work, and for efforts to improve the training of first responders in the use of geospatial technologies.

The third important characteristic of emergencies is their unpredictability, and particularly the impossibility of predicting either when—the precise timing of the event—or where—the area impacted by the event, in other words its spatial footprint. Within the impacted area of Hurricane Katrina were two states and numerous cities and counties. Literally hundreds of agencies from federal to local had some degree of responsibility for the impacted area, and numerous non-governmental organizations were also involved, from the Red Cross to universities and the volunteer GIS Corps. The need to coordinate and communicate created perhaps the greatest impediment to the effective use of geospatial technologies, and problems of an institutional nature quickly came to dominate the aftermath. Thus there is a strong argument for giving institutional issues, within the broader realm of GIStudies, the highest priority in future research.

18.3 Can Cultures Communicate?

The chapters of this book have been written largely by professionals steeped in the culture of geospatial technologies. While they may not all be advocates, and the list of authors certainly includes several skeptics, there is nevertheless a distinct tendency evident in the chapters to treat the geospatial aspects of a problem as primary. Professionals committed to GIS have often been accused of over-selling, but a more subtle issue is of major concern here, and it is the cultural divide that separates specialists in emergency management, and more broadly in homeland security, from specialists in geospatial technologies. The NRC report concludes that the two groups learn remarkably little about the other's domain in formal education, since courses in geospatial technologies often give little attention to emergency management, and courses in emergency management often make little mention of GIS.

But the divide is deeper than that, for as Bednarz and Bednarz (this volume) argue, thinking spatially is different. People who think spatially likely begin with a map, and with the perception that the Earth's surface is a continuum. Cities and states are not lists and tables, but places distributed over a landscape that varies at all levels of detail, from the coarsest to the finest. Spatial thinking places great emphasis on context, the importance of surroundings in understanding events and patterns of behavior. Spatial thinkers know about and trust Tobler's First Law of Geography, the statement that nearby things tend to be more similar than distant things. Spatial thinking may be the subject of their research, as it is for Bednarz and Bednarz (this volume); or the framework within which they develop tools; or no more than a conceptual underpinning to their use of tools. But invariably they will give primary importance to geographic space in whatever they do.

By contrast, emergency responders see the event, its victims, and their alternative courses of action as primary. If geospatial technologies are used, it is because they, like many other kinds of tools, are found to be useful—but they have no deeper level of commitment to any one class of tool. They will likely agree that geospatial technologies are important, but be unlikely to see them as in any way unique or more important than any other technology, or as a basis for organizing their workflow. In this context, the title and contents of this book appear somewhat odd. Why select these particular aspects of the subject, and what benefits result from assembling chapters on this theme rather than on any other cross-cutting theme? GIS professionals know the answer, but it is not necessarily shared by other kinds of professionals. After all, while GIS is important in all aspects of homeland security, and in many activities of the US Department of Homeland Security, it is not to date reflected at a high level in the organizational structure of that agency (though its Geospatial Management Office plays an important role), or in the topics of the various research centers it has funded to date. GIS is everywhere in the agency, but at the same time it is in danger of being nowhere.

18.4 Emerging Themes

Despite the broad coverage of topics in GISystems, GIScience, and GIStudies, there are several important areas of geospatial research that deserve mention, and that may provide challenges and opportunities for the community. In this section I review four of these. Again, I think it is important to emphasize that this is essentially a personal list, and that by describing it I hope to stimulate further discussion on areas that I have omitted.

18.4.1 Dynamics

The roots of GIS lie in the map, a paper document that once printed is difficult to update. Maps are expensive to produce, and thus tend to be directed at the most popular applications, and to emphasize features that change the least. We make maps of topography, roads, cities, soils, and vegetation, but not of the instantaneous positions of vehicles, or of flows or transactions. This is a powerful legacy for GIS, and the technology remains largely concerned with the static aspects of the Earth's surface, and difficult to adapt to its more temporal aspects.

This situation is changing, however, and the dynamics of the Earth's surface are coming more and more to dominate GIS development. While there is little mention of this work in the chapters of this book, significant efforts have been made in recent years to develop specialized GIS software for dynamic simulation (e.g. PCRaster, http://pcraster.geo.uu.nl), for the analysis of travel behavior, and more generally for the representation and analysis of the entire range of geographic dynamics (Goodchild 2007).

In the area of homeland security, work on dynamics is helping to produce real-time simulations of evacuation (Cova and Johnson 2002); models of the spread of atmospheric plumes and flood surges; and a host of techniques for analyzing intelligence and making inferences about potentially dangerous events. The chapter by Castle and Longley (this volume) presents a very useful analysis of the state of this art in modeling the behavior of pedestrians, while Zhan and Chen (this volume) show how it can be applied to the analysis of evacuation.

18.4.2 Geo-Collaboration

Another research thread focuses on the use of the Internet and geospatial tools to foster and facilitate collaboration between the various actors in matters of homeland security. Nowhere is this more valuable than in emergency management, when effective communication between all of the various responders and decision-makers can often make the difference between success and failure. Responders arriving on the scene often have only partial knowledge of the event, and lack the kinds of local awareness and skills needed to be effective. Better communication is needed not only between responders and their managers, but also peer-to-peer between responders, and peer-to-peer between managers and their agencies. Work at Pennsylvania State University (http://www.geovista.psu.edu/work/projects/geocollaboration.jsp) is exploring this notion of geo-collaboration, and prototyping the kinds of tools needed to make it operational and effective.

18.4.3 Sensor Networks

A sensor network can be defined as a distributed network of devices that know their positions and that sense various aspects of their environments and report them to a central server. They might consist of video cameras or microphones in the case of surveillance, or temperature sensors in advance of a wildfire front (http://www.rfidjournal.com/magazine/article/1724), or sensors distributed in the world's oceans to provide early detection of tsunamis. Many existing networks meet this definition, including the weather stations that provide minute-by-minute monitoring of the Earth's atmosphere. Sensors may be carried on people or vehicles; in the former case one thinks of the atmospheric pollution sensors carried by children involved in research on asthma, while the latter case includes participants in research on travel behavior that have allowed their vehicles to be tracked. Clearly many applications of sensor networks are of importance to homeland security.

As sensors become cheaper, smaller, and more versatile the associated problems of computation, compilation, and distribution of their signals is emerging as a major topic for geospatial research within the domain of GIScience. Several researchers have addressed the social impacts and implications of sensor networks within GIStudies, and there is increasing interest in such issues as privacy (Curry 1998).

18.4.4 Volunteered Geographic Information

Finally, and related in many ways to sensor networks, is the topic of volunteered geographic information (VGI), a form of user-generated content that is attracting very significant attention in the geospatial research community. VGI is exemplified by a growing collection of Web sites, including Wikimapia (http://www.wikimapia.org), OpenStreetMap (http://www.openstreetmap.org), and Flickr (http://www.flickr.com), all of which allow their users to enter data regarding places on the Earth's surface, in the form of descriptions, actual street maps, and photographs respectively. Blogs and Wikis that use geo-tags to allow contributors to reference information to points on the Earth's surface are other forms of VGI. Efforts such as these are engaging thousands of comparatively unqualified amateurs in what amounts to an increasingly significant mechanism for creation of geographic information. Research is clearly needed on the social aspects of the phenomenon (What makes people do this, and are they accurate?), the technical aspects (Can techniques be devised to mine this information?), and the conceptual aspects (What do people need to know to be more effective?).

While VGI is normally associated with the creation of conventional geographic information through unconventional means, it can also play a very significant role in homeland security. One can imagine a future in which citizens within the footprint of a disaster are able to contribute what they know and can see around them to central repositories using such devices as mobile phones; and in which the resulting information is made almost instantly available to responders. Useful intelligence, in both a commercial and a military sense, can often be gleaned from the content contributed by citizens to blogs or contained in email. This is a very different world from the traditional one—in which virtually all geographic information was compiled and distributed by central government agencies—and it is growing very rapidly.

18.5 Conclusion

Reading through the chapters of this book, one is struck again and again by the importance of geospatial technologies, and by the need for a concerted effort to make them more available, and to educate people in their effective use. In the hands of a user who thinks spatially, geospatial technologies are amazingly powerful extensions of the senses, revealing things that would be impossible to obtain in any other way. Geospatial technologies replace the tedium, inaccuracies, and lack of replicability of more traditional approaches with a workflow that is readily shared and understood. Much remains to be done to enhance their specific applicability to issues of homeland security, but the chapters of this book demonstrate the vast range of opportunities that exist.

At the same time one would be foolish not to recognize the problems that can result from their misuse. Results from computers can easily appear more accurate than they really are, because of the ample precision of the machine. Users may be overconfident in the results of simulations, conveniently ignoring the uncertainties that are inherent in any model of the world. The logical precision of a GIS may be at odds with the much more casual and ambiguous way in which humans interact with the planet, giving different names to the same place, making maps that sometimes reveal as much about the agendas of their makers as they do about the real world they are intended to represent. Geospatial technologies are not readily adapted to representing the different views and perceptions people have of their surroundings, or the ways some cultures manage and conceptualize space.

In the final analysis, however, it is impossible to avoid the fact that geospatial technologies have an enormous amount to offer homeland security, and that their greater and more effective use, along the lines presented in this book, cannot but help make the world a safer and more secure place.

References

Church, R. L. & Scaparra, M. P. (2007). Protecting critical assets: The *r*-interdiction median problem with fortification. *Geographical Analysis*, 39, 129–146

Cova, T. J. & Johnson, J. P. (2002). Microsimulation of neighborhood evacuations in the urban-wildland interface. *Environment and Planning A*, 34, 2211–2229

Curry, M. R. (1998). *Digital places: Living with geographic information technologies*. (New York: Routledge)

Geospatial Information and Technology Association (GITA) (2007). *Building a business case for geospatial information technology: A practitioner's guide to financial strategic analysis*. (Aurora, CO: Geospatial Information and Technology Association)

Goodchild, M. F. (1992). Geographical information science. *International Journal of Geographical Information Systems*, 6, 31–45

Goodchild, M. F. (2007). Representation and computation of geographic dynamics. (In K. Hornsby & M. Yuan (Eds.), *Understanding the dynamics of geographic domains*. Boca Raton: CRC) (in press)

Grubesic, T. H. & Murray, A. T. (2006). Vital nodes, interconnected infrastructures and the geographies of network survivability. *Annals of the Association of American Geographers*, 96, 64–83

Longley, P. A., Goodchild, M. F., Maguire, D. J. & Rhind, D. W. (2005). *Geographic information systems and science*. Second Edition. (New York: Wiley)

National Research Council (NRC) (2003). *IT roadmap to a geospatial future*. (Washington, DC: National Academies Press)

National Research Council (NRC) (2007). *Successful response starts with a map: Improving geospatial support for disaster management*. (Washington, DC: National Academies Press)

Index

The GeoJournal Library

21. V.I. Ilyichev and V.V. Anikiev (eds.): *Oceanic and Anthropogenic Controls of Life in the Pacific Ocean.* 1992 ISBN 0-7923-1854-4
22. A.K. Dutt and F.J. Costa (eds.): *Perspectives on Planning and Urban Development in Belgium.* 1992 ISBN 0-7923-1885-4
23. J. Portugali: *Implicate Relations.* Society and Space in the Israeli-Palestinian Conflict. 1993 ISBN 0-7923-1886-2
24. M.J.C. de Lepper, H.J. Scholten and R.M. Stern (eds.): *The Added Value of Geographical Information Systems in Public and Environmental Health.* 1995 ISBN 0-7923-1887-0
25. J.P. Dorian, P.A. Minakir and V.T. Borisovich (eds.): *CIS Energy and Minerals Development.* Prospects, Problems and Opportunities for International Cooperation. 1993 ISBN 0-7923-2323-8
26. P.P. Wong (ed.): *Tourism vs Environment: The Case for Coastal Areas.* 1993 ISBN 0-7923-2404-8
27. G.B. Benko and U. Strohmayer (eds.): *Geography, History and Social Sciences.* 1995 ISBN 0-7923-2543-5
28. A. Faludi and A. der Valk: *Rule and Order. Dutch Planning Doctrine in the Twentieth Century.* 1994 ISBN 0-7923-2619-9
29. B.C. Hewitson and R.G. Crane (eds.): *Neural Nets: Applications in Geography.* 1994 ISBN 0-7923-2746-2
30. A.K. Dutt, F.J. Costa, S. Aggarwal and A.G. Noble (eds.): *The Asian City: Processes of Development, Characteristics and Planning.* 1994 ISBN 0-7923-3135-4
31. R. Laulajainen and H.A. Stafford: *Corporate Geography. Business Location Principles and Cases.* 1995 ISBN 0-7923-3326-8
32. J. Portugali (ed.): *The Construction of Cognitive Maps.* 1996 ISBN 0-7923-3949-5
33. E. Biagini: *Northern Ireland and Beyond.* Social and Geographical Issues. 1996 ISBN 0-7923-4046-9
34. A.K. Dutt (ed.): *Southeast Asia: A Ten Nation Region.* 1996 ISBN 0-7923-4171-6
35. J. Settele, C. Margules, P. Poschlod and K. Henle (eds.): *Species Survival in Fragmented Landscapes.* 1996 ISBN 0-7923-4239-9
36. M. Yoshino, M. Domrös, A. Douguédroit, J. Paszynski and L.D. Nkemdirim (eds.): *Climates and Societies – A Climatological Perspective.* A Contribution on Global Change and Related Problems Prepared by the Commission on Climatology of the International Geographical Union. 1997 ISBN 0-7923-4324-7
37. D. Borri, A. Khakee and C. Lacirignola (eds.): *Evaluating Theory-Practice and Urban-Rural Interplay in Planning.* 1997 ISBN 0-7923-4326-3
38. J.A.A. Jones, C. Liu, M-K.Woo and H-T. Kung (eds.): *Regional Hydrological Response to Climate Change.* 1996 ISBN 0-7923-4329-8
39. R. Lloyd: *Spatial Cognition.* Geographic Environments. 1997 ISBN 0-7923-4375-1
40. I. Lyons Murphy: *The Danube: A River Basin in Transition.* 1997 ISBN 0-7923-4558-4
41. H.J. Bruins and H. Lithwick (eds.): *The Arid Frontier.* Interactive Management of Environment and Development. 1998 ISBN 0-7923-4227-5
42. G. Lipshitz: *Country on the Move: Migration to and within Israel, 1948–1995.* 1998 ISBN 0-7923-4850-8
43. S. Musterd, W. Ostendorf and M. Breebaart: *Multi-Ethnic Metropolis: Patterns and Policies.* 1998 ISBN 0-7923-4854-0
44. B.K. Maloney (ed.): *Human Activities and the Tropical Rainforest.* Past, Present and Possible Future. 1998 ISBN 0-7923-4858-3

The GeoJournal Library

---

45. H. van der Wusten (ed.): *The Urban University and its Identity.* Roots, Location, Roles. 1998 ISBN 0-7923-4870-2
46. J. Kalvoda and C.L. Rosenfeld (eds.): *Geomorphological Hazards in High Mountain Areas.* 1998 ISBN 0-7923-4961-X
47. N. Lichfield, A. Barbanente, D. Borri, A. Khakee and A. Prat (eds.): *Evaluation in Planning.* Facing the Challenge of Complexity. 1998 ISBN 0-7923-4870-2
48. A. Buttimer and L. Wallin (eds.): *Nature and Identity in Cross-Cultural Perspective.* 1999 ISBN 0-7923-5651-9
49. A. Vallega: *Fundamentals of Integrated Coastal Management.* 1999 ISBN 0-7923-5875-9
50. D. Rumley: *The Geopolitics of Australia's Regional Relations.* 1999 ISBN 0-7923-5916-X
51. H. Stevens: *The Institutional Position of Seaports.* An International Comparison. 1999 ISBN 0-7923-5979-8
52. H. Lithwick and Y. Gradus (eds.): *Developing Frontier Cities.* Global Perspectives–Regional Contexts. 2000 ISBN 0-7923-6061-3
53. H. Knippenberg and J. Markusse (eds.): *Nationalising and Denationalising European Border Regions, 1800–2000.* Views from Geography and History. 2000 ISBN 0-7923-6066-4
54. R. Gerber and G.K. Chuan (eds.): *Fieldwork in Geography: Reflections, Perspectives and Actions.* 2000 ISBN 0-7923-6329-9
55. M. Dobry (ed.): *Democratic and Capitalist Transitions in Eastern Europe.* Lessons for the Social Sciences. 2000 ISBN 0-7923-6331-0
56. Y. Murayama: *Japanese Urban System.* 2000 ISBN 0-7923-6600-X
57. D. Zheng, Q. Zhang and S. Wu (eds.): *Mountain Geoecology and Sustainable Development of the Tibetan Plateau.* 2000 ISBN 0-7923-6688-3
58. A.J. Conacher (ed.): *Land Degradation.* Papers selected from Contributions to the Sixth Meeting of the International Geographical Union's Commission on Land Degradation and Desertification, Perth, Western Australia, 20–28 September 1999. 2001 ISBN 0-7923-6770-7
59. S. Conti and P. Giaccaria: *Local Development and Competitiveness.* 2001 ISBN 0-7923-6829-0
60. P. Miao (ed.): *Public Places in Asia Pacific Cities.* Current Issues and Strategies. 2001 ISBN 0-7923-7083-X
61. N. Maiellaro (ed.): *Towards Sustainable Buiding.* 2001 ISBN 1-4020-0012-X
62. G.S. Dunbar (ed.): *Geography: Discipline, Profession and Subject since 1870.* An International Survey. 2001 ISBN 1-4020-0019-7
63. J. Stillwell and H.J. Scholten (eds.): *Land Use Simulation for Europe.* 2001 ISBN 1-4020-0213-0
64. P. Doyle and M.R. Bennett (eds.): *Fields of Battle.* Terrain in Military History. 2002 ISBN 1-4020-0433-8
65. C.M. Hall and A.M. Williams (eds.): *Tourism and Migration.* NewRelationships between Production and Consumption. 2002 ISBN 1-4020-0454-0
66. I.R. Bowler, C.R. Bryant and C. Cocklin (eds.): *The Sustainability of Rural Systems.* Geographical Interpretations. 2002 ISBN 1-4020-0513-X
67. O. Yiftachel, J. Little, D. Hedgcock and I. Alexander (eds.): *The Power of Planning.* Spaces of Control and Transformation. 2001 ISBN Hb; 1-4020-0533-4 ISBN Pb; 1-4020-0534-2

The GeoJournal Library

68. K. Hewitt, M.-L. Byrne, M. English and G. Young (eds.): *Landscapes of Transition.* Landform Assemblages and Transformations in Cold Regions. 2002 ISBN 1-4020-0663-2
69. M. Romanos and C. Auffrey (eds.): *Managing Intermediate Size Cities.* Sustainable Development in a Growth Region of Thailand. 2002 ISBN 1-4020-0818-X
70. B. Boots, A. Okabe and R. Thomas (eds.): *Modelling Geographical Systems.* Statistical and Computational Applications. 2003 ISBN 1-4020-0821-X
71. R. Gerber and M. Williams (eds.): *Geography, Culture and Education.* 2002 ISBN 1-4020-0878-3
72. D. Felsenstein, E.W. Schamp and A. Shachar (eds.): *Emerging Nodes in the Global Economy: Frankfurt and Tel Aviv Compared.* 2002 ISBN 1-4020-0924-0
73. R. Gerber (ed.): *International Handbook on Geographical Education.* 2003 ISBN 1-4020-1019-2
74. M. de Jong, K. Lalenis and V. Mamadouh (eds.): *The Theory and Practice of Institutional Transplantation.* Experiences with the Transfer of Policy Institutions. 2002 ISBN 1-4020-1049-4
75. A.K. Dutt, A.G. Noble, G. Venugopal and S. Subbiah (eds.): *Challenges to Asian Urbanization in the 21st Century.* 2003 ISBN 1-4020-1576-3
76. I. Baud, J. Post and C. Furedy (eds.): *Solid Waste Management and Recycling.* Actors, Partnerships and Policies in Hyderabad, India and Nairobi, Kenya. 2004 ISBN 1-4020-1975-0
77. A. Bailly and L.J. Gibson (eds.): *Applied Geography.* A World Perspective. 2004 ISBN 1-4020-2441-X
78. H.D. Smith (ed.): *The Oceans: Key Issues in Marine Affairs.* 2004 ISBN 1-4020-2746-X
79. M. Ramutsindela: *Parks and People in Postcolonial Societies.* Experiences in Southern Africa. 2004 ISBN 1-4020-2542-4
80. R.A. Boschma and R.C. Kloosterman (eds.): *Learning from Clusters.* A Critical Assessment from an Economic-Geographical Perspective. 2005 ISBN 1-4020-3671-X
81. G. Humphrys and M. Williams (eds.): *Presenting and Representing Environments.* 2005 ISBN 1-4020-3813-5
82. D. Rumley, V.L. Forbes and C. Griffin (eds.): *Australia's Arc of Instability.* The Political and Cultural Dynamics of Regional Security. 2006 ISBN 1-4020-3825-9
83. R. Schneider-Sliwa (ed.): *Cities in Transition.* Globalization, Political Change and Urban Development. 2006 ISBN 1-4020-3866-6
84. B.G.V. Robert (ed.): *Dynamic Trip Modelling.* From Shopping Centres to the Internet Series. 2006 ISBN: 1-4020-4345-7
85. L. John and W. Michael (eds.): *Geographical Education in a Changing World.* Past Experience, Current Trends and Future Challenges Series. 2006 ISBN 1-4020-4806-8
86. G.D. Jay and R. Neil (eds.): *Enterprising Worlds.* A Geographic Perspective on Economics, Environments & Ethics Series. 2007 ISBN 1-4020-5225-1
87. Y.K.W. Albert and H.G. Brent (eds.): *Spatial Database Systems.* Design, Implementation and Project Management Series. 2006 ISBN 1-4020-5391-6
88. H.J. Miller (ed.): *Societies and Cities in the Age of Instant Access.* 2007. ISBN 1-4020-5426-6
89. J.L. Wescoat, Jr. and D.M. Johnston (eds.): *Political Economies of Landscape Change.* Places of Integrative Power. 2008 ISBN 1-4020-5848-6

The GeoJournal Library

90. E. Koomen, J. Stillwell, A. Bakema and H.J. Scholten (eds.): *Modelling Land-Use Change*. Progress and Applications. 2007 ISBN 1-4020-5647-8
91. E. Razin, M. Dijst and C. Vázquez (eds.): *Employment Deconcentration in European Metropolitan Areas*. Market Forces versus Planning Regulations. 2007 ISBN 1-4020-5761-X
92. K. Stanilov (ed.): *The Post-Socialist City*. Urban Form and Space Transformations in Central and Eastern Europe. 2007 ISBN 978-1-4020-6052-6
93. P. Ache, H.T. Andersen, Th. Maloutas, M. Raco and T. Tasan-Kok (eds.): *Cities between Competitiveness and Cohesion*. Discourses, Realities and Implementation. 2008 ISBN 978-1-4020-8240-5
94. D.Z. Sui (ed.): *Geospatial Technologies and Homeland Security*. Research Frontiers and Future Challenges. 2008 ISBN 978-1-4020-8339-6